fn 315 cause
71 LA as metropolis
91-2 dreams of humble hollers —

35—

122 fn book —
       'd us nat'l
       march

2/94
ETG

fn 458 - chis 4—+5
       xerox

.01—
C's "never many" ~

xerox
not chapter
for class?

+ use greenman
       y 9's
              American
              talk m
              American cultures

fn 450 cause
   5-7   review of field —
          bipolar model resistance/assim
          frames two distinct "opposing cultures" [7]

55 + Ellis Island

62

58-9 list of INS

9   on immigrant groups mexico

turst thru
205

64 migration
65 patterns to LA —
   most immigrants
   spend 5-10
   years elsewhere
   before coming
   to LA

√ 11 m category of remaining in nat'l A + context ①   Ⓓ

√ 13-4

38 + Turner ① in context of immig Ⓓ/models

249-50 * pre
  activism 30's/40's
  + immig vs native
  identity

123-4

fn mann
   26
   27

271

* √ 200 + of Baich novel   home—
                          ruing   [+ fn 215
199—   wrote]
201 m   home

BECOMING
MEXICAN
AMERICAN

BECOMING
MEXICAN
AMERICAN

Ethnicity, Culture, and Identity in Chicano Los Angeles, 1900–1945

George J. Sánchez

New York   Oxford
OXFORD UNIVERSITY PRESS
1993

Oxford University Press

Oxford   New York   Toronto
Delhi   Bombay   Calcutta   Madras   Karachi
Kuala Lumpur   Singapore   Hong Kong   Tokyo
Nairobi   Dar es Salaam   Cape Town
Melbourne   Auckland   Madrid

and associated companies in
Berlin   Ibadan

Copyright © 1993 by George J. Sánchez

Published by Oxford University Press, Inc.
200 Madison Avenue, New York, New York 10016

Oxford is a registered trademark of Oxford University Press

Library of Congress Cataloging-in-Publication Data
Sánchez, George J.
Becoming Mexican American : ethnicity, culture, and identity in
Chicano Los Angeles, 1900–1945 / George J. Sánchez.
p.   cm.   Includes bibliographical references and index.
ISBN 0-19-506990-0
1. Mexican Americans—California—Los Angeles—Social conditions.
2. Mexican Americans—California—Los Angeles—Ethnic identity.
3. Los Angeles (Calif.)—Social conditions.
4. Ethnicity— California—Los Angeles.
I. Title.
F869.L89M57   1993
305.868'72079494'09041—dc20
92-45570

2 4 6 8 9 7 5 3 1

Printed in the United States of America
on acid-free paper

This book is dedicated to my loving and supportive parents,
JORGE R. SÁNCHEZ AND NINFA SÁNCHEZ
A cruzar la frontera, han criado una nueva generación.

Acknowledgments

I began this study in order to get to know my parents. Immigrants to the United States from Mexico in the 1950s, they lived in a world I always felt close to, but never fully understood. Born in the United States, I never experienced the personal challenges of moving from one country to another, changing one's citizenship, and committing one's life to an adopted nation. As I grew older, I came to realize that my own life as their son had been clearly and intentionally demarcated by the courageous decisions that my parents made years before. Their lives have been full of sacrifices made on behalf of their children. Both my father and mother provided loving guidance to each of us and demonstrated how to pass on values from one generation to another. In addition, they nurtured in my sister, brother, and myself a fierce independence of thought and action which, while certainly making parenting more demanding, also enabled us to live richer lives. For providing me with the moorings needed to live my present life, this study is dedicated to them.

Writing this book has been a continuous process of crossing intellectual and personal borders. I had the privilege of beginning it under mentors who were as committed to me as to the completion of the project and who made this journey ever more challenging. Albert Camarillo guided my entire graduate career, nurturing this project in its infancy with unbridled compassion and decency. He is a role model for me in every sense of the term. Estelle Freedman consistently raised my level of analysis by challenging me to ask difficult questions. She taught me to communicate my answers in direct and forceful ways. Other current and former faculty at Stanford University, including Carl Degler, Don Fehrenbacher, and Michael Kazin, were critical to the development of my abilities at crucial times in my graduate career.

The difficult transition of turning the dissertation into a book was aided immeasurably by several historians who shared their insights, enthusiasm, and time, reading various versions of the manuscript and offer-

ing advice and critique. David Gutierrez pored over several versions and
generously shared his wisdom and wit with me throughout the entire
process. Camille Guerin-Gonzales became a significant intellectual and
personal confidante, and our friendship has grown as we shared thoughts
about the exciting new directions in Chicano history. Bruce Schulman
scrutinized an early version of the manuscript and helped me rethink
several significant issues. William Chafe suggested ways I could connect
this work to larger issues in American history and demonstrated enormous enthusiasm and support. Robin D. G. Kelley halted his busy
schedule to critique a difficult chapter in record time, and his contribution and encouragement made the book significantly stronger. Patricia
Nelson Limerick, Vicki Ruiz, George Lipsitz, Pedro Castillo, Valerie
Matsumoto, and José Moya also read parts of the manuscript at different
stages and prodded me to sharpen my vision and clarify my words.

From the beginning, I was committed to crossing the boundaries
separating the academic disciplines, a decision that took me out of the
safe confines of strict historical inquiry and into anthropology, sociology,
ethnomusicology, cultural studies, and even literary criticism. Like all
temporary sojourners, I was careful in these new lands. I have tried to be
respectful of the knowledge I gained. All along the way, I was aided by
other scholars who helped me negotiate difficult intellectual terrain. In
particular, my colleagues in Chicana and Chicano Studies at UCLA offered helpful criticism and intellectual sustenance. Vilma Ortiz will always remain a special friend because of the academic and administrative
battles we waged together and the intellectual community we attempted
to build. Sonia Saldivar-Hull, Rosalinda Fregoso, Clara Lomas, José
Monleon, Eddie Telles, Ray Rocco, Richard Chabran, David Lopez,
Danny Solorzano, Kris Gutierrez, Edit Villarreal, Raymund Paredes,
Guillermo Hernandez, and Steve Loza each helped me think about my
work from a different perspective, and their insights are embedded in
this effort, whether they know it or not.

I have also been gratified by the profound way my work has been
influenced by daily interaction with graduate students who have shared
their perspectives with me in formal and informal settings. Ernesto Chá-
vez deserves special mention for introducing me to various uncharted
corners of Los Angeles and UCLA, and for taking time from his own
work to share sources and stories. Margo McBane was a colleague from
the start, and I learned as much from her as I taught her in our interaction. Dana Leventhal and Paula Scott each helped me complete the dissertation, and John Nieto-Phillips, Jim Pearson and Bob Myers, Jr., provided vital assistance in finishing the book. The community created by
the aforementioned students and by Linda Maram, Jeffrey Rangel, Omar
Valerio Jimenez, Irma Valdivia, Miroslava Chavez, Catherine Pet, Steven
Reyes, Liliana Urrutia, Lorena Chambers, Jaime Cardenas, Sheila Gar-
dette, Anthony Macias, Maria Elena Fernandez, Monica Russel y Rodriguez, and Susan Terri Gomez helped sustain me on a personal and

intellectual basis, and each of these individuals directly contributed to the project and to my learning by sharing a document, an insight, or a criticism at important junctures in my thinking. Their own work will surely transform the writing of history and Chicano Studies in the future. Indeed, they are the future of our profession and the intellectual heirs to my peculiar fascination with culture, ethnicity and identity.

I am proud to have been a direct beneficiary of affirmative action admissions policies and targeted minority and need-based scholarships throughout my academic career—at Harvard University, Stanford University, and as a professor at UCLA. To all the skeptics regarding the efficacy of these policies, I hope that this book will offer evidence that such programs do work. I would not have been able to make my contribution to research, writing, and teaching without the opportunities provided by affirmative action programs and those individuals that coordinate them.

nice move

The Danforth Foundation, through its Dorothy Danforth Compton Fellowship program, allowed me to begin my graduate work at Stanford and provided me with a crucial year of support at the dissertation level. The Ford Foundation, through its Fellowship Program for Minorities, funded much of the dissertation stage, then awarded me a postdoctoral fellowship to allow me to transform the manuscript into a book. I especially want to thank Dr. John B. Ervin, former Vice President of the Danforth Foundation, Sheila Biddle, Program Officer of the Ford Foundation, and Barbara Montgomery of the National Research Council for setting up conferences and networks which have proven crucial to my own intellectual and emotional rejuvenation at critical stages in my career. Edward Escobar and Noé P. Lozano at Stanford University, and Harriet Moss at UCLA helped administer these programs at the university level, and their efforts are richly appreciated.

As this project evolved, I crossed geographical borders throughout the United States and Mexico to conduct primary research. Along the way, helpful archivists and librarians made my searches more fruitful and less time-consuming. The staff at the southern California regional branch of the National Archives at Laguna Niguel, particularly Diane Nixon, Suzanne Dewberry, and Fred Klose, helped make my many months there productive and enjoyable. Roberto Trujillo of Stanford University, Anne Caiger of the UCLA Department of Special Collections, and Richard Chabran of the UCLA Chicano Studies Research Library went out of their way to track down obscure sources which have enriched this study. Licenciada Patricia Galeana de Valadez of the Archivo Historico de la Secretaria de Relaciones Exteriores helped me navigate through their extensive collection in Mexico City, as did Bob Lazar of the ILGWU Archives in New York and Father Francis J. Weber of the Los Angeles Chancery Archives. I could not have put together the photographs for this book without the assistance of Carolyn Kozo Cole of the Los Angeles Public Library and Dace Taube of the USC Regional Cultural

History Collection. To each of these librarians and archivists, a heartfelt "thank you."

Several funding agencies and organizations paid for travel and research; others provided relatively quiet time to write and reflect away from the demands of teaching. The Program on Mexico at UCLA supported a crucial trip to Mexico City's Archivo General de la Nacion and the Archivo Historico de la Secretaria de Relaciones Exteriores, while the UCLA Institute of American Cultures funded much-needed research assistance. Other support was provided by the UCLA Center for the Study of Women and the UCLA Academic Senate. Two Faculty Career Development Awards facilitated additional leave time. Judith Glass and Jessica Ross each helped keep me comfortable and sheltered during a year of intense writing at the Institute for the Study of Women and Men at the University of Southern California. The UCLA Chicano Studies Research Center generously provided office space and an intellectual community, while facilitating grants from the Inter-University Program in Latino Research and the University of California which allowed me to finish the book.

Many individuals opened their homes to me during my research. They provided not only shelter and sustenance, but intellectual stimulation as well. I want to thank Ralph and May Ziskin, Rosalind Markell, Matt Mallow and Ellen Chesler, Arturo Vargas, Penny Russell, Julie Reuben, Helen Lara-Cea, Camille Guerin-Gonzales, David Gutierrez, and my parents, Jorge and Ninfa Sánchez, for their generosity and companionship. I am especially indebted to Beatriz R. Palomarez, not only for welcoming me into her home one afternoon but also for sharing with me her recollections, mementos, and writings about her early life and that of her father, Zeferino Ramírez.

Sheldon Meyer of Oxford University Press has been the most gracious and supportive editor an author could desire. I have enjoyed our conversations immensely over the years. Stephanie Sakson was the skillful copy editor who strengthened the manuscript and patiently awaited my final revisions. Karen Wolny kept in constant contact with me in the difficult final stages of putting the book together. She especially helped me organize the illustrations for the final copy.

My family has been supportive of this effort for so many years that it is difficult to find the right words to thank them. Along with my parents, my sister Martha Dominguez and my brother Carlos Sánchez have always given me encouragement and expressed their pride in me. My mother-in-law, Rosalind Markell, contributed unendingly and cheerfully to child care and household management. She even graciously vacated her house several times for weekend retreats. While working on this book, I watched Alison Morantz grow up to become a young scholar herself. I am proud of her accomplishments and the friendship we have nurtured together. Jessica Morantz has crossed her own borders during the same period, and I wish her the very best life has to offer as she confidently makes her way in the world. Both Alison and Jessica put up with much change when I entered their lives. I thank them for their patience, compassion, and good humor. Adam Max Sánchez is almost at the age when he can read this book on his own. I hope he takes up the challenge in a few years. I often saw my hopes for the future reflected back to me in his eyes; now that the book is published we are freer to build that future together.

I could never have completed this project without the strength, talents, and efforts of my wife, Regina Morantz-Sánchez. No one has had a more profound intellectual impact on this book. I learned tremendously from our shared conversations on history and ethnicity over the years. She not only provided a keen editorial eye throughout, but she served as a constant intellectual companion, often taking time from her own work to listen and respond to my ideas. Her insights forced me to clarify my thinking and improve my writing. Moreover, she never let me underestimate the significance of this project. In addition, she gave me the freedom to struggle with my ideas by going beyond the call of duty, not always peacefully, but always with love. I doubt that I can ever repay her fully for her sacrifices, but I will certainly spend my life trying. I only hope that I can become the partner and soulmate that she has already become for me.

Los Angeles G.J.S.
March 1993

Contents

PART FOUR. AMBIVALENT AMERICANISM

List of Illustrations

List of Tables

BECOMING
MEXICAN
AMERICAN

Introduction

When Zeferino Ramírez stood up in front of his fellow residents of Belvedere, an unincorporated area east of the city of Los Angeles, on June 12, 1927, he had long since been recognized as a leader among the Mexican immigrants there.[1] That Sunday night the meeting was to focus on a community crisis, and it was no surprise that Ramírez had been asked to preside over the discussion. At issue was a plan to incorporate the area into a full-fledged municipality, a move that would certainly make Mexican settlement more difficult in the district. At least three plans for incorporation had been submitted to Los Angeles County officials within the year by real estate and manufacturing interests in Belvedere. Their strategy was to increase the taxes of local residents to pay for city services, thereby forcing the largely working-class community to sell their property in a depressed market. The area in dispute could thereby be resold to middle-class Anglo Americans, forcing up the estate values in neighboring communities and making a tidy profit for real estate companies.[2]

Ramírez had seen this kind of discrimination before. He had come to Los Angeles during the decade of the Mexican Revolution and was unable to secure employment because of his nationality, even though he had gained valuable skills in the mines of northern Mexico and Arizona. Unable to find work, he was forced to live in an insect-infested room with fellow Mexicans. Returning to Mexico briefly to bring over his wife and children, he finally managed to save enough to return to Los Angeles and buy a small home in Belvedere. For seven years he worked as a highway laborer. In the mid-1920s, he started a business of his own, opening an undertaking establishment after serving as an apprentice in an Anglo-run mortuary. As one of Belvedere's first Mexican businessmen, he quickly earned the respect of his neighbors. Always taking pride in his Mexican nationality, he spearheaded efforts to establish a Mexican school in Belvedere.[3]

Yet this story was more complicated than one of ethnic leadership

3

against conniving Americans out to deprive Mexican residents of their land. The tensions endured by many Mexican immigrant families were exhibited in Ramírez's as well. Part of the reason he had been so active in establishing a school was that he had begun to worry when his own sons spoke English at home. Although his daughters had attended an American school and were now music teachers, he continued to prohibit them from going out alone, attending dances, or viewing American movies. His oldest daughter, however, controlled the business finances and "didn't leave much money within his reach." Ramírez himself had undergone a Protestant conversion after reaching Los Angeles, and he served as a lay preacher of a Methodist church on Brooklyn Avenue. The family's clothes and meals were largely American-style, yet they tended to decorate their home with both pictures of American subjects and patriotic Mexican portraits. Later, Ramírez would visit Mexico City with the intent of moving his business, only to conclude that "everything there [was] still very backward and very disorganized."[4]

Despite these obvious signs of cultural adaptation, when some Anglo American supporters at the meeting counseled Belvedere's Mexicans to apply for naturalization, Ramírez was among the first to balk at the suggestion. These officials, among them municipal judges in the Belvedere area, argued that those Mexican immigrants without first papers would be unable to vote against incorporation, if the issue ever appeared in an election. But Ramírez joined *La Opinión* and the Mexican consulate representative in warning against this advice. To him, the negation of Mexican citizenship would have been a larger crime than that being perpetrated on the residents of Belvedere.[5]

How does one make sense of the contradictory aspects of this story? What does Zeferino Ramírez's life, and his reaction to the prospect of changing his citizenship, tell us about the cultural adaptation of Mexican immigrants to the United States in the early twentieth century? Historians and social scientists have certainly struggled with the subject of the cultural adjustment of immigrants before. More than thirty years ago, David M. Potter noted the significance of issues surrounding immigrant acculturation in an essay about the American character. He argued that because virtually all Americans descended from immigrants, they were compulsively preoccupied with the question of national identity. Americans feel deprived of an organic connection to the past, especially when confronted with their diverse religious, linguistic, and political heritage. The result has been an obsessive fixation on the elusive tenets of "Americanism."[6]

Oscar Handlin had anticipated Potter's assessment of the impact of migration and acculturation on the American character when he described in *The Uprooted* his gradual realization that writing a history of immigration was, in fact, writing about all of American history. Handlin's account emphasized the loss of European peasant roots, the arduous journey across the Atlantic, and the painful, yet to him inevitable, process of absorption into American society. Yet, as effective as he was in demonstrating the importance of the immigrant experience, Handlin mistakenly assumed that every immigrant shared the same process of cultural adaptation. Indeed, his thesis regarding the disintegration of peasant cultures in the New World was extended even beyond his European subjects to include African Americans and Latinos.[7]

In the last three decades, revisionist historians have argued that Handlin's tendency to collapse all groups' migration experiences into one story belittles the diversity of such events and distorts history. Moreover, they have taken direct aim at the notion that much was lost in the transition to life in the United States. They have pointed instead to the retention of Old World culture by many different groups of Americans and to its persistence and resiliency.[8] Unlike Handlin and his contemporaries, the revisionist historians, Howard Rabinowitz has observed:

> view such [cultural] persistence quite favorably and treat it as something that should have happened. In other words, the new social history not only finds evidence of strong ethnic and racial identification, but usually celebrates it, as well, though often expressing dismay at the frequent examples of inequality among groups.[9]

Groups of European immigrants that Handlin had depicted as thoroughly assimilated were freshly revisioned as retaining important ties to their Old World culture even after living for years in the United States. As one reviewer of this literature has noted, "since the 1960s . . . ethnicity has largely been transformed from a heathenish liability into a sacred asset, from a trait to be overcome in a conversion and rebirth experience to a very desirable identity feature to be achieved through yet another regeneration."[10] This "new ethnicity" paradigm, in fact, came to dominate historical writing about immigration in the 1970s and 1980s. So as to emphasize the continuity of culture, John Bodnar called his 1985 synthesis of the recent literature on immigration to the United States *The Transplanted* so as to leave no doubt that most immigrants had maintained their cultural roots, even when planted in different soil.[11]

Historians of the Mexican American experience entered this debate at the height of the revisionist reinterpretation. Highly influenced by the struggles for national liberation in the Third World and the civil rights movements of both African Americans and Chicanos in the United States, particularly the calls for cultural nationalism, the Chicano historians who created the field in the late 1960s and early 1970s focused their attention on the retention of Mexican culture throughout the history of the American Southwest. Although Anglo American social scientists had earlier identified and bemoaned the retention of Mexican traits among the immigrant population, Chicano scholars now took the residues of Mexican culture as positive evidence of the existence of a distinct people who could, and often did, identify themselves in opposition to the majority Anglo society.[12]

Typical of this interpretation were the writings of Juan Gómez-Quiñones, considered the dean of Chicano history in the early 1970s. In particular, Gómez-Quiñones offered his 1977 essay, "On Culture," as "a call to debate on culture academically and politically."[13] Taking an approach largely derived from Amilear Cabral and Antonio Gramsci, Gómez-Quiñones focused his attention on cultural resistance as the "negation of assimilation," downplaying any cultural act that was not "an act of struggle" against dominant and dominating values. For him, as for Handlin, there existed two cultural poles: "Mexicano" (or Old World ethnic for Handlin) "versus Anglo United States." Chicanos stood as a subculture between these two poles where "culture and identity is a safehouse and thus provides strategic and tactical elasticity vis-a-vis the dominant society." In this admittedly polemical essay, Gómez-Quiñones laid the basis for others' writings on Chicano culture when he argued that "to acculturate is not merely to exercise a cultural preference but to go to the other side."[14]

Ironically, though positioning himself in opposition to assimilation, Gómez-Quiñones in the end accepted the terms of the cultural debate set out by Handlin and others, terms that held immigrant assimilation to be inevitable. In the view of both historians, cultural adaptation occurred in a linear fashion with strict distinctions made between "traditional" and "modern" cultures.[15] Both Mexican and American cultures were depicted largely as static, impermeable, and always in opposition, with individuals constantly pushed or pulled in one direction or the other. Moreover, Chicano culture was belittled on the one hand as primarily a way-station on the inevitable path toward assimilation into the American mainstream or on the other hand as representing the "U-turn" on the return road to a pronounced Mexican nationalist stance. Largely accepting this model of bipolar cultural opposites, most Chicano historians set out to document the cultural persistence of things Mexican in new American surroundings.

Not surprisingly, most historical writings on Chicanos that were influenced by this nationalist position emphasized cultural continuity in almost all geographical settings. Cultural retention was framed not necessarily as a conscious act of resistance, but rather as a condition of the racial and class separation inflicted on the Mexican population. For Richard Griswold del Castillo, for example, the barrio, though circumscribing social and economic possibilities for nineteenth-century Mexican residents of Los Angeles, made cultural survival possible. "Proximity of residence reinforced the language, religion, and social habits of Chicanos," observed Griswold del Castillo in his account, "and thus insured the continuation of their distinctive culture." Albert Camarillo added that for the twentieth-century residents of Santa Barbara, "facilitating continued contact among Mexicanos, the [mutualista] organizations helped perpetuate Mexican culture, language, and cohesiveness in an otherwise foreign society."[16]

Chicano historians have shared much with scholars working in the fields of African American, Native American, and Asian American history. These historians too have emphasized the importance of cultural persistence, primarily by examining how minority groups faced conflict within the host culture. Investigations of antebellum black life, for example, have been interested less in the intrusion of white culture into slave quarters than in the richness of a distinct African American culture "from sundown to sunup."[17] Asian American historians have increasingly turned away from descriptions of their past as simply that of immigrants to interpretations which stress "the development of common cultural and psychological experiences in America that were rooted in, but not limited to, the cultures of the original ethnic communities."[18] Central to each of these discussions has been the argument that the histories of racial minorities in the United States, and therefore the possibilities of their assimilation into American life, are qualitatively different from the experiences of European immigrants.[19]

Unfortunately, few historians of Chicanos writing in the 1970s and early 1980s took Gómez-Quiñones's "call to debate on culture" to heart. Instead of exploring and debating the issues he raised about the symbolic and transformative significance of culture, Chicano historians focused their work on developing important alternative theoretical approaches— such as internal colonialism, the process of barrioization, or the dual labor market theory—to explain the constraints on assimilation. All of these theories emphasized the impact of race and class. For example, the concept of a dual or segmented labor market has been used to explain the disadvantaged position of Mexican workers in the Southwest. This structural approach argues that while those in the primary sector enjoy relatively decent wages, labor conditions, job security, and union membership, those in the secondary sector, including racial minority groups such as Mexicans, are relegated to low-paying, "dead-end" jobs.[20] Such work has yielded important insights. These include the understanding that Chicanos' uneven participation in American society cannot be explained simply by citing racial intolerance. Hence, the complex intersection of racial and class oppression in the American Southwest became the centerpiece of scholarship in Chicano history and Chicano studies in the 1970s and 1980s.[21] For these scholars, culture was simply a reflection of socio-economic conditions.

Though noting the severity of racial and class discrimination, some Chicano historians have also highlighted cultural change in the Mexican origin population, particularly among the children of immigrants and within the middle class. But these changes continue to be placed within a bipolar model of opposing cultures. Mario García, for example, has described a variety of experiences in El Paso, Texas, divided by age, nativity, and generation, which together resulted in "a Mexican border culture, neither completely Mexican nor American, but one revealing contrasting attractions and pressures between both cultures."[22] Arnoldo De

León also confirmed the diversity of cultural positions among nineteenth
century Tejanos all along the Rio Grande, recognizing that "some became *agringados,* some bicultural; others stayed Mexican."[23]

While both authors recognize that change did occur on the northern
side of the border, Mexican American culture remained a tenuous site of
cultural exchange, always a prelude to the attractions of a "purely" Mexican or "purely" American stance. Even in their descriptions of the inter-
mediate subculture, as Roger Rouse has observed, they noted "an or-
dered synthesis of old and new" which treated contradictions as
temporary features that were certain to disappear with the passage of
time and generations.[24] As anthropologist Renato Rosaldo has pointed
out, this "classic concept of culture seeks out the 'Mexican' or the 'Anglo-
American,' and grants little space to the mundane disturbances that so
often erupt during border crossings."[25]

Recently, however, new perspectives offered by scholars working in
the field of cultural studies force us to reexamine such static assumptions.
Across a variety of disciplines, the very language used to describe the
particularistic experiences of individuals—culture, ethnicity, identity,
gender, and race—has been challenged.[26] In particular, any notion that
individuals have occupied one undifferentiated cultural position—such as
"Mexican," "American," or "Chicano"—has been abandoned in favor of
the possibility of multiple identities and contradictory positions. Moreover, the strictly nationalist position of early Chicano historians has been
questioned, not only by cultural theorists exploring the complicated historical allegiances in the ethnic past but also by Chicana feminists who
claim that a single standard of ethnicity largely left women out of historical constructions.[27]

"Culture," of course, has been one of the most hotly debated terms
across disciplines throughout the twentieth century. If one looks at the
changing language in anthropology, for example, a discipline that has
claimed "culture" as its primary focus, one can see how volatile definitions themselves can become.[28] Once hoping to understand the "most
complex whole," anthropologists have lately come to recognize the myriad contradictions inherent in cultural systems. Recent work by ethnographers who have drawn from textual analysis and cultural studies perspectives has gone further in questioning the ability of anthropologists to lay
claim to being "scientists of culture":

> Cultures are not scientific "objects" (assuming such things exist, even in the
> natural sciences). Culture, and our views of "it," are produced historically,
> and are actively contested. There is no whole picture that can be "filled in,"
> since the perception and filling of a gap lead to the awareness of other gaps.
> . . . If "culture" is not an object to be described, neither is it a unified
> corpus of symbols and meanings that can be definitively interpreted. Culture
> is contested, temporal, and emergent. Representation and explanation—both
> by insiders and outsiders—is implicated in this emergence.[29]

It is not surprising that issues in Chicano culture should have such resonance for scholars reexamining our understanding of how culture and ethnicity work in the post-industrial age. For radical scholars, "culture is not what preexists this historic overlay [of capitalism], but is how humans whose lives are structurally defined by institutionally enacted capitalist principles respond to them in their everyday life and experience."[30] Since Mexican migrants move between two countries—one highly industrialized and the other severely impoverished—they have been among the first to experience what some have called the "postmodern condition." At least one anthropologist has recently suggested that we look toward the life experiences of these transnationals to understand our "confusing world, a world of crisscrossed economies, intersecting systems of meaning, and fragmented identities."[31]

Recently, Chicano scholars in art, literature, and anthropology have begun to develop notions of "trans-creation" to describe the process of cultural formation among Chicanos and other Latinos in the United States.[32] The movement between Mexican and American cultures is not so much a world of confusion, but rather a place of opportunity and innovation. In Los Angeles, living in this cultural "borderlands" can also lend itself to adaptations drawn from African American or other ethnic peoples, depending on the time period, the local community, and the level and nature of contact. As Gloria Anzaldúa has described this "mestiza" worldview, what often develops is "a tolerance of contradictions, a tolerance for ambiguity."[33] Mexicans, long accustomed to cultural blending and creation, continue this custom in the United States, now incorporating aspects of the "others" they find in a multicultural setting like Los Angeles. To be Chicano, in effect, is to be betwixt and between.

Yet, is there still a "there" there? For earlier scholars, the comforting presence of Mexico was all that was needed in order to project a unified cultural concept to which Chicanos could turn. Yet "Mexico," maybe even more so than other nations, was a national community that had to be "imagined" to exist, particularly given its racial and regional diversity. Not only was culture never static in Mexico, nor U.S. influence ever far removed in shaping its contours, but the construction of a Mexican national identity was never more ferociously pursued than in the aftermath of the Mexican Revolution—at the very moment thousands of Mexicans were making their way north.[34]

Thus, my own study necessarily begins with an examination of the rural villages and burgeoning towns of Mexico. Mexico during the early twentieth century was undergoing fundamental socioeconomic upheaval. Partly because of this tumult, hundreds of thousands of Mexicans crossed the border into the United States, usually seeking temporary residence and better wages in order to help their families survive through difficult times. Yet the very meaning of crossing the border was undergoing a transformation during this period; moving north signified a momentous occasion. Though back-and-forth migration continued, increasingly durable settlement north of the border was a result of tightening immigration restriction. Los Angeles attracted an ever larger number of these migratory sojourners, drawn to the city because of its employment opportunities and vibrant Mexican community.

In the United States, new "traditions" had to be invented and older customs discarded or radically transformed at the same time that Mexicans in Mexico were creating "traditions" to cement national identity. The early twentieth century was certainly a period tailor-made for the invention of traditions on both sides of the border. As Eric Hobsbawm has noted:

> Inventing traditions, it is assumed here, is essentially a process of formalization and ritualization, characterized by reference to the past, if only by imposing repetition. . . . There is probably no time and place with which historians are concerned which has not seen the "invention" of tradition in this sense. However, we should expect it to occur more frequently when a rapid transformation of society weakens or destroys the social patterns for which "old" traditions had been designed, producing new ones to which they were not applicable, or when such old traditions and their institutional carriers and promulgators no longer prove sufficiently adaptable and flexible, or are otherwise eliminated: in short, when there are sufficiently large and rapid changes on the demand or the supply side.[35]

Mexicans in the United States not only had to draw for their new traditions upon their memories of a Mexico now irreversibly transformed, but they also had to make sense of a new world north of the border that was undergoing rapid transformation. Nowhere was this more evident than in Los Angeles, where demographic upheaval meant that most residents were newcomers little versed in the culture of the region they now inhabited. Political and social power, however, was concentrated in the hands of a small group of Anglo American newcomers. This power enabled them to mold dominant "traditions" onto the cultural landscape of California.[36] The growth of the Chicano community in Los Angeles created a "problem" for Anglo American residents, one which resulted in public efforts to alter cultural loyalties among Mexican immigrants. American officials launched programs to teach these newcomers idealized versions of American practices, customs, and values. The Mexican government, for its part, worked through its consulate office to instill loyalty to Mexico, trying to persuade citizens to return to their native country. Neither of these efforts had their intended effect, but both ironically served to stimulate the process of self-recognition as ethnic Americans among the immigrant population.

These efforts reflect attempts by sovereign states to control the ethnic identity of a people in turbulent social and economic times in order to bring cohesion to their respective countries by implementing what one scholar as called the "fictive ethnicity" of the nation-state. Anglo Americans in Los Angeles, as cultural newcomers themselves, were anxious to impose coherence on a region full of diversity. They acted as surrogates for the state apparatus in an attempt to minimize ethnic difference and tie these "other" recent arrivals to American society. Their goal was to alter outmoded values and customs without necessarily encouraging social mobility. Mexican government representatives, along with members of the expatriate community anxious to promote nationalist sentiment, engaged in a similar campaign to keep emigrants' loyalties linked to an emerging Mexican nation. Perhaps even more accurately, they hoped to create "Mexicans" out of former mestizo villagers.[37] In both cases, Mexican immigrants in Los Angeles appeared to these competing nationalists as "clean slates," or culturally empty vessels. Renato Rosaldo has noted the way ethnographers, accustomed to viewing cultures as self-enclosed, have similarly treated immigrants:

> [T]he borders between nations, classes, and cultures were endowed with a curious kind of hybrid invisibility. They seemed to be a little of this and a little of that, and not quite one or the other. Movements between such seemingly fixed entities as nations or social classes were relegated to the analytical dustbin of cultural invisibility. Immigrants and socially mobile individuals appeared culturally invisible because they were no longer what they once were and not yet what they could become.[38]

But Mexican immigrants played their own part in this drama. Constrained by their lack of economic and political stature, they drew strength from the networks of family members and fellow countrymen who lived nearby. Through the daily struggle to survive in an oftentimes hostile environment, these newcomers constructed a world for themselves, shaped both by their memories of their past lives and by the reality of their present situation. During the 1920s, many Mexican immigrants gradually changed their orientation from that of temporary sojourner to permanent resident. Much of this process occurred within the family context, as individual migrants married and raised families in Los Angeles. Cultural adaptations marked the transition to a Mexican American lifestyle. Catholic religious practice, for example, increasingly narrowed to the province of women, and became less a community function and more a set of rituals performed at home. The secular entertainment industry of Los Angeles reshaped Mexican musical traditions, as immigrants made Los Angeles a center for Chicano cultural life in the United States. The search for stability encouraged Mexican immigrants to settle in particular barrios and assume new roles within their families, at work, and as American consumers.

Ethnicity, therefore, was not a fixed set of customs surviving from life in Mexico, but rather a collective identity that emerged from daily experience in the United States. As such, ethnicity arose not only from interaction with fellow Mexicans and Mexican Americans but also through dialogue and debate with the larger cultural world encountered in Los Angeles. Whether accommodation, resistance, or indifference marked an individual's stance toward American culture, everyone reacted to living in the United States. For those who chose to stay, their cultural adaptations would have lifelong implications. For over time, as Mexican immigrants acclimated themselves to life north of the border, they did not remain Mexicans simply living in the United States, they became Mexican Americans. They assumed a new ethnic identity, a cultural orientation which accepted the possibilities of a future in their new land. As anthropologist Michael M. J. Fischer observes about ethnicity:

> Ethnicity is not something that is simply passed on from generation to generation, taught and learned; it is something dynamic, often unsuccessfully repressed or avoided. It can be potent even when not consciously taught; it is something that institutionalized teaching easily makes chauvinist, sterile, and superficial, something that emerges in full—often liberating—flower only through struggle.[39]

The struggle which forged a Mexican American identity was powerfully rooted in the decade of the 1930s. The onset of the Great Depression forced many Chicano residents to reconsider their decision to remain in the United States. Moreover, the deportation and repatriation campaigns launched against Mexicans in Los Angeles profoundly disrupted the cultural centeredness of the community. Los Angeles lost one-third of its Mexican residents, and those who remained were made keenly aware of the fragility of their social position. The sons and daughters of the immigrant generation, entering adulthood during the late 1930s and early 1940s, became acutely sensitive to America's lack of tolerance. Hence, many became more active in American unions and struggles for civil rights. They found themselves profoundly affected by a generation of labor leaders, both Chicano and Anglo, who dedicated their lives to the fight for social equality in the era of the New Deal.

The repatriation campaign provided a symbolic break with the past for those who remained, while participation in American unions and other struggles for civic equality in the late 1930s and early 1940s created the context for a new identity. Rather than culture serving as a substitute for politics, it became a way to enter the political arena. As George Lipsitz has observed, culture can be a "means of reshaping individual and collective practice for specified interests . . . as long as individuals perceive their interests as unfilled, culture retains an oppositional potential."[40] Ironically, it was not the search for Mexican nationalism which engendered political radicalism for large numbers of Mexicans and Mexican Americans in the 1930s, but the forging of a new identity as ethnic Americans. "Cultural identity," as Stuart Hall has observed:

> is a matter of "becoming" as well as of "being." It belongs to the future as much as to the past. It is not something which already exists, transcending place, time, history and culture. Cultural identities come from somewhere, have histories. But, like everything which is historical, they undergo constant

transformation. Far from being eternally fixed in some essentialized past, they are subject to the continuous "play" of history, culture and power. Far from being grounded in a mere "recovery" of the past, which is waiting to be found, and which, when found, will secure our sense of ourselves for eternity, identities are the names we give to the different ways we are positioned by, and position ourselves within, the narrative of the past.[41]

This new cultural identity was forged within the context of a hostile, racist environment which sought to deny Mexican Americans a claim to being "Americans." The so-called "Zoot Suit Riots" in Los Angeles in 1943 were only the most outward manifestation of the racism they experienced. As a result, parents and children alike forged an ambivalent Americanism—one distinguished by a duality in cultural practices and a marked adaptability in the face of discrimination. Central to my thesis is the argument that Mexican American cultural adaptation occurred without substantial economic mobility, particularly since it was rooted in the context of the Great Depression. This book is in part a study of how cultural change can take place *without* social mobility. Previous studies of Chicanos in Los Angeles have been helpful in understanding the forces which militated against upward mobility among Mexican Americans.[42] But, as we shall see, these earlier works have neglected to tell the fascinating story of cultural invention which must also be included in any portrait of working class life in these years.

My own study of Mexican immigrants to Los Angeles between 1900 and 1945 focuses on the related questions of cultural adaptation and ethnic identity, utilizing new perspectives from other disciplines and from cultural studies. I argue that the emphasis in Chicano history on bipolar models that have stressed either cultural continuity or gradual acculturation has short-circuited a full exploration of the complex process of cultural adaptation. This problem is particularly regrettable when one realizes that Mexicans have been among the most numerous and significant of immigrant groups in the United States during the twentieth century. They number well over half of the total current Latino population of 30 million—a population destined to become America's largest minority group within the next thirty years.

Los Angeles was selected as the site of my research for an obvious reason: by 1928 it had the largest Mexican population of any city in the United States. Additionally, understanding California's role in twentieth-century American culture is crucial. Latinos have had and will continue to have a profound effect on that culture, especially in Los Angeles. Currently, Los Angeles International Airport welcomes more immigrants than any other port of entry in American history. Public mythology, however, still reveres Ellis Island and the Statue of Liberty and looks toward Europe. Historical writing on immigration in the United States surely suffers from this severe regional imbalance; most studies still focus on the Northeast and selected cities of the Old Northwest. The fact that the American Southwest has been the locus of one of the most profound and complex interactions between variant cultures in American history is repeatedly overlooked.[43] Obviously, mainstream immigrant history has much to learn from historians whose focus is the American Southwest.

The narrative that follows, therefore, is my attempt to unravel the many layers of cultural adjustment among Mexican immigrants to Los Angeles in the early twentieth century. As I attempt to tell this important story, I am humbled by the words of George Lipsitz, who reminds historians of our limitations:

> We need to understand the past in order to make informed moral choices about the present, to connect our personal histories to a larger collective history. But that larger history can never be fully comprehended; the complexities and pluralities of the past always resist definitive evaluation and summary. Reconstructing the infinitely complex experiences of the past through the paltry bits of evidence about it available to historians inevitably renders some aspects of the past as incommunicable.[44]

Zeferino Ramírez's refusal to give up his Mexican nationality when to do so might have aided him in facing an obvious external threat has led me to ask a multitude of questions. In answering them, I hope to offer some insights into the complex process of cultural adaptation, while gaining a better understanding of the difficult choices made by those who came before us.

CROSSING BORDERS

"Cultures" do not hold still for their portraits.

—James Clifford, 1986[1]

The psychology of the average Mexican alien unskilled worker from
Mexico is that when he enters in any manner into the United States
that he is only upon a visit to an unknown portion of his own country. He is independent and does not consider he is an immigrant
alien, but rather in what is termed the United States by right of
birth and possession, the country of his forebearers, the territory
acquired after the war of 1846 and the territory subsequently ceded
in 1856 known as the Gasden Purchase. To him there is no real or
imaginary line.

—Report to the U.S. Secretary of Labor, 1922[2]

It must have been about ten at night,
the train began to whistle.
I heard my mother say, "There comes that ungrateful train
that is going to take my son."

"Goodbye my beloved mother, give me your blessings.
I am going to a foreign land, where there is no revolution."

Run, run little train, let's leave the station.
I don't want to see my mother cry for her beloved son,
for the son of her heart.

Finally the bell rang, the train whistled twice.
"Don't cry my buddies, for you'll make me cry as well."

Right away we passed Jalisco, my, how fast the train ran.
La Piedad, then Irapuato, Silado, then La Chona,
and Aguas Calientes as well.

We arrived at Juarez at last, there I ran into trouble.
"Where are you going, where do you come from?
How much money do you have to enter this nation?"

"Gentlemen, I have money so that I can emigrate."
"Your money isn't worth anything, we have to bathe you."

Oh, my beloved countrymen, I am just telling you this,
That I was tempted to go right back across.

At last I crossed the border, and left with a labor agent.
And there, dear countrymen, was much that I endured.

—From the corrido, "El Deportado" (ca. 1930)[3]

CHAPTER 1

Farewell Homeland

Carlos Almazán was born into a world on the brink of monumental change. The occasion of his birth around 1890 was undoubtedly celebrated throughout the estate near Zamora, Michoacán, where the Almazán family had resided for as long as anyone could remember. Neighbors congratulated his parents for having another strong boy, one that would, as he grew older, certainly improve the family's economic situation. The Almazáns made their living from the land, and in a late nineteenth-century Mexican community dependent on agriculture, every healthy child proved indispensable to the family's economic subsistence.[1]

Unfortunately, tragedy soon struck—Carlos's father died. Señora Almazán had no choice but to carry on the farm work by herself with young sons. Although she struggled to maintain ownership of the land, the small farm gradually slipped from her hands. Like many others in the region, she became a sharecropper. As Carlos and his brothers grew older, they learned to plant corn and other grains, using old plows that had been passed from generation to generation. Farm work completely occupied their lives. With the help of her sons, who had been propelled by misfortune into early manhood, Señora Almazán gradually managed to stabilize her economic situation after the difficult decade following her husband's death.

But Carlos, now a teenager, grew restless. Tired of the backbreaking work in the fields, he decided to go to Mexico City. Such a decision would have been improbable only a few years before. Zamora and its surrounding communities had been relatively isolated until a newly constructed railroad connected the region to the nation's capital. This transportation network allowed Carlos to leave with high hopes and seek his fortune in the city. It was not long before Carlos made a promising start selling meat and other foods on the streets. He soon married and had children. Yet his prosperity was short-lived. How could he have predicted that Mexico would soon be embroiled in a revolution that would leave him bankrupt? Defeated, Carlos and his family returned to Zamora, but not for long. At the urging of his older brother, Carlos once again made a momentous decision regarding his future: he boarded a train for the north, leaving his homeland for the United States in 1920.

Carlos became part of a massive movement of individuals and families who crossed the Mexican border to the United States in the first three decades of the twentieth century. Approximately one and a half million Mexicans migrated northward between 1900 and 1930, most settling in the Southwest. This process eventually made Mexico one of the largest single sources of immigration to the United States. For Mexico, the migration resulted in the loss of about 10 percent of its total population by 1930.[2]

Most scholars who have analyzed this movement north have focused almost exclusively on the socio-economic factors involved in this migration.[3] This chapter will review those issues, but will also put into context the larger cultural questions raised by such a massive movement of people between two nations with unique histories. The railroads not only led to economic growth in Mexico and the American Southwest, they also facilitated the transmission of cultural values and practices between the two countries.

By concentrating on cultural transformations occurring in Mexican villages, this chapter will also examine the beliefs and traditions that immigrants to the north brought with them. The structure of authority in the village, the rise of Mexican nationalism, and the adaptations in familial customs in this period all played a role in defining the outlook of Mexican immigrants. Finally, this chapter will explore the very decision to migrate itself, one which was clearly driven by economic considerations but also culturally conditioned. This examination will stress that the culture Mexican migrants brought with them, rather than being a product of a stagnant "traditional" society, was instead a vibrant, rather complicated amalgamation of rural and urban mores, developed in Mexican villages during half a century of changing cultural practices.

Recent scholarship has made clear that migration to California is not merely a twentieth-century phenomenon. Ever since Mexico had lost its northern territories in the aftermath of the Mexican-American War, there had been movement of Mexicans into the United States. With the discovery of gold in 1848, perhaps as many as 20,000 experienced miners rushed to California from Sonora and Zacatecas, only to be driven out of the mines by the early 1850s. Yet despite the many returnees, this migration probably still signified a larger movement north to California than any other during the entire Spanish (1771–1821) and Mexican (1821–48) eras.

While Mexicans drifted across the border during the remainder of the nineteenth century, most located in the mining towns of southern Arizona or the ranches and farms of south Texas, where they were within easy reach of their homeland. These two states alone accounted for over

80 percent of the 103,393 Mexican-born residents of the United States in 1900. Despite the heavy gold-rush migration to California, the 1900 census reported only 8,086 inhabitants of that state who were born in Mexico. This figure is striking when compared with the approximately eight to ten thousand noted at the time of the signing of the Treaty of Guadalupe Hidalgo in 1848.[4] Although migration had occurred in the last half of the nineteenth century, it paled in comparison to the mass exodus of Mexicans in the first thirty years of the twentieth. This time California, which in particular had received relatively few Mexican immigrants before 1900, experienced a dramatic rise in the proportion of new settlers.

No historian of the period disputes the notion that the American Southwest held strong economic attractions—often characterized as pull factors—for such immigrants. The mining industry in Arizona and New Mexico had encouraged Mexicans to cross the border even in the late nineteenth century. After 1900 the growth of mines in these states, as well as in Colorado and Oklahoma, induced more workers to flock to the area. Mexicans also played a crucial role in the construction and maintenance of southwestern railroad networks. In addition, railroad work provided the transportation by which job seekers moved from site to site throughout the Southwest.

However, it was the expansion of agriculture which created the most pronounced demand for labor, particularly in California. Irrigation revolutionized California farming, allowing arid land to be converted into vast new farms. By 1929, California became the largest producer of fruits and vegetables in the Southwest, a region generating 40 percent of the total United States output. Meanwhile, Mexicans rapidly replaced the Japanese as the major component of the agricultural labor force.

Although certainly paid less than Anglo Americans, a Mexican worker could earn a wage in any of these three industries far above the 12 cents a day paid on several of the rural haciendas of central Mexico. For example, clearing land in Texas paid 50 cents a day, while miners earned well over $2.00 per day. Most railroad and agricultural laborers were paid between $1 to $2 a day.[5]

The demand for labor created by the expansion of southwestern industry in the early twentieth century was compounded by the curtailment of Asian and European immigration; the Chinese Exclusion Act of 1882, the 1907–8 Gentlemen's Agreement with Japan, and, finally, the Immigration Acts of 1917, 1921, and 1924 all effectively limited other sources of cheap labor.[6] Employers began to look longingly toward Mexico as a source of labor for their steadily increasing needs. Not surprisingly, immigration restrictions directed against Mexicans were at first consistently deferred under pressure by southwestern employers and then, when finally enacted, were mostly ignored by officials at the border. American administrators, in effect, allowed migrants to avoid the head tax or literacy test—instituted in 1917—by maintaining sparsely monitored checkpoints even after the establishment of the border patrol in 1924.

More characteristic of prevailing American attitudes toward Mexican immigration before the 1930s, however, was the elaborate network of employment agencies and labor recruiters stationed in border towns such as El Paso. These networks provided the workers for the railroads, factories, and farms throughout the West. Although the contract labor provision of American immigration law strictly prohibited the hiring of foreign workers before their emigration, agents often traveled undisturbed to the interior of Mexico and to towns along the border to search out likely candidates.[7]

The pull factors represented by a burgeoning southwestern economy and a federal government willing to allow undocumented migration through a policy of benign neglect were factors which contributed to mass migration across the border during the early years of the twentieth century. But there were complicated "push" factors as well. Changes in the Mexican economy under the thirty-five-year dictatorship of Porfirio Díaz were perhaps even more important than American industrial development in bringing Mexicans to the United States.

The Díaz administration followed a land policy which encouraged the growth of large haciendas at the expense of small farmers and communally owned lands, or *ejidos*. While the more productive haciendas grew significant quantities of sugar and coffee for export, thousands of rural poor were left landless. Previously independent peasants were forced into debt peonage or into joining the growing migratory labor stream. At the same time, the shift to export crops severely decreased the production of maize, the staple food in the Mexican diet. Along with other governmental policies, this decline in production boosted the cost of living. Simultaneously, wages fell because of the labor surplus created by both the land policy and the population boom of the late nineteenth century. By the time of the Revolution of 1910, these factors had combined to bring the rural masses of Mexico to the brink of starvation.

Although the violence and economic disruption brought about by the revolution did not alone cause Mexican emigration to the United States, they certainly played a crucial role in stimulating movement. While *campesinos* crossed the border fleeing for their personal safety, hacienda owners often fled for fear of reprisals from their employees. Warring factions also destroyed farmland and railroads, bringing much of the economy to a halt. Unemployment rose along with inflation, forcing many to leave Mexico simply to survive. For other agricultural workers, revolution severed the bonds of debt peonage, emancipating workers from their haciendas and freeing them to move north.[8]

Migration occurred before, during, and after the revolution, but became practical only after the development of a railroad transportation network in Mexico linking the populous central states with the northern border. Indeed, a clearer understanding of the role of the railroads in

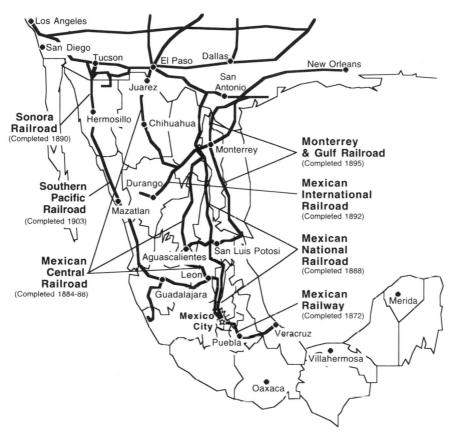

Map 1 Main Railroad Lines in Mexico, 1910

Sources: Lawrence A. Cardoso, *Mexican Emigration to the United States, 1897-1931: Socio-Economic Patterns* (Tucson: Univ. of Arizona Press, 1980), p. 15; Paul J. Vanderwood, *Disorder and Progress: Bandits, Police, and Mexican Development* (Lincoln: Univ. of Nebraska Press, 1981), p. 74.

creating the exodus reveals that Mexican immigration to the United States cannot be viewed simplistically in push-pull terms. The process was a great deal more complex. The creation of the Mexican railway system was both a product of and had consequences for not just one, but both sides of the border. Accordingly, classifying the factors contributing to emigration into "American" and "Mexican" ones can mask the unique relationship between these two neighbors—a relationship shared by no other country that has contributed masses of immigrants to American society.

Railroad development in Mexico occurred almost entirely during the *Porfiriato*—the reign of dictator Porfirio Díaz from 1876 to 1910. The year before he took office, Mexico had a mere 663 kilometers of railway lines in service. In contrast, when Díaz resigned in 1910, Mexico's railroad network stretched 19,748 kilometers, representing a thirtyfold increase in thirty-five years. More important, the rapid expansion of the rail system linked central Mexico to the northern border with completion of the two major railroads of the nation, the Mexican Central and the Nacional (see Map 1).[9]

Díaz hoped that the erection of a national railroad system would unify the nation and modernize its economy. Consequently, he aggressively encouraged foreign investment, since Mexico itself lacked the capital needed to finance construction on such a grand scale. By 1911, American investments in Mexican railroads totaled over $644 million, 61.7 percent of the total capital in support of the system. This amount more than doubled the total American investment in any other Mexican industry. Many influential Americans were involved in railroad promotion in Mexico, including such prominent figures as former President Ulysses S. Grant and the utopian reformer Albert Kimsey-Owen.[10]

Quickly, however, the same financial magnates that controlled the Southern Pacific, Santa Fe, and other railroads in the American Southwest became the major shareholders in Mexican railroads. Financiers J. Pierpont Morgan, Jay Gould, Collis P. Huntington, Thomas Nickerson, and Thomas A. Scott dominated investment in railroads on both sides of the border. As one journalist predicted upon seeing the Southern Pacific begin to build an extension into Sinaloa, "Mountainous Tepic . . . and aristocratic, languor-loving Jalisco . . . are about to be swung upon the railway chain that Uncle Sam swings from his belt." Rather than creating the strong, independent economy that he had hoped for, Díaz unintentionally made Mexico an economic appendage of the United States. By 1911, the United States received more of Mexico's trade than all European nations combined, and between one-fourth and two-fifths of all American foreign investments went to Mexico.[11]

On a local level, the coming of the railroad drove up land prices and led to increased concentration of land ownership among the *hacendados*. Communal village property was expropriated and small landowners found it extremely difficult to hold onto their plots. It is probable that the railroad that eventually took Carlos Almazán away from Zamora played an important role in forcing his mother into sharecropping. Porfirian officials encouraged this new concentration of landownership because they believed that only large-scale agricultural production would lead to "progress." Villages which had existed for centuries with systems of widespread land ownership saw this traditional form of equilibrium disintegrate in favor of a highly stratified society.[12]

The railroads also introduced new ideas and material goods, producing other profound changes in Mexican society. Some Mexican intellectuals referred to this uprooting of culture as the "Americanization" of Mexico. Yet despite such dissent, most of the Mexican elite welcomed the sudden appearance of American and European goods into the middle-class markets of Mexico City and other regional centers. Many claimed that this process would lift the largely Indian nation out of its "traditional backwardness" and pull Mexico into the twentieth century.

Such attitudes were reflected in legislation passed by several states
banning the *pantalón,* the traditional baggy cotton pants worn by the
men of the central plateau. Euro-American style slacks, made readily
available by the railroads for purchase throughout the region, became the
preferred fashion. Significantly, beer replaced tequila as the most popular
beverage in northern Mexico. American railroad workers on Mexican
lines introduced baseball in urban centers throughout central and northern Mexico, while D. S. Spaulding opened sporting-goods stores in Mexico City in 1888. As baseball became the most popular sport in the
northern state of Sonora, the loss of spectators forced bullrings in the
region to close.[13]

Of course, this penetration into Mexico of Euro-American culture
was uneven. Many villages remained relatively isolated, especially if they
were bypassed by the railroad lines. Those communities located directly
in the path of the railroads retained established ways of doing things
even as they instituted innovations. One American commentator, for example, pointed to the irony of seeing women and burros carrying heavy
loads to markets on the very roads which followed the paths of the trains.
In Tepoztlán, a modern mill stood idle as women continued to grind
their own corn, despite the additional time and effort the task required.[14]

On the other hand, sewing machines rapidly appeared in the modest
homes and isolated villages of central Mexico. The machine itself became
a status symbol among villagers, while it provided the necessary tool for
an alternative source of income through clothing manufacturing. According to Ernesto Galarza, who grew up in the isolated mountain village of Jalcocotán, Nayarit, the sewing machine was so esteemed that
only his mother was allowed to touch it. However, all could marvel at
the "remarkable piece of machinery"—this product of modern civilization—to see "the treadle see-saw dizzily, the belt whip around the balance wheel, the thread jerk and snake from the spool, and the needle stab
the cloth with incredible speed."[15]

The railroad, however, could as easily close off certain options for
income as open others up. In northwestern Michoacan, the coming of
the railroad to one agrarian town in 1899 displaced many different kinds
of workers. Muleteers, who had previously provided the backbone of the
local transportation network, were the first to be driven from the scene.
Wheat stopped being milled locally since it became cheaper to ship it to
Mexico City, Toluca, or Irapuato—larger urban centers—for processing,
thereby displacing other laborers. Traditional handicraft workers and
other artisans who had provided most goods for local consumption saw
their modest fortunes evaporate as new city-made goods were introduced
into town. Finally, as the railroad and increased irrigation made more
intensive agriculture possible and profitable, the demand for sharecrop-
pers diminished.[16]

The Mexican immigrant during this period most often came from
these unsettled communities exhibiting both customary modes of
thought and behavior and recently arrived examples of machinery and

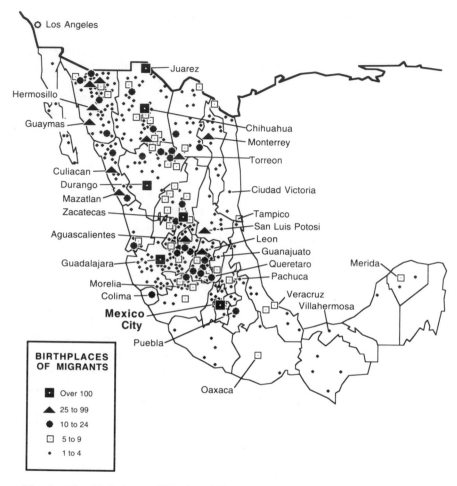

Map 2 The Birthplaces of Mexican Migrants to Los Angeles in the Early Twentieth Century
Source: Naturalization Documents, National Archives, Laguna Niguel, California.

culture. Map 2 shows that most were born in areas within a day's walk from the railroad lines, while many saw the trains pass right through their own villages. Although a significant group came from urban centers such as Mexico City or Guadalajara, areas which were heavily influenced by Euro-American culture, most were representatives of smaller towns and villages in transition. A large group of immigrants came from the northern border states, particularly Chihuahua and Sonora, which maintained constant, ample contact with their American neighbors. Few came from areas, such as the tropical southern states of Oaxaca and Chiapas, that were relatively isolated from the profound changes taking place in Mexican society.

To describe adequately the culture that immigrants from Mexico brought with them to the United States, careful attention must be paid to these communities in transition. Fortunately, the work of pioneer anthropologists such as Robert Redfield and Manuel Gamio provide us with descriptions of early twentieth-century village life in central Mexico. More recently, historians of Mexico such as Luis González and Carlos B. Gil have written "microhistories" of several municipalities, works which resemble the community studies that have been written over the last two decades by American social historians. Studies by other social scientists working in Mexican villages in the post–World War II era, such as Erich Fromm and George M. Foster, can, if used selectively and carefully, provide additional insights into the social and cultural life of earlier migrants to the United States.[17]

Too often Mexican culture during this important period of change has been portrayed as static and "traditional." This description is rooted in the work of social scientists, particularly anthropologists, who ventured from the academies of the United States or Mexico City in search of the traditional Mexican countryside—the antithesis of modernity and industrial society. Their picture of Mexican culture among the peasantry, therefore, was usually set in sharp contrast to the society of the observer.[18] In truth, substantial interaction with urban and industrial society characterized much of rural Mexico during this period. A few anthropologists in the period recognized this situation and incorporated this interaction into their models and descriptions. Robert Redfield, for example, developed the anthropological definition of "acculturation" based on his fieldwork in the Mexican countryside with villagers in contact with modern society.[19] Rather than simply abandoning a world characterized by cultural systems passed from generation to generation, Mexicans who eventually came to the United States in the early twentieth century were products of a vibrant, rapidly changing society, one which was coming to terms with what in the future would be both modern *and* Mexican.

One can accurately describe Mexico during the Porfiriato as a nation of villages, albeit villages undergoing profound social and economic change. Mexico in 1910 had only 68 cities over 10,000 in population, while the United States contained 601 in the same year. Outside of Mexico City and Guadalajara, no other city in Mexico had over 100,000 inhabitants. Like other predominantly agricultural nations, a great number of smaller centers with several hundred or a few thousand people dotted the Mexican landscape. Unlike the United States, the largely agricultural Mexican population almost never lived on isolated farms, but rather congregated in small towns. Even the urban areas in Mexico, defined as communities with a population of over 2000, resembled overgrown villages.[20]

In almost every village and barrio, the plaza acted as the community's center. Here one found the local church and the marketplace, surrounded by the shops and dwellings of those who provided services and leadership. Built into the weekly schedule were periods of relaxation from the drudgery of work in which both men and women used the plaza to do their shopping, attend religious worship, or simply meet with friends. At these times, villagers shared their feelings about community issues and hopes for and fears of the future. Ironically, community affiliation among Mexicans and the role of the plaza were strengthened as railroads solidified the importance of central markets and urban centers of exchange. Even individuals who lived outside of the town proper periodically came into the plaza to exchange their goods, using the opportunity to reestablish ties to the villagers. Such centers acted as a "focus of culture," encouraging the display both of a refined social etiquette and of agrarian folkways. As migrants left Mexico for the north, they attempted to recreate these centers of communication.[21]

Within each village, authority was divided among a variety of individuals. Traditionally, elderly men were respected for their wisdom and contribution to the community over many years. In Jalcocotán, Nayarit, Don Cleofas, the oldest person in the pueblo whom everyone called "Tata," provided the village with its history, passing down orally to the younger generations the stories of *jalcocotecanos* fighting Spanish conquistadors and French soldiers. Moreover, these men were usually patriarchs of a large constellation of families living in the village and surrounding communities. Older women, like Doña Eduvijes of Jalcocotán, often gained the respect of the younger generations with their tales of the supernatural. Others, like midwives, played important roles in life cycle events, thereby exercising a certain community-wide respect.[22]

Respect, however, did not necessarily translate into power, either economic or social, in Porfirian Mexican society. In particular, women who had been restricted by the edicts of the Catholic Church and the dynamics of a communal family economy now found themselves increasingly separated from the emerging cash-focused economic order. Unlike the description of increased female power in New Mexican village communities provided by historian Sarah Deutsch, women in Mexican villages along the railroad route quickly lost whatever standing they had possessed because the entire village was absorbed into a network where mobility became fundamental for economic survival. This transformation led to changes in community-held values, changes which stressed access to outside information and economic opportunity, while women's place continued to be circumscribed to household and village activities. Women continued to be active in productive work and providing village stability, but this very activity was rapidly devalued in the new mobile order.[23]

In Naranja, Michoacán, for example, women along with men had previously engaged in part-time weaving of straw mats and hats to supplement a subsistence economy of intensive agriculture and fishing. Once Porfirian hacienda agriculture was instituted, which used a developed railroad network to get grain to other parts of Mexico, women were forced to weave "with ever greater intensity" during all waking hours of the day, "and even while walking." Every opportunity to make additional income had to be intensified since subsistence agriculture rapidly disintegrated as an option for survival. Moreover, younger and female children replaced the men alongside women weavers, since husbands and older male children were now looking for day labor. The practice of women aiding the effort of fishing or farming at crucial times in the harvest cycle diminished, except if the entire family was engaged in sharecropping. Tortilla-making had previously engaged two older women for one to three hours a day because of the time required to grind the corn and pat out the tortillas; increasingly, poverty began to be judged by the inability of women to provide enough tortillas for their families because of lack of time and the unavailability of maize.[24]

With changes in the economic and political order, novel influences intruded on older lifestyles. Accentuated class divisions increased the leverage of the wealthier sectors of many communities. Despite general social intimacy in all relationships in Mascota, Jalisco, Carlos Gil found strict social protocol practiced when addressing someone of higher economic status. For example, Rosendo, the indentured plantation worker, always referred to Ascensión, the merchant, or Ponciano, the landowner, "as *usted,* invariably prefixing their first names with the title don." "Rosendo never became don Rosendo to anyone" and was invariably addressed with the informal *tú.*[25]

The gradual advent of literacy complicated these social dynamics. Those who provided the previously isolated, often illiterate, villagers access to the larger Mexican nation and the outside world also became important local figures. In the town of San José de Gracia in Michoacán, the influence of the schoolmasters expanded as villagers were divided into those who had access to the printed page and those who did not. The middle-class merchant who traveled to distant markets earned much respect from the village by bringing back news and goods from the city. The mule driver, or *arriero,* who ventured into the mountainous region of Mascota after having passed through populous Guadalajara, ancient Ameca, or strategic Puerto Vallarta returned with important information, especially during revolutionary periods. Along with merchants and muleteers, rancheros, artisans, state employees, and teachers were the largest beneficiaries of increased schooling during the Porfiriato. For communities largely dependent on word-of-mouth communication, these individuals were critical intermediaries linking together a mixed literate/illiterate network of discourse.[26]

Porfirio Díaz's educational policies, however, had uneven consequences throughout Mexico. Although the number of public primary schools doubled and enrollments tripled during the Porfiriato, 68 percent of all Mexican adults still could not read in 1910. Moreover, gender differences in literacy expanded in the period, with only 13 percent of women in 1910 capable of both reading and writing, compared with a

33 percent male literacy rate. The gap between female and male literacy fifteen years earlier had been only 7 percent. Outside of the Federal District, most of the central plateau of Mexico, especially in the rural villages, experienced only minimal advances in schooling and literacy. In Mascota, Jalisco, educational institutions existed "despite enormous odds and only at the behest of enlightened municipal authorities or private parties who supported them." The Catholic Church continued to organize most schooling, since it was not until 1921 that a constitutional amendment permitted the federal government to provide public education for all its citizens. A comprehensive rural school program was not established until 1934.[27]

On the other hand, states in northern Mexico spent the most on primary schooling, enrolled the highest percentage of school-age children, and achieved the highest rates of adult literacy: an average of 45 percent, as compared with 27 percent in the center and 14 percent in the impoverished south. This advancement was made possible because this region was relatively commercially prosperous and revenue-rich, thereby allowing a greater expenditure on education. Public education dominated in the north, representing 90 percent of primary schools in Sonora, for example, as compared with one-half to two-thirds in the north-central states. Moreover, the gender gap in literacy in the north was much smaller than that which existed farther south. According to the 1910 census, the gap between male and female literacy in both Sonora and Chihuahua was less than 3 percent; in the central state of Michoacan the gap widened to 5 percent; while next door in Guanajuato an 8 percent gap existed. These differences between regions would prove vital to making a distinction between migrants to Los Angeles from the north and those from the central states of Mexico.[28]

Despite the changes occurring in the countryside, the local priest continued to stand out as the most important authority figure in most Mexican villages of the period, particularly in central Mexico. He represented to the populace the elaborate institutional network of the Catholic Church, beginning with the bishop to whose jurisdiction he immediately answered. In San José de Gracia, "Padre Othón was the highest authority" who "presided over all social ceremonies; he prescribed them, embellished them, and saw that children and adults alike took part in them." According to an anthropologist studying another village in Michoacán, the priest's support in the community was the result of his role as intercessor between God and the villagers. Armed with the threat of witholding the favor of God, the local priest was able to coerce individuals to take certain actions in decisions far removed from purely religious matters.[29]

In contrast, government officials were usually seen as outsiders, interlopers in community affairs, and interested only in milking the village of its resources through taxation and graft. Barely tolerated by the villagers, few officials emanated from the communities they served and most listened more readily to their superiors in the state capitals and Mexico City than to local sentiment. There existed a general distrust of government and its representatives, and most villagers were happy to have minimal contact with people holding any formal office. Even when a certain town was transformed by the national government into a political entity of importance, such as San José de Gracia's promotion from a *ranchería* to the seat of a *tenencia*, tension often continued to exist between local political leaders and higher authorities.[30] According to Luis González, villagers in that area felt "if there were so many taxes . . . if there were no jobs . . . if the courts were corrupt . . . it was all the fault of the authorities, who had no fear of God."[31]

This situation suggests that Porfirio Díaz's plan to unify the nation and promote patriotism was far from successful. While the railroads did link many important regional centers with the nation's capital, many communities remained isolated and uninterested in the national politic. The people of Tepoztlán, living only fifty miles from Mexico City, did not celebrate the anniversary of the signing of the Mexican constitution because most had no notion of what the national patriotic fiesta meant. Their definition of patriotism extended only to the borders of the mountain-walled valleys of northern Morelos. In contrast, however, in San José de Gracia political arguments did arise. Some undoubtedly were fueled by articles in *El País,* the national Catholic newspaper, or by discussions among the seminary students who had traveled outside the immediate surroundings.[32] About the budding nationalism of San José's villagers, Luis González has written:

> On the eve of the revolution their lives were beginning to be affected by nationalistic sentiments, an interest in politics, an awareness of the outside world, curiosity about new inventions, and the desire to make money. Whether they liked it or not, the social elite were coming to realize, to feel, and to welcome the fact that they were inscribed in the diocese of Zamora, the district of Jiquilpan, the state of Michoacán, and the Republic of Mexico. The better-informed citizens know who Porfirio Díaz, Aristeo Mercado, and the prefects of Jiquilpan were; but the majority were unaware of the move toward nationalism, or even toward regionalization.[33]

Most Mexicans who came to the United States in this period were from areas experiencing this maturing, yet uneven, national sentiment. Yet most remained full of distrust of central authority.

Another factor that limited the spread of Mexican nationalism was racial diversity. At the end of the Porfiriato, Mexico still had two million Indians who did not speak Spanish, a group cast aside by Díaz's program of unifying the nation. The Spanish elite, concentrated in the urban centers, remained distant and impervious to the needs of the masses. The largest segment of the Mexican population were the mesti-

Table 1. Racial Complexion of Mexican Immigrants to Los Angeles

Complexion	Number	Percentage
Dark (including brown, brunette, olive, and ruddy)	1,608	82.9
Medium (including medium dark and medium fair)	184	9.5
Light/fair (including light brown and white)	148	7.6
Total	1,940	100.0

Source: Declarations of Intention, Naturalization Records, National Archives, Laguna Niguel, California.

zos, those containing both Spanish and Indian blood, although exactly where "Indian" ended and "mestizo" began was as often a function of social definition as it was a boundary set by genetic configuration. The source of most of the emigration to the United States was clearly mestizo/Indian, particularly from areas in Mexico that were deemed by anthropologists to be dominated by "acculturated Indians"—that is, where Indians spoke Spanish and blended native and European practices. An indicator of this racial makeup is the complexion of immigrants to Los Angeles as subjectively recorded by Immigration Bureau officials and presented in Table 1. This evidence suggests that the vast majority were either mestizo or Indian.[34]

Social commentators in Mexico often argued that the predominance of this mestizo/acculturated Indian in Mexican society softened racial attitudes in favor of public tolerance. For example, a citizen of Arandas, undoubtedly a light-skinned Spaniard, compared attitudes toward race in Mexico with that in the United States: "There is a more universal spirit here—more a spirit of social distinction and class than of race." Mexican history had unquestionably produced more of a class than caste society by 1910, yet racialist thinking drawing on the works of Charles Darwin and Herbert Spencer influenced liberal thinkers during the Porfiriato at the end of the nineteenth century. The Mexican Revolution turned these ideas on their heads, but did not do away with either racial antagonism in the villages nor racial control by non-Indians in the capital. In fact, color consciousness permeated Mexican culture of the period. As we shall see in later chapters, the attitudes that the largely Spanish elite held toward the mestizo/Indian would shape government policies both toward villages in Mexico and toward immigrants in the United States.[35]

The one institution that all Mexicans valued and the only one that could compete in importance with the village and the church was the family. The strength of the Mexican family was rooted in the formal social bonds that held together individual members, beginning with the rites of courtship and the ritual of marriage. Even these sacred practices, however, were undergoing tremendous change brought about by the rapid economic and social transformation of Mexico. Importantly, fewer parents were able to arrange marriages, since adolescents were sent off at early ages to earn money, resulting in the loosening of parental control over their behavior.[36] Thus, geographic mobility greatly increased the threat of erosion of parental lines of authority within families. In my analysis of case records of Mexicans who immigrated to Los Angeles before 1940, only 28.6 percent of those marriages occurring in Mexico were between individuals from the same place of birth. Of course, even less parental authority prevailed when young Mexican adults chose their marriage partners after migrating to the United States.[37]

This transformation from parental control of marriage partners to personal choice was indirectly linked to the penetration of capitalist markets and the increased mobility of young adults. Ramón Gutiérrez, in a stunning analysis of courtship and marriage in colonial New Mexican villages, described this development in the nineteenth century:

> The displacement of persons through rural landlessness and the creation of wage laborers and petty producers for the market broke down the hierarchical authority relationships between a father and his children and allowed greater personal choice in partner selection at marriage. Freedom from the moral constraints of a village economy through migration for employment and the loosening of patriarchal control due to new material exigencies, allowed persons to behave more as individuals and to choose their spouses on the basis of love.[38]

Nevertheless, in some villages great care was taken by all to institute practices which legitimized parental authority in new ways. George Foster, for example, found that as late as the 1950s in the village of Tzintzuntzan, the act of *robo*, or elopement, was the preferred method of engagement. Rather than the actual "robbing" of a young woman against her wishes, this custom was part of a highly stylized ritual enacted by the young couple to circumvent direct control by the parents of their children's choice of a mate. At the same time, *robo* was followed by a formal visit of the young man's father to the home of the young woman to *hacer las paces* ("to make peace formally"), apologize for his son's conduct, and legitimize the union. Though the woman's father feigned the proper offended dignity, he rarely objected. Foster suggests that this ritual was made possible by both the lack of real property to pass on that would necessitate more judicious marriage alliances and the desire to preserve a hierarchical family structure.[39]

In many ways, there was nothing new in this maintenance of the social ideals of parental authority in the face of countless demonstrations of individual will by young people in love. Gutiérrez found similar discrepancies in court records as far back as the early eighteenth century between the formal norms established by the state and the church and the actual practices of New Mexicans. What had developed over time was more standard rituals which took the loss of parental authority fully into account. The power of the ideal, therefore, despite tremendous socio-economic changes, was manifested in continued marriage rituals which incorporated supposed "traditional values" as they obviously transformed them.[40]

In this era, couples chose to maintain earlier village practices concerning engagement and matrimony even if their courtship had resulted from the new mobility of the era. Carlos Gil described in detail the formal engagement of twenty-three-year-old farmhand Rosendo Peña to twenty-two-year-old Trinidad Peña (no previous relation) in 1908 in Mascota, Jalisco. Both families walked two hours to the church notary office in town from their respective homes on the hacienda to arrange a *presentación,* or espousal, with the local pastor and the church notary. Forms were signed by the parents noting their approval of the union, even though Rosendo himself had initiated the relationship with Trinidad on his own while working as a laborer at a neighboring hacienda. Witnesses confirmed that the betrothed were upright Catholics. Plans for the three-day wedding festivities were carefully described and announced to the entire community.[41]

In Porfirian Mexico, marriage remained a momentous event in the life of a villager. It involved the creation of expectations for economic survival for the new family, and anxiety over its effect on the families of the spouses. In this agrarian society, romance had to be tempered with economic reality. Most marriages in Mascota occurred at the end of the wheat harvest in April or once the corn was gathered in November. Moreover, in contrast to the pervasive myth of widespread teenage marriage in rural Mexico, the seriousness of marriage was reflected in more mature unions. Gil reports that the median ages of grooms in Mascota in the late nineteenth and early twentieth century was between twenty-five and twenty-six years, while for brides it was between nineteen and twenty-one years.[42]

It does appear that the Mexican family at the turn of the century was undergoing a transformation from a highly patriarchal, stable institution with a strict separation of the sexes to a more adaptive and insecure structure forced to conform to increased geographic mobility and economic dislocation. The nature of the relations between the sexes has been a controversial topic in descriptions of family life in Latin America. In particular, the debate concerning to what extent *machismo* permeated male-female dynamics in the Mexican family has yet to be settled.[43]

Although the ideal man might still be expected to possess the characteristics of "physical courage, cleverness, *machismo,* integrity, wealth in money and land and cattle, health, vigor, and, finally, 'manliness,' " in actuality, men of the peon class had a difficult time providing for their families and maintaining their own well-being under the Porfirian economy. Erich Fromm has called this kind of society the "undermined patriarchy," in which the attitude of toughness within *machismo* is actually a façade that masks a more pronounced sense of powerlessness. These tendencies which were just beginning to affect Porfirian society were more marked in fluid, mobile environments than in communities that retained their hold over acceptable behavior. The more cavalier and destructive aspects of *machismo,* ranging from alcoholism to wife beating, seemed to be a result of the breakdown of community control, a situation which better characterized the growing urban centers and the northern border communities.[44]

For women, the one ideal most valued was female chastity before marriage, which historically had been linked to the social class system. Unlike the notion of "female passionlessness" in Victorian America, women in Mexico were seen as particularly vulnerable creatures whose honor had to be protected by men. Chastity functioned as less a notion of female responsibility and more as a function of community and social class mores. The rigidity of this code of behavior, however, was differentiated by one's position in the social hierarchy, with the aristocracy most concerned with preserving female purity. Working-class women, and those of Indian/mestizo background, were less restricted by this ideal, yet their variance from it was clearly used as evidence of their inferiority. During the nineteenth century, the disparity between the ideal of female purity, along with other familial ideals, and the day-to-day reality grew larger because of increased mobility of the population and the breakdown in effective community control of behavior.[45]

Sex for procreation in marriage, however, was expected and encouraged. Young brides could envision spending much of the next twenty to thirty years bearing children and being primarily responsible for raising them. Most women in Mascota, Jalisco, for example, bore five or more children, with some giving birth to many more. High mortality, particularly for those under the age of five, accompanied high fertility, however. Children under the age of six made up nearly half of those that died in Mascota during the Porfiriato. Seen from another perspective, in this village, approximately one-quarter of all infants did not live beyond the first five years of life, many succumbing to epidemics of smallpox, stomach disorders, or fevers. Despite the prevalence of infant mortality, rural Mexican society in the Porfirian age consisted mostly of youngsters because of the high birth rate and low life expectancies of adults. Nearly 50 percent of the population in Mascota in 1895 was under the age of twenty.[46]

During the revolutionary period, wartime violence and economic upheaval did not escape children. Adolescents were particularly vulnerable to its repercussions. Stories of young females victimized by rape abound in the literature of the period. Left to fend for themselves because fathers and older brothers were off fighting in the war or working in distant fields, girls could be attacked by rebel or government troops or by local

Table 2. Age of Migrants at Time of First Crossing

	Male		Female		Total	
Age	*Number*	*Percent*	*Number*	*Percent*	*Number*	*Percent*
Child		30.1		37.7		32.1
Under 13	581		261		842	
Adolescent		22.6		19.2		21.8
13 to 15	184		67		251	
16 to 18	253		66		319	
Young adult		22.4		20.7		21.9
19 to 21	235		72		307	
22 to 24	197		71		268	
Adult		24.9		22.4		24.2
25 to 29	187		68		255	
30 to 39	214		68		282	
40 to 49	67		17		84	
Over 49	12		2		14	
Total	1,930	100.0	692	100.0	2,622	100.0

Source: Naturalization Records, National Archives, Laguna Niguel, California.

men. Oscar Lewis described the experience of Guadalupe, the maternal aunt of the Sánchez children, as an adolescent in revolutionary Guanajuato:

> When Guadalupe was thirteen, Fidencio, a man of thirty-two, broke into the house when her parents were away and carried her off at knife point. He lived on the other side of the street and had made advances to her ever since she was nine. He took her to a cave and raped her. She bled profusely and he brought her to his mother's house in Hidalgo. Guadalupe stayed in bed unattended for fifteen days until the hemorrhaging stopped. Her father found her and whipped her so badly she had to be in bed another two weeks. He told her he didn't like girls who were deflowered because they weren't "worth anything anymore," and he forced her to marry Fidencio in church.[47]

Young boys were dragged into military service, sometimes by federal troops and sometimes by opposition forces. When not obligated to fight, they often became the sole supporters of their families as fathers and older brothers were pressed into service. Not surprisingly in this age of violence, many children of both sexes were orphaned when their parents became victims of the revolution.

Tragedies and responsibilities, therefore, were not new to adolescents. Already full participants in a Mexican family's social and economic well-being, older male teenagers were well prepared to contemplate a step as radical as leaving the village to migrate north. Many undoubtedly welcomed the chance to escape the drudgery, the violence, or the boredom that had characterized their adolescence. Data from the sample of Los Angeles migrants indicates that 43.7 percent of the individuals who left Mexico between 1900 and 1930 crossed the border between the ages of thirteen and twenty-four (see Table 2).[48]

The characteristics of the Mexican family are important to the history of immigration because the decision to send a member north was usually a family one. Many older sons sought extra income in Mexican urban centers, like the capital or Guadalajara, in the prosperous mines or agricultural fields of the northern states, or as track laborers in Mexico's growing railroad network. Others crossed the border to work in the mines, railroads, and agricultural fields of the American Southwest and sent earnings home. Manuel Gamio was able to trace the dynamics of Mexican immigration to the United States by analyzing money orders forwarded to families by immigrant wage earners in various parts of the United States. In July 1926 alone, 12,321 such orders were sent totaling 592,065 pesos ($296,033), an average value of 48.05 pesos ($24) per draft. The amounts ranged from 0.52 pesos (26 cents) to 207.25 pesos ($103.63) per note. The money received was often used to purchase or to retain land under the threat of dispossession. Other families relied on the extra cash simply to survive the ravages of inflation, crop failure, or revolutionary destruction, all of which had rapidly decreased their purchasing power.[49]

While many migrants were single, fathers also left their families and ventured north. The lack of elder sons or the economic strains placed upon newlyweds often made this decision necessary. In Arandas, Jalisco, Paul Taylor reported "prolonged separations of husbands and families, sometimes commencing almost immediately after marriage and lasting for years, and entailing emotional distress and other inevitable hardships." Though some wives and children were abandoned, Taylor found that the vast majority were supported by remittances from the emigrants.[50]

Women rarely emigrated alone. Like most other immigrant groups in the United States, when women did migrate it was usually with other family members or to join a husband who had settled in the United States.[51] Older daughters were expected to stay close to home until they married, so few were seen crossing the border by themselves. Villages throughout the states of Jalisco, Michoacán, and Guanajuato had a comparatively high ratio of women to men, as husbands and brothers left for the United States or other parts of Mexico. According to the census of 1921, overall these three states had a female to male ratio of 106 to 100.[52]

When individuals or families did decide to migrate directly to the United States from their villages in central Mexico (more will be said about this process in following chapters), they often had to overcome widespread negative attitudes toward the United States. One merchant-ranchero in Arandas told Taylor that "down in their hearts the Mexicans do not like the Americans, collectively . . . the United States took more than half of this country [Texas, California, etc.]. But [with intense emotion] I tell you, it will be Mexico again, not now, but in hundreds, or a thousand years." Another young man who was herding burros, when asked whether he was interested in migrating north, replied: "No, I don't wish to go; it is too far. This is *mi tierra*."[53]

Given assurances that money could be made in the United States, however, many families facing economic uncertainty in Mexico chose to send family members north in hopes of improving their situation. Such a decision was surely made easier by the belief that the migration would be temporary. The same railroads which took migrants north could just as easily bring them back. Few saw their initial departure as permanent. Even during the revolution after 1910, most hoped that the end of hostilities would allow a return to Mexico.[54] Mexican immigrants did not find their life in the United States altogether foreign, for they usually traveled and worked in the American Southwest, a region that had once been part of Mexico and still retained a strong tradition of things Mexican. Moreover, they had the advantage that the two countries were contiguous. Working on the rail lines or in the agricultural fields with other Mexicans on the northern side of the border could not have appeared so very different from doing the same a little farther south.

After the first migrants left their communities, others contemplating migration to the United States saw the positive effect migration had on their own villages. Taylor reported that Mexican workers in the United States sent on the average 58,071 pesos ($29,036) per year to Arandas, Jalisco, between 1922 and 1931 via postal money orders. In addition, almost 90 percent of the registered letters arriving from the United States contained either bank drafts or currency, and the amounts sent in this fashion were probably greater than the funds transferred in money orders. This extra cash was vital for temporarily raising the standard of living for Arandas families. While some were able to purchase land, others bought goods, including American-made radios or sewing machines, making their lives more comfortable. Undoubtedly, the economic distinction between families who could rely on this extra income and those who did not encouraged many to participate in the emigration process.[55]

Not all of the ramifications were positive, however. Prolonged separation of husbands and wives heightened the chances of marital strife and infidelity, not to mention the possibility of permanent abandonment. Moreover, community life was affected by the absence of many of the most industrious young men from the village. Whether the departure of so many individuals made it more difficult for Mexico to recover from the ravages of war and advance into the twentieth century as a democratic nation has yet to be studied.

We must also remember that the vast majority of Mexicans did *not* migrate to the United States during this era. Even the highest estimates of migratory behavior indicate that 90 percent of Mexico's population did not leave Mexico. Most Mexican migrants opted to improve their family's situation by trying their luck in Mexican cities and larger towns. Arandas sent large numbers of migrants to Guadalajara, León, Piedra Gorda, La Piedad, Pénjamo, Guanajuato, Atotonilco, Venta del Astillero, Hacienda del Plan, and Zapopan.[56] Mexico City grew tremendously during the early twentieth century as a result of internal migration from villages throughout central Mexico. Still others went north to Torreón, Monterrey, Chihuahua, or Ciudad Juárez without ever deciding to cross the border.

Many remained in their native villages and weathered whatever storms approached. For better or for worse, village culture in Mexico was rapidly changing during the Porfirian and revolutionary periods. Few areas remained isolated from the social and economic pressures altering the countryside. Sometimes the overwhelming feeling of powerlessness created by such a transformation found its expression in the subconscious of those who were left behind. One man's dream in a small village in Morelos eloquently reveals the inner struggles which change posed for him and his neighbors: "I dreamt that I was in bed in my house with all my family, all in bed, when I saw a train, an engine that came over all of us. On seeing the engine, I jumped from the bed, yelling to the one driving that he stop his machine and not crush us all."[57] Though this man remained in Mexico, others elected to migrate to the United States and confront the challenges associated with emigration. Yet what would become of them as they ventured northward could only dimly be imagined.

CHAPTER 2

Across the Dividing Line

The most prevalent image of the American West is, of course, the frontier—an image fixed in American history by Frederick Jackson Turner,
but popularized since by a host of writers. The frontier has always pro-
jected one myopic vision, that of the East looking West, civilization looking toward chaos, Europe looking toward the rest of the world. It casts
the Euro-American as conqueror of both nature and foreign peoples,
sometimes depicted as "savages," and speaks to the belief that the young
American country would know no bounds in fulfilling its destiny to become the world's leading nation. It serves as a continuation of the story
of migration to the New World, depicting the movement west as a destiny just as manifest as the momentous undertaking of crossing the Atlantic was a mission of redemption.

A concept of the border has had no comparable chroniclers among
American historians for obvious reasons. The international border suggests limitations, boundaries over which American power and might
have little or no control. It implies a dual vision, that of two nations
looking at each other over a strip of land they hold in common. It ac-
knowledges that at least two distinct peoples meet in this region, neither
having the certain destiny of cultural and military superiority, and with
conflict being an ever-present historical possibility. While "frontier"
evokes an image of expansive potentialities, "border" speaks to what is
real and limiting between nations and peoples.

The border, however, is also a social construct and has a distinct
history. Simply demarking a line in the desert or a point on a river which
designates the jurisdiction of two governments does not address the social and cultural significance assigned to that spot. It fails to account
for the complex cultural and economic relationships that intertwine two
countries when they share a common border. Moreover, the relationship
between the United States and Mexico is further complicated by the fact
that the northern side of this legal boundary was once held by the Re-

38

public of Mexico. As movement across this boundary increases, both sides have a vested interest in "creating" and "recreating" the border to suit the new social and economic realities of the region. The first four decades of this century saw the border socially invented, in its modern version, to meet the needs of both governments.[1]

This chapter will address the impact of the social construction of the border in the early twentieth century upon the migrants who crossed it, particularly focusing on the border crossing at El Paso–Juárez, the major entry point of those on their way to Los Angeles in this period. First, the means by which migrants left their hometowns and arrived at the border will be discussed. The effects of this burgeoning mobility on border communities were massive, creating both labor conduits for direct migration to the U.S. and new and expanded settlements that would serve as future sending points for migrants north. Finally, the changes which marked the actual border crossing from 1910 to 1924 will be analyzed, with particular attention paid to the transformation of the process in the minds of Mexicans who chose to cross this line of demarcation.

The massive migration across the United States–Mexican border in the early twentieth century did not occur simply because of individual decision-making and vague notions of economic opportunity to the north. Rather, it began as a highly organized movement to provide the American Southwest with substantial labor from Mexico's populous central plateau area. Because the climate and topography of the central plateau was most suitable for agriculture, Mexico's population had been concentrated here since before the Spanish conquest, as the arid north and tropical south provided less opportunity for successful farming. Government policies during the nineteenth century also encouraged the concentration of urban populations in relatively few centers, particularly around Mexico City and Guadalajara. The centrality of village life kept most Mexicans rooted in the areas where they were born.

The linking of the nation by the railroad network, coupled with economic policies during the Porfiriato which discouraged small-scale agriculture, created the impetus for the movement of people from central Mexico. The specific recruitment patterns of American railroad companies, however, set the mass migration of the early twentieth century in motion. Though contrary to American law (the Alien Contract Labor Law of 1885), recruitment in the Mexican interior mobilized individuals ready to take advantage of new opportunities. Once started, migration took on a life of its own, departing from its original pattern and evolving throughout the twentieth century in unprecedented fashion.

Initial movement often resulted from overt attempts to contract labor in Mexico. In 1910 the supervising inspector for the Immigration Service in El Paso, F. W. Berkshire, was forced to admit that "the contract labor law has been flagrantly and openly violated in the past and that Mexican immigration was largely solicited a few years ago." He believed, however, that the practice of actively recruiting laborers within Mexico had been checked by the Immigration Service since 1908.[2] A thorough report prepared in 1910 by Inspector Frank R. Stone described one method by which these early migrants were recruited, a tactic confirmed by all the railroad conductors in the area:

> On the Guadalajara Division of the Mexican National Railway, running from Guadalajara to Irapuato, in the state of Jalisco, my investigation discloses the fact that it was a common occurence [sic] for a labor contractor from the United States to stand on the rear platform of a North-bound train and as it passed through the various villages, at the depots of which were gathered a great many laborers employed on the adjacent haciendas, exhort these laborers to come to the United States, depicting the conditions obtaining and the comparitively [sic] high wages paid there; and this agent would later collect such peons as desired to come to the United States shipping them out in large gangs, paying their transportation to Juárez; even furnishing their bridgetoll over the Rio Grande to El Paso; giving them instructions regarding the responses they should make to questions asked them by our officers.[3]

In Mexico, an overabundance of labor was a distinct economic advantage for employers. It allowed them to keep wages low and workers pacified. Any threat to this labor supply was quickly rebuffed. Consequently, most of the states and many of the municipalities of west central Mexico established regulations prohibiting the contracting of laborers within their jurisdiction. These regulations were enforced to prohibit laborers from leaving for the United States and for work in other states in Mexico. For example, the mayor of the city of Guanajuato, an important hacendado himself, went so far as to have contracted peons physically thrown off trains and placed in jail.[4]

Of course, these measures could not put an end to the movement north. Even the actions of the Guanajuato mayor were circumvented by laborers who went to Silao, the next railroad station, and there purchased tickets to Juárez. Rather, once the practice of contracting laborers stopped, the initial recruitment seemed sufficient to create momentum for increased movement north. El Paso supervisor Berkshire observed that "it does not appear that it is necessary for such [recruitment] tactics to be resorted to at the present time, as the Mexican aliens who have come to the United States, secured employment, and after a period returned to their homes in Mexico, have so diffused the information that wages and living conditions are so far superior in the United States to Mexico that the influx has by these natural means increased from year to year." In order to aid these "natural means," industrial and agricultural employers in the United States continued to encourage their employees to write home and pass on information to induce immigration.[5]

In fact, most villages that sent these early migrants north from central Mexico experienced the return migration that Berkshire described. For example, in 1910 Inspector Stone conducted a four-day investigation in the railroad office in Zamora, Michoacán. Among the 639 people he interviewed who were on their way to the United States, 30 percent (189) had been in the United States before, and an additional 386 individuals were accompanying them. Only 84 of the total had neither first-hand nor secondhand knowledge of the United States, and, consequently, only a relatively small number of Mexicans from a given community were migrating north with no information about their destination.[6]

Paul Taylor also observed that the attitude of those who had experience in the United States was decidedly more positive than those who had never been north. Despite some stories of police brutality or racial prejudice, "the agreeable aspects of their experience in the United States far overshadowed the disagreeable." What was noted by almost all was the standard of living—the possibility of nice clothes and automobiles, along with the pleasant public parks and comfortable housing.[7] The exchange of this form of information could only increase the likelihood that others would join the migrant stream.

Mexicans who migrated to the United States generally came from families engaged in years of creative adaptation to adversity, and were therefore keen on this sort of information. Unlike European immigrant families, whose movement into American society could best be described as chain migration, Mexican families were much more likely to be involved in a pattern of circular migration. Although most European immigrant groups also had high rates of return migration, ranging from 25 to 60 percent, only Mexicans exhibited a pattern of back-and-forth movement that would continue for years.[8] Men ventured north across the border to engage in seasonal labor, then returned south for a period of a few months or a couple of years. If economic circumstances once again necessitated extra cash, the circular pattern began anew. During World War I and up until 1921, the United States government contributed to this pattern by giving entrance visas to temporary workers in order to regulate their movement back into Mexico at the end of a season.

Invariably, adult men formed the bulk of the initial migrants from each village and town in Mexico. In 1910, for example, immigration officials in El Paso reported 35,886 Mexican aliens admitted, almost all of them obstensibly for temporary sojourns of less than one year. Women over the age of fifteen numbered 2,442 (6.8%) of the total, and most of them were probably spouses of male immigrants or the grown children of migrating parents.[9] Relatively few single women unattached to a family unit ventured north during this period. The only significant exceptions to this pattern were nuns fleeing religious persecution in Mexico during the 1920s and high school and college students, usually from the wealthier families of Mexico's urban centers.[10] Family migration, which grew in importance during the Mexican Revolution and the 1920s, was the context in which most women came to Los Angeles.

Consequently, any discussion of immigration from Mexico during

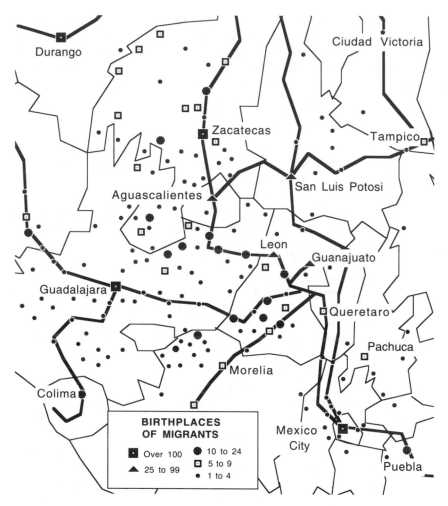

Map 3 Birthplaces of Mexican Migrants from the Bajío Region of Mexico
Source: Naturalization Documents, National Archives, Laguna Niguel, California.

these years must begin with the migration of male laborers. Family migration, described more thoroughly in Chapter 6, was a very important, but secondary process. Although these two forms of migration significantly overlapped chronologically, male migration usually occurred first, both at the level of the individual family and village.

Male migration fell into three distinct patterns. First, young single men came who hoped to relieve their family's dire economic situation in Mexico. Married men formed a second group of migrants who desperately needed to help their families by working in the United States and sending money home to their wives and children. Finally, some males arrived in the United States in family groups, either as children or as heads of households who brought their entire families with them.

Many who made it to the United States had originally intended to move only within Mexico. The country's growing urban centers attracted former rural peasants who sought various employment as a way to earn extra income. For example, from 1900 to 1930, the populations in cities such as Guadalajara and Aguascalientes grew over 175 percent, as nearby campesinos tried their luck in the city. Several cities rapidly became centers of commerce and manufacturing on the west central plateau, largely because they benefited from the direct connections to the capital and the north via rail lines. Other towns in the Bajío region bypassed by major rail connections lost population, and their markets were limited to the immediate region (see Map 3). León fell from being the nation's fourth largest city in 1900 to eighth place in 1921; Guanajuato, eighth in 1900, dropped to 27th; Querétaro fell from 13th to 19th. These centers were also greatly affected by the disruption of economic activities during the Mexican Revolution. The flow of refugees from the countryside particularly aided the growth of Mexico City, as it accounted for 60 percent of the nation's urban growth from 1910 to 1921. The capital consistently attracted rural villagers from throughout central Mexico, and its population multiplied threefold between 1900 and 1930.[11]

Others left haciendas and small villages to earn extra cash by working on irrigation projects, in mines or factories, or with the railroads. In particular, railroads used the enticement of higher wages to attract workers, paying between 25 and 75 cents per day, depending on the availability of labor. As a result, hacendados found it necessary to raise the pay scale to keep campesinos from abandoning the fields altogether.[12] For many migrants, these floating labor communities served as points of transition from a familiar, highly structured, rural, local environment to a more geographically mobile and urban atmosphere. No wonder American Protestant missionaries, searching for vulnerable locales where the hold of Catholicism over the populace could be broken, established their first congregations close to the new railroad depots, mines, and textile factories of the Bajío region. It was among laborers experiencing such dislocating change that the message of personal responsibility and individual discipline appeared to be most attractive.[13]

For a certain percentage of those who moved to Mexican cities or began work on Mexican railways, that first move signaled the beginning of a pattern of mobility which eventually led them to the United States. For example, Julian Ruiz, a resident of the small town of Calvillo, decided at age seventeen to move thirty miles to the city of Aguascalientes. Julian had quit school at the age of fourteen, and could scarcely read and write, so Calvillo presented few opportunities for him besides the two or three pesos a week he received from his storeowner brother. As a night clerk in an Aguascalientes hotel, Ruiz frequently encountered tourists and businessmen passing through town who extolled the virtues of life in the United States. After five years in Aguascalientes, he continued his migration, which ended in southern California. Using his sisters' savings

Table 3. The Migrant Journey of Adult Male Mexicans

	Number	Percent
Migration directly from place of birth	487	41.8
Migration from state of birth (but not same birthplace)	151	13.0
Migration from region of birth (but not same birth state)	67	5.7
Migration from regions north of region of birth	262	22.5
Migration from regions south of region of birth	87	7.5
Unknown	111	9.5
Total	1,165	100.0

Source: Naturalization Records, National Archives, Laguna Niguel, California.

for the trip, Julian Ruiz set out in 1923 for a border crossing at El Paso and eventual life as an agricultural laborer in California's citrus fields.[14]

Ruiz's pattern of migration north was typical. An analysis of naturalization records for adult male migrants to Los Angeles revealed that up to 58.2 percent of the Mexican immigrants surveyed began their journey from home by making an interim stop in a Mexican town or city (see Table 3). In addition to a general movement toward urban areas, Mexican internal migration was characterized by a pronounced movement to the northern border states. Much of the railroad construction took place in the north, as did the increasing mining activity of the early twentieth century. Northern Mexican employers, facing a shortage of labor, consistently offered higher wages to draw workers from the south. At the same time, these industries were faced with competition from employers across the border, a fact that also contributed to higher wages and continued labor shortages. Many Mexican laborers whose original intention was to earn the higher wages on the Mexican frontier found themselves, after a few years, ready to seek even higher pay by venturing across the border.

In fact, most of the Mexicans who came to Los Angeles during the late nineteenth century and first decade of the twentieth were residents of a contiguous border region. During this period, Mexican communities from Tijuana to Piedras Negras were transformed from isolated outposts to important ports of exchange for trade and labor between Mexico and the United States. For example, El Paso del Norte, renamed Ciudad Juárez in 1888, remained unimpressive as late as 1884. One scholar describes its physical appearance that year:

Table 4. Population Growth in the Cities of the Border, 1900–1940

Mexico-U.S. pairs	1900	1910	1920	1930	1940
Ciudad Juárez, Chih.	8,218	10,621	19,457	39,669	48,881
El Paso, Texas	15,906	39,279	77,560	102,421	96,810
Nogales, Sonora	2,738	3,117	13,445	14,061	13,866
Nogales, Arizona	–	3,514	5,199	6,006	5,135
Nuevo Laredo, Tam.	6,548	8,143	14,998	21,636	28,872
Laredo, Texas	13,429	14,855	22,710	32,618	39,274
Mexicali, B.C.	–	462	6,782	14,842	18,775
Calexico, Calif.	–	797	6,223	6,299	5,415
Piedras Negras, Coah.	7,888	8,518	6,941	15,878	15,663
Eagle Pass, Texas	–	3,536	5,765	5,059	6,459
Tijuana, B.C.	242	733	1,028	8,384	16,486
San Ysidro, Calif.	–	–	–	–	–
Matamoros, Tam.	8,347	7,390	9,215	9,733	15,699
Brownsville, Texas	6,305	10,517	11,791	22,021	22,083

Source: Oscar J. Martínez, *Border Boom Town: Ciudad Juárez since 1848* (Austin: Univ. of Texas Press, 1975), 161.

Its main avenue was crossed by nine smaller streets on which were located adobe homes, vineyards, orchards, and empty lots. The business sector consisted of three main establishments which sold clothes, groceries, drugs, hardware goods, and other items, in addition to small shops which specialized in foods and meats. A small hotel, described by a contemporary observer as "dirty and unhealthy," and the usual artisan shops also formed part of the landscape.[15]

Within a few years, the town had grown tremendously. Once again, the railroads played a major role in the expansion of Juárez and other border communities. (See Table 4 for population growth in the early twentieth century.) Railroad lines first linked the Mexican interior to Matamoros in 1882, Nogales and Nuevo Laredo in 1888, and Piedras Negras in 1892. Often, however, the connection of these communities to the American side occurred earlier than its rail link to the Mexican interior and remained more important from an economic standpoint. The first railroad to reach the El Paso–Juárez area, for example, was the Southern Pacific in 1881. This link to Los Angeles was followed by the Atchinson-Topeka-Santa Fe line originating to the north and the Texas-Pacific and the Galveston-Harrisburg-San Antonio from the east. All three of these lines were completed by 1883. Construction of the Mexican Central and Mexican National Railways proceeded simultaneously from the border area southward and from the interior northward. Thus, residents of Juárez could travel directly to California via the railroad well before they were linked to Mexico City in March 1884.[16]

Table 5. Origins of Adult Male Mexican Immigrants

Mexican states	From naturalization records[a]		From money orders from California[b]	
	Number	Percent	Rank	Percent
Chihuahua	160	14.4	7	4.7
Jalisco	126	11.3	2	21.1
Sonora	119	10.7	10	1.7
Zacatecas	115	10.4	4	8.5
Distrito Federal	92	8.3	6	5.0
Guanajuato	88	7.9	1	22.9
Durango	72	6.5	5	7.3
Sinaloa	53	4.8	9	2.6
Coahuila	39	3.5	11	1.6
Michoacán	36	3.3	3	16.0
Aguascalientes	35	3.2	8	2.7
Baja California	25	2.3	13	1.0
San Luis Potosí	22	2.0	16	0.5
Veracruz	21	1.9	17*	0.3
Nuevo León	19	1.7	19	0.2
Puebla	15	1.4	20*	0.2
Colima	10	0.9	12	1.1
Hidalgo*	9	0.8	20*	0.2
Nayarit*	9	0.8	14	0.9
Oaxaca	8	0.7	23	0.1
Tamaulipas	7	0.6	15	0.6
Guerrero	6	0.5	17*	0.3
Chiapas*	5	0.4	25	0.1
México*	5	0.4	24	0.1
Yucatán*	5	0.4	27*	0.0
Querétaro	4	0.3	22	0.2
Campeche*	2	0.2	27*	0.0
Tlaxcala	2	0.2	27*	0.0
Morelos*	1	0.1	26	0.1
Quintano Roo	1	0.1	27*	0.0
Tabasco	0	0.0	27*	0.0
Total	1111	100.0	—	100.0

[a] RG 21, National Archives, Laguna Niguel, Calif.

[b] Adapted from postal money orders from California during January 1927. See Manuel Gamio, *Mexican Immigration to the United States: A Study of Human Migration and Adjustment* (1930, rpt., New York: Dover, 1971), Table X, p. 17.

*Tie

The railroads did not provide the only stimulus to economic development. The policies of the Mexican government helped as well. Since the east-west lines which brought products to and from the east and west coasts were all on the American side of the international boundary, each of the landlocked border towns on the Mexican side found themselves at a great disadvantage. Together, they were utterly dependent on their American counterparts for outlets to potential markets, and remained only auxiliary partners in a burgeoning new international trade complex. In January 1885, after considerable debate, the Mexican government decided to extend the free trade privilege, previously instituted only in the state of Tamaulipas, along the entire length of the border for a distance of twenty kilometers from the boundary line. Merchants capitalized on this new situation by shipping goods from around the world through the United States without paying custom duties, and consequently they sold items in the *Zona Libre* at greatly reduced prices. With this new commercial incentive, the growth of Mexican border towns was accelerated, attracting workers from the Mexican interior and merchants from the American side. In one year, the value of exports transported through El Paso del Norte (Juárez) increased from $2.5 million to $8.7 million. Similar advances occurred throughout the border area.[17]

Many of the families of migrants who settled permanently in the United States first took up residence in the Mexican northern states during this period of economic boom. During the Porfiriato, the states of Coahuila, Durango, Nuevo León, Chihuahua, and Sonora consistently drew large numbers of migrants, while México, San Luis Potosí, Jalisco, Guanajuato, and Zacatecas were among the states yielding the greatest number of migrants.[18] This population redistribution laid the groundwork for twentieth-century migration from the Mexican border states. Sons and daughters of such internal migrants, and sometimes the migrants themselves, gradually ventured onto American soil when conditions seemed opportune. Table 5 indicates that permanent settlers in the United States (as represented by those who sought naturalization) often had been born in the northern states of Chihuahua and Sonora, while other migrant-sending states produced both permanent settlers and temporary immigrants.[19]

Rapidly changing economic conditions often served as impetus for further movement. In 1891, responding to heavy criticism from the United States and from competing regions in Mexico, the Mexican government amended the Free Trade Zone legislation by imposing heavy duties on goods manufactured within the zone and shipped to the interior. This move resulted in conditions which proved so restrictive that trade between the frontier and the rest of Mexico was effectively cut off. Moreover, the government imposed first a 10 percent and then an 18 percent tariff on all foreign goods entering the zone. To add to the north's economic woes in the 1890s, the worldwide depreciation in the value of silver devalued the peso from 92 cents to the American dollar in 1890 to 40 cents in 1897. All of these conditions contributed to a huge increase in the cost of living at the border, prompting many to cross over to the United States. The Juárez area, which had reached a population of roughly 29,000 in the Free Zone period, saw its numbers dwindle to

Table 6. Ports of Entry of Adult Male Migrants to Los Angeles

U.S. port	Mexican bordertown	Number	Percent
Land ports			
El Paso, Texas	Ciudad Juárez, Chih.	679	58.3
Nogales, Arizona	Nogales, Sonora	196	16.8
Laredo, Texas	Nuevo Laredo, Tam.	88	7.5
Calexico, Calif.	Mexicali, B.C.	31	2.7
Naco, Arizona	Naco, Sonora	30	2.6
Eagle Pass, Texas	Piedras Negras, Coah.	24	2.1
Douglas, Arizona	Agua Prieta, Sonora	23	2.0
San Ysidro, Calif.	Tijuana, B.C.	15	1.3
Columbus, N.M.	Las Palomas, Chih.	4	0.3
Other land ports	–	13	1.1
Sea Ports			
San Diego, Calif.	–	17	1.5
San Francisco, Calif.	–	14	1.2
New York, N.Y.	–	10	0.9
Galveston, Texas	–	6	0.5
New Orleans, La.	–	3	0.2
Other sea ports	–	2	0.1
Unknown or unlisted	–	10	0.9
Total		1,165	100.0

Source: Naturalization Records, National Archives, Laguna Niguel, California.

8,218 by 1900. In Piedras Negras the population declined by 5,000, and Nogales and Matamoros experienced similar losses.[20]

Despite this turn of events, the north continued to attract workers from the interior of Mexico because of its higher wages. The chronic shortage of labor and the keen competition for workers with employers from the American Southwest forced many companies to offer wages consistently higher than the rest of the nation.[21] By 1900, common laborers earned an average of 23 cents (U.S.) a day in the interior, compared with 88 cents in Juárez, while some made $1.00 or $1.50 a day near the border. Northern railroad companies provided the best pay, with the Mexican Central the leader in this regard.[22]

Once in the north, however, workers soon realized that the high cost of living and erratic job opportunities ate away at their wages, and many soon learned to reevaluate their decision. Particularly after the turn of the century, this unstable economic situation forced many to seek better wages, stable working conditions, and lower living costs across the border. Ciudad Juárez, and to a lesser extent Nogales, were thereby transformed into major labor conduits, initially luring workers north but often becoming simply stepping stones for further migration into the United States. While each of the border communities grew steadily during the first four decades of the twentieth century, they also assumed an equally important role as temporary depots for international migration. At any given moment between 1900 and 1940, Ciudad Juárez, Nogales, or Nuevo Laredo contained a large proportion of newly arrived residents who did not intend to stay (see Table 6). Some made their way north across the international border, while others returned to the interior after time in the United States.

These were frontier towns in every sense of the word. Along with individuals seeking greater opportunities for themselves and their families were adventurers of all stripes looking for more immediate gratification. Some migrants settled permanently at the border, using the hopes and dreams of others moving through town to their advantage.

Those who decided to seek work in the United States, however, did not have to cut their ties with their homeland. Most men venturing north believed that their stay in the United States was temporary, and indeed many returned to their homes in Mexico. Lax enforcement of immigration restrictions at the border, the concentration of Mexican workers in seasonal employment, and the liberal policies of railroad companies toward transporting workers back and forth combined to make it easy for individuals to see United States employment as an extension of their work experience in Mexico. According to Immigration Service records, for example, one man crossed the border on a work permit for ten straight years, returning every winter to his family.[23]

This type of migration had a mushrooming effect, especially on villagers throughout northern and central Mexico. After the first few men went north to work, they returned with knowledge that made it easier for others to follow in their footsteps. Moreover, what began as a primarily border phenomenon was soon transformed into an option for peasants deep in the Mexican interior. By the time of the Mexican Revolution, most villages on the north central plateau had begun the process of circular migration to the north. With the violence of the civil war and the increasing labor needs of American employers during World War I, more villagers went directly to the United States for employment, by-passing Mexico's urban centers and the northern states altogether. Many ingenious methods were devised to facilitate this movement.

Getting to the border was often the first obstacle facing men seeking work in the United States. Laborers used a variety of means to pay for their trip north, reported to be 27 Mexican pesos in 1910 from Irapuato, the center of the agricultural and mining district of the west central plateau. Many saved for months to raise cash for the trip, or dipped into whatever savings were available. Others obtained funds for passage from their relatives, sometimes taking up a collection from each family member. A substantial group borrowed money. Inspector Stone told of a host of merchants in cities and towns in west central Mexico who loaned money at usurious rates to campesinos going north. Rarely was security furnished against these loans, as merchants relied for collateral on the borrower's honesty and the likelihood that he would return.[24]

One example of the way in which groups of men from the interior migrated together is found in the transcript of a Board of Special Inquiry held by the Supervising Inspector of Immigration in El Paso on May 13, 1910. Nine residents of Nochistlán and Llano Grande were stopped from crossing the border by officials who believed that they had been contracted for work. They were all headed for the Santa Fe & Topeka railroad to hire on as extra gang workers on the line from El Paso to Stafford, Kansas. A pass for one trip to Stafford had been sent by a foreman to a worker who had labored at the camp for sixteen months before he returned to Mexico to visit his family. Only two of the nine had worked at Stafford before, and these were the only men allowed to cross. Some had worked previously in Morenci, Arizona, Santa Ana, California, and Linden, Kansas, while others were coming north for the first time. Those who could not afford passage to the border borrowed money from their parents or from others in the group. Each had left either parents or a wife at home, and all expected to return to Mexico within the year. Though the news of a free pass had encouraged them to leave as a group, each admitted having a long-standing desire to come.[25]

Not often present in accounts of international migration is a description of the actual border crossing. The experiences of Mexican immigrants differed greatly from those of millions of other immigrants from Europe and elsewhere. Of course, what it means to cross our borders has changed over time, and thus immigrant experiences have also changed. In the first four decades of the twentieth century, crossing over from Mexico was transformed from a casual and easy task—with perhaps few questions asked by officials—to a tense and formal ritual full of suspicion.

Internal reports from the Immigration and Naturalization Service provide detailed accounts of procedures used for inspection at the El Paso border. In the early period, the primary concern of immigration officials was the entry of Chinese who evaded the 1882 Chinese Exclusion Act by entering from Mexico. There was little attention paid to Mexicans. In late 1906, for example, an inspector crossed the border incognito to observe this process. "On the night of the 27th," he reported,

> [I] got on the car at Juárez. When the bridge was reached the Customs officers boarded the car, the Immigration officer following them. He looked around, asked a man ahead of me in Spanish *"¿De dónde viene usted?"* (Where do you come from). He asked me the same question and I answered, *"Yo soy Mexicano"* (I am a Mexican), and he passed on.[26]

Two days later, Inspector Seraphic crossed again, noticing how little attention was paid to the movement of Mexicans across the border. While the immigration official on duty at the Stanton Street Bridge sat in his office, hundreds of Mexicans passed back and forth without inspection. After nearly an hour of observation, Seraphic reported that the official had made no attempt to leave his seat and sat inside by the stove reading a newspaper. Mike Romo, a longtime El Paso resident, remembered a similar situation on the Santa Fe Bridge. ". . . One evening I was coming along about 6:00 or 7:00. There was a man [there]; he was the only one at the bridge, on a soap box. Evidently he was sleeping; he had his head down. Anyway, I guess he heard me walking. He lifted up [his head] and looked at me, and then down it [went] again. They didn't bother about anything!"[27]

Charles Armijo, who crossed the border in December 1910 as his family fled Villa's troops in central Chihuahua, explained that upon arriving in Juárez, "well, we just came over. There were no restrictions then about Mexicans coming over. They were free to come in and go out without any passport, without anything else. Everybody was allowed to go back and forth whenever they wanted. . . . And we came over on the streetcar." "All you had to do coming from México, if you were a Mexican citizen," recalled Cleofas Calleros, "was to report at the immigration office on the American side—give your name, the place of your birth, and where you were going to." Conrado Mendoza also remembered crossing during the Mexican Revolution: "All one had to do was get on the electric trolley, or on the electric streetcar, and cross over to the United States, and no one told you anything." The electric streetcar itself had only recently replaced mule-drawn trolleys in 1908 as the most used public transportation to get across the bridges spanning the Rio Grande. These mule-drawn vehicles had carried passengers across the El Paso Street and Stanton Street bridges since 1877.[28]

The laxity with which American officials patrolled the border crossing in 1910 was not due to an absence of immigration statutes on the books. Although no quotas were applied to Mexico until 1965, there were restrictions against border crossings by those deemed morally suspect, diseased, engaged in contract labor, and "likely to become a public charge." Any of these categories, particularly the "LPC" provision, could have barred most entrants from Mexico in this period. But the presence of a strong border culture in which passage had been largely unregulated—this area, after all, had been known as "the northern pass"—mitigated against stringent enforcement of these regulations. Instead, civil servants working at the border concentrated their efforts on the surreptious entry of the Chinese and patrolled against criminal activity. Thus the economic function of the border passage took firm root relatively unencumbered.

Both American officials and entering aliens understood that it was the labor needs of the American Southwest that defined Mexican migration to the United States and not laws drawn up in Washington. Making this point emphatically were representatives of labor recruitment agencies, who stood directly outside the buildings of the Immigration Service. These agencies, called *reenganches* by Mexicans, had operated out of El Paso since 1882, when the city was firmly connected to the large railroad network in the American Southwest. F. W. Berkshire, supervising inspector at El Paso, acknowledged the situation when he reported to his superior in Washington:

> We can exclude practically all of the Mexican aliens of the laboring class who apply for admission at this port as persons likely to become a public charge, for the reason that they are without funds, relatives or friends in the United States, and have no fixed destination; at the same time we know that any able-bodied man who may be admitted can immediately secure transportation to a point on the railroad where employment will be furnished him.[29]

Knowing that labor agencies immediately employed the immigrants, the Immigration Service established a policy of admitting all such aliens. As Berkshire put it, the Bureau had "to recognize the fact that this [El Paso] is a labor market and will unquestionably continue to be such." The importance of these labor agencies in distributing Mexican labor in the United States is confirmed by the fact that in 1910 some 43,548 alien Mexicans were shipped from El Paso alone to points throughout the Southwest and Midwest. This compares with a total Mexican-born population of 221,915 in the United States, according to the 1910 census—little more than five times the number shipped in one year! The law prohibiting those likely to become public charges, intended for immigrants from Europe and Asia who traveled long distances across oceans before being allowed to secure employment, was circumvented on the border to meet the demands of the labor market. Immigration officials remained acutely sensitive to the needs of American employers and condoned what they felt they could not prevent.[30]

The Immigration Service could hardly have been unaware of the implications of their actions. Outside their offices, labor "rustlers" conducted their business openly, to a point that Berkshire felt that the noise produced "when they begin to ply their wares" was "almost intolerable."[31] Another El Paso official described the transfer process in this fashion:

> As they [the aliens] applied and were admitted, they would be taken to the basement of the Immigration Service building and held until detailed individual examinations were completed. When a group could be released, guards would escort the men outside to the rear of the building and line them up. There agents representing the railroads and the ranches would make speeches about the delightful quarters, good pay and fine food they would have if they went to work for their company. When the promising was over, the agents would shout, "This way for the Santa Fe," "This way for the Southern Pacific," and so on, the men following the agent they thought offered the best or most benefits.[32]

Despite the compromised behavior of the Immigration Service, its relationship with the labor recruitment agencies remained problematic. While ignoring the "public charge" clause of the immigration laws, the Service did attempt to enforce provisions against contract labor. Labor agencies and employers were prohibited by law from soliticing for labor on the Mexican side of the border. But reports consistently poured into immigration offices describing attempts by recruiters to gain advantage over competitors in El Paso by soliciting in Juárez or on trains bound for Juárez. Thus, aliens arriving in El Paso were often predisposed to signing up with one particular labor agency, since promises had already been made.

The special relationship between these labor agencies and the Immigration Service was exposed in a series of proposals advanced and executed by Supervisor Berkshire of El Paso from 1909 to 1913. In a letter to the Commissioner-General of Immigration on November 1, 1909, Berkshire suggested that all labor agencies be housed in one building under the control of a joint representative that would insure that no solicitation occurred in Mexico and that no illegal aliens be shipped out for railroad work. In defense of the plan, Berkshire argued that the proposal was a compromise, for he admitted agents could not be eliminated entirely. He acknowledged, however, that the idea was originally advanced by the three "reputable" labor firms, the Holmes Supply Company, L. H. Manning & Company, and the J. E. Hutt Construction (later the Hanlin Supply) Company, who sought only to perfect "some arrangement whereby they might secure, in a perfectly proper and legitimate manner, their proportionate share of the Mexican laborers admitted at this port."[33]

Although obstensibly put forward to preserve law and order, these proposals were tinged with behind-the-scenes machinations which revealed the favoritism and discriminatory attitudes of local immigration officials. Mexican laborers who crossed the border were much more likely to take up the offers of labor agencies run by Spanish-surnamed individuals, most of whom had been in operation for many years and who had built up a reputation with the returning migrants and their companions. Zarate & Avina, for example, was the largest labor agency in El Paso. It was so successful that the three Anglo-run agencies in town were constantly trying to buy the company out. Although Berkshire called the Mexican American agencies "unquestionably irresponsible," other evidence indicated that, to the contrary, it was the newer Anglo-run agencies which had established commissaries for their "recruits," charged exorbitant rates, and, in effect, served only to exploit further the Mexican workers. According to the Mexican Americans involved in labor recruitment, Berkshire moved in the same social circles as their Anglo American competitors, and often enjoyed New Mexican hunting and camping trips with company executives.[34]

Berkshire's own attitude toward the immigrant was generally paternalistic and founded on his conviction that "the Mexican peon is childlike and travels with a party from the same locality as himself." He felt that the immigrants' tendencies to trust only those agencies with Mexican names gave such companies an unfair advantage. Consequently he prohibited all agencies from using their own names when recruiting. Instead they were allowed only to present themselves as representatives of specific railroad lines. It did not trouble him that the major railroad companies used Anglo-run firms almost exclusively.[35]

It was not long before Berkshire's superiors in Washington began to have reservations about the El Paso office's entangling alliances with labor agencies and southwestern employers. One investigator felt that Berkshire had "simply become a party to an agreement whereby those interested are permitted to secure, with his approval, a class of laborers against the admission of which the law is directly aimed"—the poor and destitute paupers. An assistant solicitor warned that "the proposed plan would practically have the effect of making the acceptance by these indigent peons of the offers of employment the test of their right to enter and remain in the United States."[36]

In late 1912, the Office of the Solicitor ruled that Berkshire's plan to develop an official relationship with the city's Anglo-run labor agencies was in violation of immigration law. Berkshire continued to push variations of the plan for several months, however. Finally, Berkshire admitted defeat and decided that, given this finding, little more could be done to strengthen enforcement of the contract labor law.[37] Though resigned to the lack of formal arrangements, border officials continued to maintain informal ties to labor recruiters and southwestern employers for many years to come. Given that the movement of Mexicans across the border was not primarily a legal question, but rather an economic one, it is not surprising that this relationship continued.

The unraveling of the Mexican Revolution, however, heightened the sense of tension along the border. Since Juárez and other border towns were often sites of intense revolutionary conflict, Americans were warned not to cross into Mexico, and normal border traffic was often disrupted. Immigration officials sometimes took out their frustrations with the situation on Mexicans wanting to cross into the United States. One resident of El Paso, Harry C. Carr, finally complained to his senator:

> I remember another night that an old Mexican woman came across the bridge. The custom house is located at one side of the road. She walked along the other side. One of the custom officials yelled for her to come back. When she stood before him, he yelled, "What have we got this house here for?" She said she didn't know. "Well, we'll learn you what it is here for," he said. She was finally allowed to go after having been brutally insulted before four or five admiring inspectors.[38]

Misunderstandings because of language were a frequent source of anxiety. Many American officials did not speak Spanish, and this ignorance only aggravated the situation, as evidenced by another account by Carr:

> On the morning of June 29, there was a street car strike and traffic was delayed. When a car finally did come over from Juárez it was crowded. In

order to better examine the crowd, the officials found it necessary to have some of the people leave the car. This is the way they made the request: "Git out of here." The Mexicans did not know what he meant and hesitated. "Jesus Christ," he yelled at the top of his voice. "Can't you God Damn fools git out when I tell you to?" With that he reached into the crowd and began yanking them out as though they were bales of rags.[39]

Carr himself worried about the ramifications of such treatment. He thought that "it does not seem right for a great government like the United States to allow its petty officials to bully these humble peons and to inspire them with a hatred for us that will live for generations."[40] Though his complaints were investigated, officials at the border denied the allegations and no further action was taken.

At other times, however, the lack of Spanish-language ability among early immigration inspectors provided a ready-made resource for ridicule of Anglo-Americans by Mexican immigrants and Mexican Americans. One Tejano who worked for the Border Patrol as a translator from 1924 to 1930 recalled an incident involving an Anglo officer who believed he could speak Spanish well:

> He walked up to one fellow and asked:
> —¿Cómo se llama yo? (What is *my* name?)
> —Pues quién sabe, señor. (Well, who knows, sir.)
> And then he turned to me and said:
> —How stupid can these people be, they don't even know their own names.[41]

In 1917, the United States Congress passed an immigration act which, for the first time, placed substantive restrictions on European immigration and on those who entered from Mexico. These included a literacy test, a medical examination, a head tax, and the institution of an investigation procedure into the likelihood that the individual would become a public charge. Although an official exemption was extended to Mexicans until 1921, the Immigration Service formalized its procedure in El Paso, borrowing many of the techniques it used for European immigrants at coastal ports such as Ellis Island. These new requirements compelled the government to expand its personnel on the international bridge and construct additional facilities for the inspection of aliens. A report from the inspector in charge of the main crossing at El Paso described this new procedure:

> On arriving at the American side of the Santa Fe Street Bridge, aliens are first inspected and, if necessary, vaccinated by the Public Health Service. . . . The majority of the second class arrivals are also bathed and deloused and their clothing and baggage fumigated by that Service. This occurs in the Public Health building between the boundary and this office. After the delousing and fumigating process is completed the aliens are discharged into a courtyard or "patio" which is entirely surrounded by the buildings of this and the Public Health Service. Fumigating, delousing and vaccination is

commenced at about 7 or 8 a.m., dependent upon number of arrivals and other circumstances.[42]

Particular attention was paid to the medical condition of arriving aliens, since one of the earliest distinctions made between "desirable" and "undesirable" arrivals revolved around their health. From early in the history of the American Republic, immigrants were blamed for spreading contagious diseases throughout the land. Furthermore, the feeble or disabled were barred, since they could not contribute to the labor needs of the United States. The process by which this distinction was made clear to the new arrivals was direct. All newcomers were separated into a special room where they were carefully and systematically examined by medical practicioners hired by the Immigration Service.[43]

Another procedure borrowed directly from Ellis Island was the separation of "first class" passengers from the "laboring classes" by a rather subjective visual analysis. It was only the "second class" migrants (the concept of steerage passengers borrowed from ocean travel) who experienced the rather humiliating medical examination. Wealthier looking immigrants were usually quickly inspected without having to disembark the trains they were riding, and only those in obvious bad health were subject to closer scrutiny. If apparently healthly, immigrants were allowed to pass even when the medical examiner was not available. Moreover, the literacy test was not given to "members of the learned profession or high officials" to avoid embarrassing incidents. In December 1923, the number of immigrants arriving daily in this fashion from Mexico was estimated at usually 5 or 6, and rarely exceeded 10 or 12. Most were professionals, officials, and returning residents from El Paso.[44]

What Mexicans who crossed the border after 1917 at El Paso seemed to remember most vividly were the baths they were forced to endure each time they crossed. The implementation of the medical provisions of the 1917 Act coincided with an outbreak of an influenza epidemic in the border region, thereby solidifying the Anglo-American public's connection of immigration with disease. The baths were maintained at least through the 1920s and became one of the most humiliating aspects of the border crossing. Migrants would be forced to remove their clothes and bathe while their clothes were washed and dried. It was easy to distinguish people who had recently crossed the border because their clothes were often quite wrinkled from this process. As one migrant remembered, "they disinfected us as if we were some kind of animals that were bringing germs." Many tried to avoid the baths by taking special care to be clean and well-dressed in order to persuade officials to waive the requirement.[45]

The literacy test, implemented in 1917, required that all aliens over the age of sixteen admitted to the United States be able to read in at least one language. Besides six test cards provided by Washington, immigration officials in El Paso prepared twelve additional ones containing excerpts from the Proverbs, which were translated into Spanish.[46] This literacy test was one of the final examinations administered to arrivals at the border. Recounted one official,

> Each inspector is provided with two or more of the official Spanish language test cards or typewritten copies thereof on 3" x 5" cards, and no two aliens of the same family or group are tested by the same card. The cards are changed frequently. . . . Aliens sometimes represent that they have forgotten or are without spectacles or eyeglasses. For use in such cases printed in one-inch letters by the use of a rubber stamp, the reading matter copied from the official Spanish language test cards, are sometimes used in testing such Mexican aliens. . . . A record of the inspection is made on manifest (Form 548) and the alien required to sign it. This assists to some extent in detecting illiteracy.[47]

Upon completion of a successful literacy exam, the names of the individuals were compared with lists of aliens previously excluded or deported. Aliens claiming marriage were required to show proof, an obvious hardship to those having wed under common-law practices in Mexico. Individuals who could not produce such documentation went before a special board of inquiry that attempted to ascertain whether the "alleged relationship" indeed existed or whether the couple was "immoral" and should be rejected for "moral turpitude." Finally, head tax receipts were endorsed, and all aliens not required to appear before the board of special inquiry were discharged from the station.[48]

A head tax in the amount of $8 was imposed in 1917. In El Paso, the agent of the transportation company, usually the El Paso Electric Railway Company, collected the money. The 1924 Act added the payment of a $10 fee to secure a visa from the nearest American consul prior to departure. During the first full year of the 1917 Act, 5,745 Mexicans were turned away at the border for being "unwilling or unable to pay the head tax." After the 1924 Act was implemented, legal immigration dropped from 90,000 during 1924 to 32,378. The increased financial barriers to immigration encouraged workers to enter the country illegally, and the rise in illegal arrivals after the imposition of these restrictions probably matched, if not surpassed, the decline in legal immigration.[49]

The extensive complicity between employers and government officials which had developed since the late nineteenth century was further exposed by the 1917 Immigration Act. The type of labor available for Mexican men in the United States did not require literacy nor the possession of money at the border. Up to 1917, the majority of men turned away were debarred only for medical reasons, a fact which didn't trouble employers since their value as laborers was already questionable. On the other hand, women and children might often be barred because their value as railroad laborers was minimal. With the new provisions of the

1917 Act, border officials instituted "boards of special inquiry" to investigate the circumstances surrounding the immigration of children traveling to the United States without their parents and women traveling alone. Both of these situations, unlike those confronting adult men, were seen as circumstances which made a potential immigrant "likely to become a public charge." Thus, officials further gendered the border crossing, making the female "condition" grounds alone for suspicion.[50]

The coming of the 1917 Immigration Act, however, did complicate matters for male migrants because of the new expense. As early as 1913, however, officials realized that the restrictions would lead to increased violations of the law. And they were right. For the first time, significant numbers of aliens illegally crossed into the United States to avoid the head tax. The Immigration Service also realized early on that labor recruiters and southwestern employers cared very little how their prospective employees had made it to the United States.[51]

To prevent the indiscriminate hiring of undocumented workers at El Paso, the Immigration Service began to demand a head tax receipt from every alien at Union Depot. This procedure caused difficulties for some, but many simply boarded freight trains outside the city bound for inland areas of employment, particularly California, where enforcement of immigration laws in this period was almost nonexistent. According to two Labor Department investigators, train officials did not molest these riders, since "the railroad supply agents figure that upon their arrival in California, they will secure their percentage of track work and if they do not, the alien unskilled worker will secure employment in the many seasonal activities." These conditions led the chief inspector in El Paso to report that supervision of the border was so lax "that practically any alien desirous of entering the United States and possessed of ordinary intelligence and persistence could readily find the means of so doing without fear of detection."[52]

The new immigration laws were rarely conceived with the realities of the border in mind. In Washington, politicians focused primarily on restricting European immigration. Their lack of interest in the Southwest is demonstrated by the fact that the Border Patrol was not established until 1924. Before then, enforcement of the immigration laws was a shared function of the Customs and Labor departments and focused primarily on ports of entry established in border towns. Outside of these towns, there were few clear demarcations along the border, making surreptitious entry relatively easy. In addition, a limited number of civil service employees kept watch over smugglers' most traveled routes.[53] One report spelled out the difficulties:

> The inland borders of the United States are inadequately guarded, particularly the Mexican Border and unless one traverses the entire distance for more than 2,500 miles he has no conception of the vast stretches of unguarded territory and cannot realize the obstacles the immigration officers have to contend with. One travels from mountain to mountain, over valleys and across desert and plain, rocky bluffs, stretches of arid wastes, across

rivers and irrigated lands. One meets with many conditions, favorable and adverse, with sections thickly populated and in others with but a few souls for hundreds of miles. . . . [54]

Only about 40 officers patrolled the area around El Paso, the largest region of legal and illegal entry.[55] From the beginning, the Border Patrol realized that it was almost completely unable to stem the rising tide of immigration from Mexico.

Early Border Patrol officers had a great deal of latitude in applying immigration statutes, because they received so little preparation for the work itself. One of the first officers hired along the Texas-Mexico border, Wesley Stiles, was frank about his lack of training. When asked whether he was sent to a school or was told what his authority was, he answered:

> —No. That was the thing about it. None of us spoke any Spanish to speak of and somebody suggested, we'd better get you a speaking dictionary (laughter). That's the best way to learn Spanish. But anyway, we didn't have any schooling. No one knew what to do. That was the big trouble.
>
> —And you didn't know what your authority was?
>
> —Just look for aliens. The little law book that they gave us was about that thick, it wasn't an inch.[56]

Another recruit who joined in 1925, Edwin M. Reeves, summed up his training like this: "Just give you a .45 single action revolver with a web belt—and that was it."[57]

The Border Patrol, however, was crucial in defining the Mexican as "the other," the "alien," in the region. J. C. Machuca, who worked for the El Paso Department of Immigration in the late 1920s, recalled that some of the early immigration inspectors were members of the Ku Klux Klan, which was a leading organization in the El Paso region at the time. Officials would consistently denigrate those who crossed at the bridge, even if their papers were perfectly legal. Eventually crossing the border became a painful and abrupt event permeated by an atmosphere of racism and control—an event that clearly demarcated one society from another.[58]

An unintended result of the new immigration laws and the tensions they produced was to make temporary immigrants already living in the United States think twice about returning to Mexico. Many on six-month work permits planned to go back to their homes and families through El Paso in the fall. Since the Literacy Act, the head tax, and the visa fee made a future reentry prohibitive, many stayed on. Avoiding the racism one could easily encounter at the border crossing might also have played a part in this decision. The new laws insured that onetime entry from Mexico, rather than the back-and-forth migration widespread before 1917, would be much more likely in the future. Consequently, when work in the fields or on the railroads proved temporary, an increasing number of Mexicans settled in the larger cities of the American Southwest.[59]

Even the return trip through El Paso-Juárez became more problematic as restrictions were tightened for passage. One El Paso resident remembered working as a teenager directing "norteños" (those who had ventured north of the border to work) to money exchange houses and hotels for several nights' stay on their way south. This sort of guidance was necessary because "many got robbed, others were sick, some got drunk, then got robbed." Unscrupulous residents of the border area took advantage of returning migrants, who were often exhausted and disoriented, and who tended to bring their wages with them in cash. A friendly, sympathetic guide became a necessity for maneuvering through El Paso-Juárez. Teenage boys were able to make a living from tips from the migrants and from owners of establishments catering to them, but this form of "protection" only added to the returnee's burden.[60]

As the Border Patrol became more of a presence and uncomplicated passage disappeared, some migrants who had originally crossed without proper documents sought to legalize their status in the United States. Jesús Pérez, who had crossed in 1923 without documents, decided in 1928 to formalize his status because the immigration service had stepped up its activities around his hometown of Fabens, Texas. "I gathered my coins and I came, arriving exactly on September 1, 1928 to the immigration station on the other side. Right away I took care of this business in seven days. On the 7th of September they gave me my passport to cross into the United States and until this day here I am, yes sir."[61]

For others, this process could turn out to be a humiliating ordeal. Catalina Aranda recalled assisting a friend who wanted to legalize his status. "He came here very young, but after many years he went to Mexico and then he returned. . . . And then he wanted to fix his passport, so I took him. And then the Americans there laughed at him, because they asked him when he had crossed." After much ridicule and confusion regarding his story, the immigration official continued:

—Well, didn't anyone mark your passage when you crossed?

—No, there wasn't anyone here.

—Yes, this office is full of officials dressed in green uniforms.

—Well yes, but they said nothing to me and I just crossed.

This gentleman was able to secure a valid passport only after engaging a lawyer.[62] Things had changed so rapidly along the border that many newcomers to immigration work, like this boastful American official, knew little about procedures that had only recently been discarded or modified.

Yet even after the establishment of the Border Patrol in 1924, the arbitrariness of enforcement continued throughout the 1920s. Even more important than the regularized procedure was the cultural dimension which clearly showed Mexicans who was in control of the border pass. Those Mexicans who had been long-term border residents could continue to cross in a casual fashion if, and only if, they were granted this special privilege by some Anglo benefactor. Angel Oaxaca remembers carrying a special letter during the 1920s written by a prominent El Paso physician, a Dr. Gallagher, which allowed him to cross the border at any time, day or night.[63]

Another long-term resident who lived on the Mexican side of the border and had been crossing regularly since at least 1912 remembered an encounter he had with Chief Immigration Officer Pierce in 1926, after more strigent restrictions had been put in place. Epitacio Armendáriz recalled returning from a job in Santa Rita, New Mexico, with a local passport which prohibited working in the United States. Driving a shiny new truck that he had bought with his earnings, he encountered Pierce on the bridge, who asked him:

—Where are you coming from?

—Well, I went to do a little work in Santa Rita.

—No, listen, you can't work with a local passport.

—Oh yeah? Then what is a local passport for? For going around here locally.

—No, it is to enter and leave [the border], for stays of a week or eight days, or ten days, or fifteen days, but does not allow you to get a job.

—Oh, now I understand you.

—It's okay. It doesn't matter, I'm not going to take your passport away. But don't cross again like that. If you do again, well, don't tell anyone about it, don't even tell me.[64]

Even when individual exceptions were made, the new immigration statutes and their administration on the border heightened the significance of the boundary line between Mexico and the United States. Indeed, the modern version of the border was created during the first three decades of the twentieth century. It became a much more rigid line of demarcation, as the intricate economic relationship between Mexican labor and American capital was perpetuated through the labor recruitment agents. Here immigration officials, through their inspection of new arrivals and the enforcement of laws barring illegal entry, made it clear that passage across this barrier in the desert was a momentous occasion, a break from the past. The new role of the immigration inspector was duly noted by an El Paso attorney when he wrote to Washington, D.C.: "His business has brought him in contact with the poor, the ignorant, the friendless and the foreigner, over whom he has practically almost limitless power."[65] It was this power over the dreams of the individual immigrant which became increasingly evident at the border crossing.

Ironically, it was in this period of transition that the term "alien" first began to be applied to the Mexican in the Southwest. Men born hundreds of miles away from the border region—in the American East or Midwest—were suddenly given the task of enforcing laws passed by a majority of legislators who had probably never seen the area. These laws were directed against individuals who had deep roots in the region, many of whom had crossed countless times before. Though Mexicans knew that they would have to come to terms with the new reality, the irony of history was surely not lost on them. They were now interlopers on familiar land, even as their labor became increasingly crucial to its economic development and they had begun to settle their families in the United States. Mexican immigrants learned to live with the contradiction, partly because they continued to feel wholly Mexican, but mostly because they could do little to change their lot.

CHAPTER 3

Newcomers in the City of the Angels

Like that of many other Mexican immigrants, Zeferino Velázquez's path to Los Angeles was circuitous. Arriving in the city for the first time in 1919 at the age of twenty-five, Velázquez, a native of León, Guanajuato, had already been in the country for eight years. Similar to many other young men who crossed the border in the early twentieth century, the railroad, family contacts, and sustained geographic mobility characterized Velázquez's entry into American society. In 1911 he and his brother-in-law had crossed over from Ciudad Juárez to El Paso, where they had immediately contracted themselves for railroad work in Kansas. Finding work on the railroads too arduous, Velázquez next obtained a job in a Kansas City packing house. Here, he was able to impress his foreman and secure a modest raise. Perhaps it was this modicum of financial security that enabled him to marry a Mexican immigrant woman from La Piedad, Michoacán. In Kansas City, all seemed to be going well for Velázquez, whose story began as one of traditional immigrant social mobility.

Within a year, however, Velázquez's wife had died, leaving him with an infant son. Moreover, Velázquez himself had recently broken a leg at the packing plant, and was unable to collect any compensation for the damage. To add insult to injury, American officials in Kansas began pressuring him to enlist in the army, now mobilized as a result of World War I. Yet Velázquez had carefully maintained his Mexican citizenship. To avoid Kansas draft officials, he fled to California. There he secured work as an agricultural laborer for Japanese farmers in the Imperial Valley. In 1918, he returned briefly to Juárez to help his father, his sister, and her children move to the United States. By the end of that cotton picking season, the Velázquez clan, working as an economic unit, managed to save enough money to settle in Los Angeles.

Once in his new home in the Lincoln Heights barrio, Velázquez took a job as a laborer with the Los Angeles Paper Manufacturing Com-

Table 7. Migration Patterns of Adult Male Migrants to Los Angeles

Time from border crossing to L.A. residence	Decade of L.A. migration				Total	
	Before 1910	1910s	1920s	1930s	Number	Percent
Direct to L.A. (0 to 3 mos)	40.0	18.7	13.9	2.4	126	15.2
3 mos. to 1 year	13.3	6.2	8.4	7.3	67	8.1
1 to 3 years	16.7	16.1	12.7	7.3	108	13.3
3 to 5 years	3.3	13.7	11.6	7.3	95	11.5
5 to 10 years	13.3	23.6	24.0	31.7	198	23.9
10 to 15 years	10.0	13.0	12.2	26.8	108	13.0
15 to 20 years	0.0	2.5	6.9	12.2	50	6.0
20 to 25 years	0.0	1.2	5.7	2.4	37	4.5
25 to 30 years	0.0	0.6	2.2	2.4	15	1.8
Over 30 years	0.0	0.6	1.5	0.0	10	1.2
Unknown	3.3	3.7	1.0	0.0	13	1.6
Total	30	161	597	41	829	–

Source: Naturalization Records, RG 21, NA—Laguna Niguel, California.

Note: Numbers from the years before 1910 and after 1930 are too small to reach reliable conclusions about migration patterns in those periods.

pany. When employment in the city was irregular or his pay was cut, he continued to do agricultural work in the Imperial Valley. In the 1920s alone, Velázquez worked as a lemon picker, an independent vegetable planter, a sharecropper, and a gardener in Beverly Hills. He eventually remarried after 1920 to a woman from San Francisco del Rincón, Guanajuato, who he had met while working in the Imperial Valley. Although he worked consistently during the decade between 1920 and 1930, Velázquez was fired at least once in the mid-1920s on the grounds that he was not an American citizen.[1]

Velázquez's geographic mobility and variable work patterns were typical of the vast majority of Mexican immigrants who found themselves in Los Angeles during the first three decades of the twentieth century. Data taken from naturalization records of 829 adult male Mexican immigrants in Los Angeles indicates that only 15 percent came directly to the city after crossing the border (see Table 7). In fact, the majority did not arrive in Los Angeles until they had lived five or more years in other parts of the United States.[2] Thus, most Mexican immigrants to Los Angeles had initial experiences with American life elsewhere.

This chapter will explore the complicated patterns by which Mexican migrants arrived in Los Angeles in the early twentieth century. Unlike foreign newcomers to other parts of the United States, Mexicans who settled in Los Angeles were often seasoned sojourners who chose this city after careful exploration of other options. The opportunity and attraction that Los Angeles held out for these newcomers, however, was tempered by the reality of life for Mexicans living in the city. Poor housing options and unhealthful living conditions were as much a part of Mexican life in the city as other more favorable conditions. The establishment of fresh Mexican communities within the Los Angeles area, as this account will reveal, created new foundations for an emerging Mexican American culture.

Given the regional settlement patterns in Mexico and the pattern of railroad routes bringing migrants north, few Mexicans crossed into the United States at the California border. The available naturalization records indicate that less than 7 percent of migrants who eventually made their way to Los Angeles first arrived in the United States via the land ports of Calexico and San Ysidro, or the sea ports of San Diego and San Francisco. The overwhelming majority entered through Arizona or Texas, with El Paso serving as the port of entry for close to 60 percent of all immigrants who eventually settled in Los Angeles.[3]

Since immigrant workers were concerned primarily with finding work, the geographic location of such jobs was only of secondary interest. Most first chose to settle in the area immediately adjacent to their crossing, unless recruited by labor contractors who transported them to other parts of the United States. According to a fact-finding committee commissioned by Governor C. C. Young of California in the late 1920s, between 64 and 84 percent of Mexican immigrants arriving in the United States from 1909 to 1926 had originally declared Texas to be their state of "intended future permanent residence." Only 3.8 percent of the immigrants admitted in the three-year period from 1909 to 1911 declared California, although by 1924–26 that figure had risen to 17 percent.[4]

Many of those who crossed at El Paso made that city their initial home. Each of the eight railroad lines which passed through town had set up El Paso maintenance shops employing hundreds of Mexican workers. Others labored in industries linked to the extraction of metal ores, sometimes working for the largest employer of Mexicans in the city, the El Paso Smelter. El Paso's growth as a commercial and transportation center also led to jobs in construction and commerce, and the city's Mexican population approached 40,000 by 1920. Yet proximity to the border and the ever present flow of newcomers kept competition for employment high and wages relatively low. But given El Paso's role as a railroad terminus, it was relatively easy for workers to head else-where when economic conditions turned sour.[5]

In the same manner that El Paso served as a hub for the transportation and mining industries of the surrounding region, San Antonio became a major agricultural and cattle-raising axis for south and central Texas. The largest city in Texas in 1900, San Antonio contained the largest number of Mexican residents in the Southwest in 1920—nearly 41,500. The building of military bases around the city during World War I created many new opportunities in construction, maintenance, and

Table 8. Previous U.S. Residence of Adult Male Migrants to Los Angeles

Previous U.S. residence	Decade of L.A. migration				Total	
	Before 1910	1910s	1920s	1930s	Number	Percent
Direct to L.A.	40.0	18.7	13.9	2.4	126	15.2
Other California	6.7	4.3	5.2	12.2	45	5.4
Arizona	16.7	18.7	21.3	29.3	174	21.0
New Mexico	3.3	0.6	1.8	2.4	14	1.7
Texas	30.0	51.6	54.4	48.8	437	52.7
New York	0.0	1.2	0.8	0.0	7	0.8
Colorado	0.0	0.6	0.5	0.0	4	0.5
Louisiana	0.0	0.6	0.4	0.0	3	0.4
Other states*	0.0	0.0	0.7	4.9	6	0.7
Unknown	3.3	3.7	1.0	0.0	13	1.6
Total	30	161	597	41	829	–

Source: Naturalization Records, National Archives, Laguna Niguel, California.

*Includes states of Iowa, Kansas, Michigan, Nevada, Washington, and Wyoming.

transportation. Additionally, the entire region of south Texas developed rapidly as an agricultural area, requiring a large migrant labor force. Yet along with jobs picking cotton or harvesting foodstuffs came intense racial discrimination and strict segregation. Moreover, when a Mexican picker received a daily wage of $1.75 for working Texas cotton, but could get up to $2.75 in Arizona or $3.25 in California, farmers had trouble keeping their work force. In fact, growers in this region went so far as to try to prevent Mexican laborers from purchasing automobiles and prohibiting out-of-state labor recruiters in order to keep their workers both plentiful and immobile.[6]

In spite of these efforts, both San Antonio and El Paso served as primary labor markets where recruiters from companies located coast to coast ventured to find able-bodied men and, to a lesser extent, entire families. Because of their proximity to the border, many immigrants maintained family members in these cities, while working part of the year in the steel mills of Indiana, the auto plants of Michigan, or the agricultural fields of Kansas and Colorado. Because these cities served as labor "clearing houses," both exhibited an imbalance in the ratio of Mexican men to women. In 1930, for example, there were 86 men to every 100 women in El Paso, and 95 to 100 in San Antonio. Only these two cities in the entire United States experienced this demographic phenomenon.[7] Given the unique geographic considerations and the mobile nature of Mexican immigrant society, it is no wonder that most Mexicans who eventually arrived in Los Angeles had first spent time living in Texas (see Table 8).

Two other areas were starting points for a substantial group of Mexican migrants to Los Angeles. Nogales, along with the smaller towns of

Naco and Douglas, served as entry points into the state of Arizona, especially for immigrants from the neighboring state of Sonora and other areas along Mexico's west coast. From here, many sought employment in the mines at Bisbee and Morenci or in the agricultural fields of the Salt River Valley. Arizona cotton growers banded together to recruit labor directly from Mexico, sometimes moving whole families in special trains financed by the growers. According to local studies, Mexicans made up 60 percent of Arizona's smelter workers in 1911 and 43 percent of its copper miners in 1927, and they continued to work in the mining industry in large numbers though they consistently received lower wages than Anglos. The same situation prevailed for their counterparts working in the fields. However, only a short train ride away, Los Angeles beckoned seductively to those who hoped to find better jobs and higher wages.[8]

Newcomers to Los Angeles also came from other parts of California. Though a few moved from San Francisco or San Diego, most in-state migrants came from rural areas, particularly the newly fertile Imperial Valley. Like the Rio Grande Valley, the Winter Garden area of Texas, and Arizona's Salt River Valley, the growth of this major agricultural region of California was directly linked to the development of an irrigation system which allowed large farms to develop out of desert lands. Beginning with the federal Reclamation Act of 1902, combinations of private entrepreneurs and public monies brought water to the Imperial Valley region, turning the huge arid area near the Mexican border into 120,000 acres of intensive crop farms—dubbed by one historian an "immense garden." Most large-scale farmers in the southern half of the state recruited Mexican laborers. By the late 1920s, one-third of the labor force of the Imperial Valley was of Mexican origin, and a study conducted in 1928 found that over 55 percent of the farms over 80 acres preferred Mexican laborers over all others. Also spreading northward into the San Joaquin Valley, Mexicans formed the largest single ethnic group among farm workers in California as early as 1920.[9]

This migratory pattern of Mexican immigrants eventually led to employment less mobile but more varied which a city like Los Angeles could provide. Railroad work, done by maintenance or construction gangs in outlying areas, often led to Los Angeles, a major depot with large facilities for railroad repair and maintenance located near the downtown area. Often migrants took advantage of the free travel offered by the railroads to newly hired laborers assigned to distant work sites, only to jump off trains when their destinations were reached. One observer claimed that the loss of workers in every shipment of railroad laborers was close to 50 percent. Backbreaking mine work or migratory agricultural labor often followed, with the ambitious always looking for better-paying jobs. After several years' experience in America's migratory labor market, punctuated by occasional trips to Mexico to rejoin family, those immigrants searching for more stability, greater opportunities for employment, and a more congenial atmosphere increasingly looked to the urban areas of the Southwest and, to a lesser extent, the cities in the Midwest. After World War I, Los Angeles appeared to offer migrants much of what they desired.[10]

As discussed in the last chapter, changes in immigration laws and procedures after 1917 also pressured migrants to settle down in the United States. Once the literacy test and an eight-dollar fee for crossing the border was put into effect, many more laborers chose not to return home for the winter months. Even though exemptions were plentiful at the border for unskilled workers, bureaucratic delays or red tape could detain laborers in Juárez for as many as fifteen days. These delays ate away at hard-earned savings because Mexicans had to pay the daily three pesos for board and lodging, expenses usually covered by labor agencies or the railroads. Moreover, when returnees brought other family members with them, numerous complications could arise, especially if someone in the group became ill. In addition, each child under sixteen was charged a $10 visa fee. The unintended consequences of policies designed to make immigration more difficult were to encourage those already in the country to stay, thus transforming what had been a two-way process into a one-way migration. The result was an increase in the total population of Mexican immigrants in the United States.[11]

Los Angeles, of course, possessed an affable climate and a growing community of Mexicans, a fact that made the city an excellent alternative to home. Moreover, labor agencies and contractors were well established in the area by the 1920s, so that agricultural employers and the railroads were no longer compelled to recruit their workers directly from the border. Railroads, in particular, began to curtail recruitment in El Paso in favor of cities such as Los Angeles, Chicago, and Kansas City. To insure a consistent and plentiful labor supply, California agricultural interests also began to develop relationships with recruiters in Los Angeles. Employers as far away as Salt Lake City and Ogden, Utah, centered their recruitment efforts in Los Angeles.[12]

Moreover, the movement of Mexicans into Los Angeles from agricultural areas did not necessarily mean a change of occupation. Agriculture remained one of Los Angeles County's principal industries well into the 1930s. In 1938, for example, agricultural production in Los Angeles County topped $75 million, compared with less than $24 million in Imperial and Orange counties. In the World War I era, bean fields still dotted the west side of the city on both sides of Wilshire Boulevard, as did celery fields in Culver City and acres of vegetable farms in the South Bay region. Citrus groves would continue to dominate the neighboring counties of Orange and Riverside well into the post–World War II era. For many Mexican residents of Los Angeles, agricultural labor proved to be an important first job in the area, and for others was combined with industrial labor to provide a year-round income.[13]

Many farmworkers in the 1910s and 1920s could live in the city and take Red Line train cars to work in the fields. The Pacific Electric interurban railway network, an empire carefully constructed and developed by Henry Huntington, had grown to cover more than a thousand miles by 1913, stretching as far north as San Fernando, east to Riverside, south to Newport Beach, and west to the ocean. As Anglo Americans in the 1920s became more and more wedded to the private automobile, Mexicans and other ethnic minorities in the city remained heavy users of public transportation. For Mexicans accustomed to relying on the railroad for movement across the southwest, these commuter lines put more control and flexibility into the hands of residents to pursue diverse work opportunities. The combination of this public transportation network and the diverse regional economy made the city an effective base from which to engage in farm labor for a significant portion of the year.[14]

Mexicans in Los Angeles, particularly those with a few years' experience, gained considerable knowledge about the regional labor market and used this knowledge to their economic advantage. Having alternative employment opportunities beyond those offered by a labor recruiter allowed many workers to avoid the most exploitative arrangements. As one recruiter lamented: "I get tired of hearing that twenty-nine cents is 'muy barato' [too cheap]."[15] This gradual change in attitude also contributed to an increasing recruitment of new laborers directly from Mexico, since the more seasoned workers scorned the most unpleasant positions. The stereotype of the "lazy Mexican" was thus reshaped to account for laborers unwilling to take just any offer of work, no matter how underpaid or dangerous. As one frustrated Los Angeles labor recruiter put it:

> The men fresh from Old Mexico are decidedly better. After they have loafed around here and had charity they won't stay out at Kingman, Arizona, and they aren't worth a damn. The man who comes fresh from Old Mexico finds that anything he gets is better than what he had. He finds in L.A. he can get free food and he sees Mexicans who are better off than he is. They get spoiled.[16]

Another reason that migrants chose Los Angeles as a more permanent home was that it contained a large, growing Mexican community. Since the nineteenth century, the city had a longstanding tradition of influence among California's Mexican communities. Yet Mexican-born immigrants had little contact with long-term, native-born Californios. The most obvious reason for this lack of contact between older and newer groups of Mexicans is that few longstanding Spanish-speaking residents of Los Angeles were present in the city to welcome newcomers and to aid in their adjustment. Historians who have written about Mexicans during this period in Los Angeles have noted that residential instability, rather than persistence, characterized Mexican Los Angeles in the second half of the 1800s. Richard Griswold del Castillo, in his work on nineteenth-century origins of Los Angeles' Mexican barrio, found that only twelve heads of households who lived in the city in 1844 remained

there in 1880. He concluded that "migration was a main source of population change." In fact, he estimated that more than 90 percent of the
total Mexican population of Los Angeles in 1880 had migrated there
after the Mexican War of 1848. Not surprisingly, this rate of transiency
is extremely high when compared with other cities in the nineteenth
century.[17]

The pattern continued into the twentieth century. Pedro G. Castillo,
in his analysis of Los Angeles city directories, identified a persistence rate
for Mexicans of 31 percent from 1900 to 1910 and 29 percent for the
period between 1910 and 1920. Ricardo Romo found a similar rate for
Mexicanos in Los Angeles (29.4%) for the period between 1918 and
1928. This figure may be compared with persistence rates for the general
Los Angeles population of around 50 percent per decade. Stephen
Thernstrom also found consistent rates of between 50 and 60 percent for
five other cities—Boston, Omaha, San Francisco, Norristown, Pennsylva-
nia, and Waltham, Massachusetts.[18] With residents moving out and newcomers moving in, the Los Angeles Mexican community was consistently
dominated by successive waves of immigrants from Mexico. As early as
1910, Mexican-born residents outnumbered American-born citizens of
Mexican descent in Los Angeles. By 1920, the ratio of foreign-born immigrants to native-born Mexicans was approximately 2 to 1. Ten years
later, five times as many Mexican families in Los Angeles were headed by
foreign-born heads than by individuals born in the United States.[19]

Those Californios who did remain in the southern California area
often lived at a distance from the newly arrived Mexican immigrant settlements in central and east Los Angeles. Most who could afford to do
so moved to the Anglo-dominated west side of the city where they were
completely isolated from the growing Mexican community in the central
city and east of the river. One list of Californio "pioneers" gathered by
the Circulo Mexicano, a nationalist organization of the World War II
era, listed 41 individuals living in West Adams, Brentwood Heights,
Hancock Park, and Hollywood, with only three in the downtown area
and one living east of the river in San Gabriel. Even Leo Carrillo, born
in the Plaza area in 1880, was more likely to mention contact with Chinese residents than Mexican immigrants in his autobiography. Moving
to Santa Monica with his family as a youngster, Carrillo rarely had any
contact with newly arrived Mexicans to California after 1900, except for
the occasional vaquero he encountered at farflung ranches. The few families who remained in the downtown area so fully assimilated into the
Mexican immigrant community through marriage that they became
largely indistinguishable for the newly arrived.[20]

To compound the demographic change and geographic isolation,
Anglos distorted the Spanish and Mexican past of Los Angeles by developing a romanticized version of local history and the idea that
nineteenth-century Mexican/Spanish California was a lost civilization. Beginning in the late 1880s, Los Angeles promoters cultivated an image of

southern California as a simple, pastoral society. The "mission myth," as it has been called, was intended to attract tourists and settlers. As Charles Fletcher Lummis, the city's leading publicist, pointed out, "the Missions are, next to our climate and its consequences, the best capital Southern California has." Thus, this romanticized Spanish past of California, best symbolized by the restoration of Franciscan missions and the novel *Ramona,* written by easterner Helen Hunt Jackson and published in 1886, served to distort the highly stratified Spanish reign in the region. In addition, it totally glossed over the Mexican heritage and influence in the region, and the clash of cultures between Mexicans and Americans in the state. By depicting the city's Latino heritage as a quaint, but altogether disappearing element in Los Angeles culture, city officials inflicted a particular kind of obscurity onto Mexican descendants of that era by appropriating and then commercializing their history.[21]

It is not surprising that the native-born Californio element would be subsumed in the newly emerging collective consciousness of Los Angeles residents around the turn of the century. Beginning in 1880, southern California underwent what one historian has called "the most extended period of sustained growth ever experienced by any equally compact region of the United States." The demographic consequences of this growth were staggering throughout the period from 1880 to 1930. During the first six years of the new century, for example, the population of the City of Los Angeles alone jumped from 100,000 to 250,000. In the twenty-year period ending in 1930, Los Angeles grew at a rate never matched by any other American metropolis; it quadrupled its population from 319,000 to 1.24 million.[22] What was left of California's nineteenth-century Mexican life and culture was completely transformed in the face of rapid American settlement and urbanization.

By the turn of the century, the native-born element in the population had been reduced to a relatively insignificant constituency in the life of the metropolis. Thus, even though the social position of twentieth-century Mexican immigrants had been foreshadowed by that of their nineteenth-century predecessors, most newcomers remained oblivious to a fact largely irrelevant to their current predicament. Yet remnants of the Californio heyday remained tucked away in the Mexican community in the core area of the burgeoning city.

The most important of these landmarks was the central plaza of Los Angeles itself. Since its founding in 1781, the Plaza of *La Reyna de Los Angeles* ("The Queen of Angels") had served as the central gathering place for the Mexican community.[23] After being settled in its present location in the early 1820s, distinguished Californio families—like those of José Antonio Carrillo, Pío Pico, Ygnacio del Valle, and Vicente Lugo—built town houses around the square. Dominated by the Plaza Church, this communal space witnessed countless religious ceremonies, bullfights, and political spectacles.

After the American takeover in 1847, the Plaza took on a less sanctified aura, as early gold seekers turned the streets around the courtyard into dens of gambling, vice, and crime. Anglo Americans nicknamed the Plaza area "Sonoratown," in mock imitation of the birthplace of a large group of Mexican miners who resettled there after being forced to leave the northern California gold fields during the early 1850s. Town houses originally built by the elite Californios passed into the hands of Anglos in the 1860s, while a burgeoning Chinatown on the one hand, and purveyors of vice on the other, established footholds in surrounding buildings. Despite such changes, a few residents waged a continuous battle to preserve the integrity of the Plaza. Pío Pico, the last governor of Mexican California, for example, returned in 1869 to construct the city's most opulent hotel in one corner of the area. But his efforts eventually were thwarted. By the time Pico died—in near poverty—in 1894, the Plaza area had fallen into disrepair.

During the 1890s, the Plaza was used as a wholesale market for growers of vegetables and fruits, and once again became a functioning part of the economic life of the city. In this form it was a familiar sight to Mexican immigrants, who often came from Mexican communities where goods were traded in a central plaza area. But tellingly, Charles F. Lummis and his Landmarks Club spearheaded efforts to prevent the Plaza from being used in such a practical fashion. Foreshadowing the Plaza's tourist-oriented future, Lummis argued that a crass display of commercial trading belittled the mythic and symbolic role that the Plaza held for Anglo Los Angeles.[24]

As the city grew, the social, business, and cultural center of Los Angeles moved away from the Plaza area in a southward and westward direction. City Hall, built in the late 1920s, served as the centerpiece for the surrounding civic center that emerged a short distance south. Further to the south, Los Angeles' embryonic manufacturing center took hold and crowded out residential neighborhoods. Institutions springing up along Wilshire Boulevard increasingly transferred financial power west. The significant migration of newcomers from other parts of the United States began to fill out the entire Los Angeles basin, creating various regional centers and suburbs that decreased the need for any one central focus to southern California society.

Increasingly isolated from the rest of the city, the Plaza area provided shabby but welcome living quarters for many newcomers. Most migrants from Mexico first settled in and around the area prior to World War I, but the Plaza also attracted large numbers of other foreign-born, especially immigrants from Europe. Over twenty different ethnic groups were represented in the Plaza community, with Mexicans and Italians accounting for over three-fourths of the total.[25] This part of town had emerged in the late nineteenth century as Los Angeles' immigrant quarter, and it continued to serve such a function during the early twentieth century. After 1910, however, the needs of Los Angeles' industrial employers required more cheap immigrant workers than could be housed in the Plaza

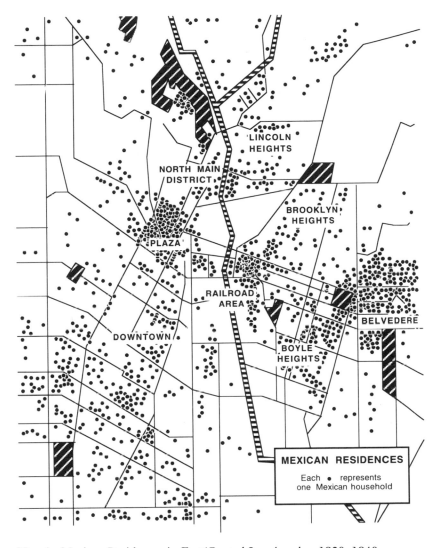

Map 4 Mexican Residences in East/Central Los Angeles, 1920–1940
Source: Naturalization Documents, National Archives, Laguna Niguel, California.

area alone. As native-born Anglo Americans increasingly settled in the expansive western parts of the city between downtown and the Pacific Ocean, immigrants were left to settle the eastern fringe as well as communities directly to the north and south of the Plaza.

Map 4 indicates just how widespread Mexican settlement in Los Angeles was in the first four decades of the twentieth century. Mexicans lived in almost every part of the city, with close to 20 percent living beyond the east and central region shown. Though there was a significant community of Mexicans in the Watts area to the south, others were

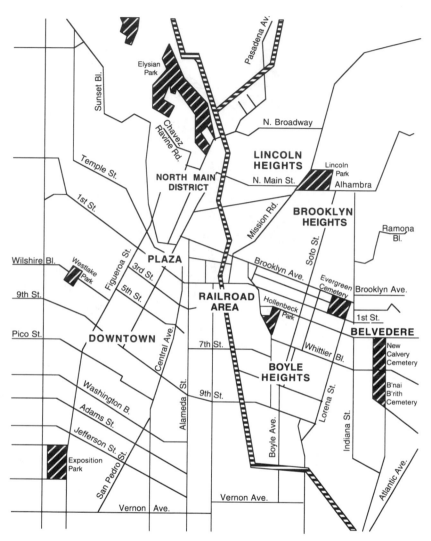

Map 5　Neighborhoods in East/Central Los Angeles, 1920–1940

spread out to the west and south in areas mostly populated by Anglo
Americans. By and large, however, the largest concentrations of Mexicans were in the Plaza area, directly to the east of the Los Angeles River,
in Boyle Heights, and outside the city limits in Belvedere. Other, less
concentrated settlements existed in the North Main Street district,
around Elysian Park, in Lincoln Heights, and between Main Street and
Central Avenue south of downtown (see Map 5 for neighborhoods).

In almost every section of Los Angeles where Mexicans lived, they
shared neighborhoods with other ethnic groups. The Plaza vicinity, of
course, was home to many different groups, and included Chinatown
and Little Tokyo, the main Japanese community. Along the river, in an

area adjacent to the Plaza area and dominated by the city's railroad repair and maintenance yards, lived other Asian immigrants, together with the largest Russian colony in Los Angeles, mostly members of the Molokan sect. Farther east in Boyle Heights, the center of the Jewish community developed along Brooklyn Avenue. At least one-third of the Jews in Los Angeles lived there during the 1920s. By 1930, nearly 10,000 Jewish households could be found in the area. Additionally, a small African American community settled in Boyle Heights, directly west of Evergreen Cemetery.

North of Brooklyn Avenue, a more prosperous and religiously orthodox Jewish section emerged in City Terrace. In other adjacent areas such as Brooklyn Heights, Jews, together with working-class Anglo Americans, formed the large majority of the population. But a smattering of Mexicans lived here, too. Farther north in Lincoln Heights, and west of the river in the North Main district, Mexicans and Italians were the two main ethnic groups, with immigrants from Italy more dominant the farther east from the river one traveled.

The most ethnically diverse communities in Los Angeles were located south of the Plaza. In the 1920s about 40 percent of the small African American community in the city lived along Central Avenue, an area that later in the century contained the largest black population in the western United States. Many of these migrants from the South replaced an earlier generation of Jewish emigrants who had lived in the area. Between Main Street and Central Avenue could be found a foreign-born population so mixed that it was difficult to claim the domination of any one group.[26] In 1926, the *Christian Advocate* printed a diagram of a city block in this area to demonstrate its ethnic complexity (see Figure 1). In his attempt to raise funds for the local Church of All Nations, Methodist minister G. Bromley Oxnam, the author of the accompanying article, was able to highlight graphically the diversity of Los Angeles during the 1920s.[27]

For Mexicans in the 1920s, Belvedere, located east of the city limits past Indiana Street, represented an extreme contrast to these mixed neighborhoods. They had begun to move to Belvedere during the late 1910s because housing was cheaper than within the city limits. There they mixed with native white residents and European immigrants. In the early 1920s, the interurban railway system completed a Belvedere line that made movement into the city for employment relatively fast and inexpensive. And by 1930, the area had developed into a community dominated by twenty to thirty thousand Mexican residents, by far the largest single concentration of Mexicans in the Los Angeles basin. Alone, it accounted for as many Mexicans as Chicago or Laredo, Texas, making Belvedere the fifth largest Mexican urban community in the nation. Out of 10 public schools in Los Angeles reportedly having over 90 percent Mexican enrollment in the school year 1927–28, six were located in Belvedere.[28]

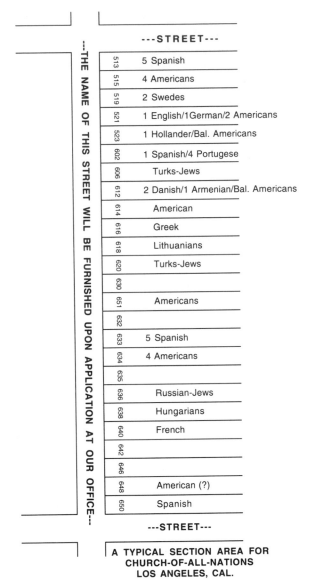

Figure 1 City Block in All Nations Church Neighborhood
Source: G. Bromley Oxnam, "Los Angeles: A City of Many Nations," *California Christian Advocate,* 18 February 1926, p. 14.

The most striking aspect of Mexican residential distribution in Los Angeles between 1900 and 1940, however, was not intense segregation; rather, it was the widespread dispersal of Mexican homes throughout central and eastern Los Angeles. As the future head of Los Angeles' Housing Commission observed in 1912, "there is no spot which we can call 'Mexican Villa' or 'Little Mexico.' "[29] Instead, the incoming native-born Anglo American population did its best to separate itself from *all* immigrants and ethnics by settling on the west side, leaving downtown and East Los Angeles to the foreign-born and blacks. Since Mexicans were the largest immigrant group in the city, they naturally made up significant portions of many of these communities. Most districts in this part of the city contained between 20 and 40 percent Mexican residents, with the area immediately surrounding the Plaza housing around 60 percent.[30]

This is not to say that racial restriction was not prevalent in Los Angeles in the interwar period; rather, this period saw the widespread use of racially restrictive covenants by real estate brokers and owners intent on keeping "undesirables" out of Anglo American Protestant communities. The "undesirables," however, included not only Mexicans but also African Americans, Asian Americans, Jews, and, at times, other foreign-born. Those intent on keeping the "others" out of their communities cared little about how these ethnics distributed themselves in central or east Los Angeles. Because of the boom nature of Los Angeles' growth in this period and the immense migration of a working population to fill the demand for cheap labor, ethnic intermixing characterized most, but not all, central and east-side communities. Rather than "a group tightly clustered residentially and socially," the Mexican community would remain a settlement of scattered communities until the demographic changes of the post–World War II era created one cohesive east-side barrio.[31]

This situation was unique in the history of Los Angeles. In the nineteenth century, Mexican residents had been increasingly segregated into the Plaza area and the southern part of the city. Barrioization had pushed nearly 70 percent of the Mexican population into these two regions by 1880, leading one historian to describe the barrio as "a well-defined enclave within the heart of the city surrounded by Anglo suburbs."[32] After 1940, segregation of Mexicans once again increased as European ethnics moved out of East Los Angeles and immigrants from Mexico took their place. By 1960, residential segregation was significantly higher than it had been in 1880.[33] In between these two periods, however, the rapid transformation of the Los Angeles economy, coupled with competition with European, African American, and Asian newcomers for housing and employment, reduced strict racial segregation. Instead, class-stratified Los Angeles exhibited a rigid residential separation between its core and eastern regions and the rest of the city. The East Los Angeles "barrio" never had a more heterogenous ethnic population than it contained during the first forty years of the twentieth century.[34]

Moreover, the rapid development and geographic mobility characteristic of Los Angeles during the early twentieth century meant that most communities underwent residential transformation continuously. The area south of downtown where the All Nations Church was located, for example, had no sooner become an ethnically mixed neighborhood in the 1920s when it was rezoned to make way for a manufacturing and retail district. Farther east outside the city limits, new meat packing, steel, and auto assembly plants eventually persuaded laborers to move out to Belvedere and other areas on the eastern fringe. The development of the city's civic center and the building of Union Station railroad terminal led to high rents which pushed residents out of the Plaza area. Like other working-class ethnics in Los Angeles, Mexicans could hardly settle down permanently in a community when control of their neighborhoods was firmly entrenched in the hands of Anglo American industrial and commercial interests. The residential preferences of immigrants were always tempered by the zoning practices and labor needs of the city's establishment.[35]

If measured in terms of housing conditions and health, the quality of life in this area of Los Angeles was generally poor. Most Mexican immigrants adapted as best they could given the poor conditions and the limited variety of housing arrangements. Single men, for example, often opted first to reside in the cheap lodging houses of the Plaza district upon arriving in the city. In 1914, rooms, often subdivided by board partitions, went for 20 cents a night. More common was the renting of a single bed, which could be had for 10 to 15 cents. In one house, 32 beds were spread about in one large room. Some houses contained stools or stands as furniture, while others consisted of only beds. Usually, one or two toilets, a wash bowl, and a tub were the only conveniences provided for upward of thirty or forty men.[36]

The house court, another alternative, was often one of the only possibilities available for newly arrived families. Some of the poorest of these residences were dubbed "cholo courts" because of the dominance of Mexican residents. These became the primary targets for investigation by the Los Angeles City Housing Commission, created in 1906. Although the city had few East Coast–style tenements, Jacob Riis compared such quarters to the worst of New York City. Barrack-like structures with thin walls separating each family, they were likened by one observer to "stalls for cattle instead of homes for human beings." Tenants shared outside toilets, water was obtained from outdoor faucets, and bathing facilities were rarely supplied. In 1912, the City Housing Commission found that in the most common house court habitation, a family of four shared two rooms, one for cooking and the other for sleeping. Most of the furniture consisted of boxes. Rents ranged from $3 to $16 per month, roughly one-fourth of the monthly income of the average family.[37]

A house court located at 742 New High Street just north of the Plaza (see Map 6) provides a perspective on housing conditions typical in the area at the time. According to a survey conducted in 1914, Folley Court extended from New High Street through to San Fernando Street, measuring a total of 170 by 44 feet. Filling this space were two long rows of habitations, each 12 by 15 feet with a doorway cut through a partition to separate a five-foot-wide kitchen from the rest of the room.

FOLLEY COURT
742 New High Street

| 1 | 2 | 3 | 4 | 5 | a | 6 | 7 | 8 | 9 | 10 | 11 |

←— 43 ft. ——→ ←——————— 170 ft. ———————→

| 12 | 13 | 14 | 15 | 16 | b | 17 | 18 | 19 | 20 | 21 | 22 |

a TOILET (MEN)
b TOILET (WOMEN)

MEXICAN RESIDENCE
⌂ Lodging House
Arrived in L.A.
• Before 1917
♦ 1917-1923
▲ 1924-1928
□ 1929-1936

Map 6   Mexican Residences in the Plaza Area
Source: Naturalization Documents, National Archives, Laguna Niguel, California.

At least fifty-five people living in the house court shared three toilets each for men and women, along with ten hydrants with sinks. Families of three to six people filled each of the nineteen occupied habitations.

The men living in Folley Court earned from $1.50 to $2.50 a day in railroad work, cement construction, bricklaying, or other common labor. Six had been in Los Angeles less than two years, although eleven spoke some English. Most residents had migrated to Los Angeles in the past five years from other parts of the United States, with many having lived in the north for some time. One family, consisting of a husband and wife and their three sons, had been in the United States for four years. Their meager furniture consisted of a bed and a small cot, a battered dresser, a stove, and a dry goods box nailed to a corner containing a few dishes and cups. Refuse lumber from a recent neighborhood fire was piled up outside the door on the muddy dirt ground.[38]

Other families took advantage of housing provided by their employers. The railroad companies, in particular, constructed and managed some of the most deplorable housing in the area. The Southern Pacific, the Santa Fe, and the Salt Lake railroads, along with the Pacific Electric interurban railway system, all owned house courts, small shacks, boxcars, and empty lots for their employees. This "generosity" came at a price, however. Workers for these companies were customarily paid as little as $1.25 a day, although they received this housing free of charge. In effect, employers were charging a 50 cents per day rental fee for space in these dilapidated section camps by reducing their employees' wages. The railroad companies made substantial profits from this scheme because the charge was generally double the average rent paid in most house courts. An additional advantage was having workers close at hand—a situation that facilitated better control. In emergencies or when needed elsewhere in southern California, laborers could be summoned immediately and shipped out at a moment's notice. Disruptive employees, of course, were doubly punished by being fired and turned out onto the streets. A 1914 Housing Commission report indicated that at least 29 percent of the employed men in central Los Angeles were railroad employees.[39]

No matter how humble the structure, a separate, detached shack provided laboring families with coveted freedom and privacy. Belvedere was a desirable area for immigrants, for example, because it was a community that had been planned for single family residences. However, because it lay beyond the city limits, developers managed to ignore city ordinances concerning sanitation and overcrowding. Often, two or three shacks were constructed on one tiny lot, leaving little unoccupied ground space. A 1928 survey of eight city blocks in the district reported 317 houses containing a population of 1,509 persons—an average of 40 houses per block and 4.8 persons per house. Light and ventilation normally remained poor and plumbing facilities substandard. In addition, unpleasant odors and pollution became standard fare in this Mexican "suburb," as gas works, soap factories, and meat packing plants briskly established themselves around Belvedere.[40]

In spite of the shortcomings, the goal for most immigrants was home ownership rather than renting. Owning gave residents a sense of security and a sense of belonging. But industrial development in the central Plaza area in the 1920s pushed prices too high to allow Mexicans to purchase homes. Other areas, like Lincoln and Boyle Heights, also offered property for sale, but that too was expensive. In these communities, homes usually were owned by Italians or Jews. In Belvedere, where over 60 percent of the shacks or bungalows were occupant-owned in 1928, Mexicans could at least afford to buy property "on time." Often payments were as low as $10 a month, half as expensive as renting in the

Plaza area. The sense of pride in one's own plot of land prompted many to keep spotless houses and yards, despite congested circumstances. Home ownership was a valued ideal among Mexicans in Los Angeles, who shared this aspiration with other working-class immigrant groups across the nation.[41]

A few families were able to improve their situation to the extent that they could afford more than two or three rooms. A description of one such home in the late 1920s provides us with an image of what a working class family could aspire to:

> The D. home is unpretentious. They have four rooms. The furnishings are better than in most Mexican peon homes. The front room is furnished as I said with a piano, several very comfortable, though inexpensive chairs, one of which is a rocker, and a bright colored rug covers the floor. It is cheap and gaudy but it is a rug. Neat, clean curtains hang at the windows. There are beds and dressers in both bedrooms, small rugs on both floors. The kitchen has its oil stove, table, and cupboard. Everything is immaculately clean, even the yard is swept until it is smooth and hard. Flowerbeds of many colorful varieties are planted between the house and the nearest neighbor's place.[42]

Unfortunately, irregular employment and extremely low wages forced most Mexican immigrants to live in conditions much worse than the D. home. A 1912 survey found 80 percent of 700 Mexican families surveyed living in habitations of one or two rooms. Endemic unemployment in the early twenties and during the Great Depression could easily wipe out the investment of more prosperous laborers who had purchased homes "on time," while forcing others into less expensive, more crowded housing. Without adequate resources or knowledge, some fell victim to unscrupulous landlords. In 1925, for example, a minister found five families being exploited a few blocks northeast of the Plaza. Two shacks on a small lot at 213 Augusta Street were found to be housing these families, including ten children, in severely overcrowded and unventilated conditions. These families paid a total of $100 a month for these accommodations, and an additional water bill of as much as $5 a month. Moreover, the owner of the property also owned a grocery store across the street from which each family was forced to purchase all their groceries and supplies at high prices.[43]

To add even greater uncertainty to an already precarious situation, periodic attempts at urban renewal forced others out of existing residences, often only to condemn them to less desirable housing situations. As early as 1912, one group of Mexican homeowners at East Seventh and Utah were evicted, probably to make way for railroad development. Between 1906 and 1912, the Housing Commission had 400 units demolished and fifty others vacated. Rarely was there any attempt to find alternative housing for displaced tenants.[44]

The driving force behind most of this residential displacement was the planned growth of business establishments in the downtown region.

Los Angeles' urban planners gave priority to large economic interests and led government officials away from improving living conditions or investing in urban parks or recreation centers. Conditions were often allowed to deteriorate in anticipation of turning a particular area into an industrial district, promising owners higher profits. As one respected sociologist observed, "they are allowed to continue in their illness-producing and death-dealing roles, because love of money by some 'Americans' is greater than the love of the Mexican's health and life. . . ."[45]

Substandard housing and overcrowded conditions, of course, bred disease and general poor health. Of particular concern was the prevalence of tuberculosis among the immigrant population. Although Mexicans comprised no more than 10 percent of the city's population, they constituted over one-fourth of the tuberculosis patients and contributed one-fourth of tuberculosis deaths in the city. In the late 1920s, for example, Mexicans accounted for over 20 percent of the patients at Olive View Sanitarium, a county institution for the tubercular. In a 1918 report about health conditions among residents in the North Main district, Dr. Gladys Patric discovered that 105 of the 331 houses studied had been infected with tuberculosis, and a majority had experienced at least one death. Another study of those afflicted by tuberculosis in Los Angeles in 1926 found that seven-eighths of the heads-of-households in Mexican families having tuberculosis were born in Mexico. Four-fifths of these, however, had been in the United States for over five years, strongly indicating to the researchers that living conditions in this country had encouraged the development of the disease. Although Mexican immigrants had suffered the ravages of war and poverty in their homeland, Los Angeles could present comparable dangers to their well-being.[46]

Particularly hard hit were Mexican babies. The Mexican infant mortality rate in 1917 was 152 per 1000 live births, three times higher than for Anglos. In Belvedere and the remaining unincorporated section of Los Angeles, the infant mortality rate for Mexicans was almost twice as high as in the city proper. Much progress was made in bringing down infant mortality in the 1920s, but the rate for Mexicans remained two to three times that of the general population in spite of these efforts.[47]

Reformers were well aware of the ravages of disease in the Mexican population. Sickness, including acute illness, tuberculosis, and physical disability, accounted for 67 percent of the Mexican relief cases at the County Charities, and Mexicans comprised almost one-quarter of all applicants. Moreover, the death or incapacitation of a breadwinner caused many needy families to apply for relief. According to reformers, disease, even when it did not lead to death, could still destroy the future opportunities of poor residents. As a Protestant pamphlet put it, "Every time a baby dies the nation loses a prospective citizen, but in every slum child who lives the nation has a probable consumptive and a possible criminal."[48]

Yet the task of awakening officials to the problems of Mexican new-comers proved very difficult given the intoxicating optimism civic leaders had toward the growth of "the last purely American city in the nation."[49] In a September 1924 Los Angeles *Times* article entitled "Where Folks Are Folks," essayist Timothy Turner went to great lengths to explain to his readers that Los Angeles was the only place left in the country truly made up of goodhearted "folks." In a city described as "more Anglo-Saxon than the mother country today," the alarming housing and health conditions affecting Mexicans and other ethnic working class groups were often trivialized by boosters caught up in promoting economic growth and Anglo American migration. As Turner described it: "Here our chief foreign element is Mexican, but it even is American in the broad sense: it belongs to the soil. It furnishes common labor with a minimum of social complexities, for the Mexican labor ebbs and flows over the border as it is needed here."[50] Having been reduced to a slice of nature, the "social complexities" of Mexican immigration and settle-ment could be ignored.

At times, however, shutting out the harsher realities could prove im-possible. Later in the fall of 1924, a dramatic occurrence revealed just how far Los Angeles had to go in providing adequate living conditions for all its residents and acknowledging the social realities in immigrant neighborhoods. An outbreak of bubonic and pneumonic plague forced city authorities to quarantine the entire Plaza/downtown area. Probably carried into California by Asian immigrants, the plague spread rapidly in the congested district. Thirty-four out of the 39 cases diagnosed were fatal. On October 31, local health authorities placed the Macy Street Mexican quarter under quarantine, and in November, the City Council passed an emergency ordinance expanding the restricted area from the junction of the Los Angeles River and North Broadway to Eighth Street at Olive. The city launched a massive program to eradicate the substantial rat population, which carried the disease to humans through infected fleas.[51]

Reformers used the outbreak of the plague to press for greater atten-tion to conditions in the city's "foreign district." But once the immediate crisis passed, the Anglo American majority lost interest. After all, how could a community that boasted of being a "City without Slums" realisti-cally come to grips with its endemic social problems? As one astute ob-server noted, "If our Chamber of Commerce would spend less money telling Eastern people we have no slums, and more money removing our slums, the City as a whole would be far better off."[52] Most Los Angeles residents, however, rarely considered Mexicans or other foreign-born part of that "whole." The sporadic attention the city's immigrant districts did get was usually inadequate to the complexity of the task. The spirit of growth and boosterism that pervaded Los Angeles in this period served only to mask the blighted lives of those individuals who had left their homes and traveled long distances to toil in the City of the Angels.

DIVIDED LOYALTIES

Our task is to break up these groups or settlements, to assimilate and amalgamate these people as part of our American race, and to implant in their children, so far as can be done, the Anglo-Saxon conception of righteousness, law and order, and popular government.

> —Ellwood P. Cubberley, Stanford University, on immigrant education, 1909[1]

You learn everything that the gringos teach you, but don't believe half of it.

> —Advice from a Mexican grandfather to his grandson attending school in the United States[2]

We have seen frequently that natives or mestizos in rural districts in Mexico have not much notion of their nationality or their country. They know their town and the region in which it is situated, and this is a "little country" for them. People of this type, as immigrants in the United States, learn immediately what their mother country means, and they think of it and speak with love of it. Indeed, it can be said that there is hardly an immigrant home where the Mexican flag is not found in a place of honor, as well as pictures of national Mexican heroes. Love of country sometimes goes so far that little altars are made for flag or hero, or both, giving patriotism thus a religious quality.

> —Mexican anthropologist Manuel Gamio, 1928[3]

Mexican education in the United States . . . seeks to reserve for the *patria* those thousands and thousands of children who either came here at an early age or were born here, the ultimate goal of which is to one day, when the conditions of our country improve, reincorporate them as factors in real progress; for, they will carry with them the advantage of having two languages and the experience of two social mediums which have marked differences which, once compared and culling from them, could produce a level of superior life.

> —Editorial in *La Opinión*, June 21, 1930[4]

CHAPTER 4

Americanization and
the Mexican Immigrant

Any Mexican who ventured into Los Angeles in the first three decades
of the twentieth century must have been immediately struck by the sights
and sounds of this burgeoning metropolis. Buildings several stories high
dominated the downtown skyline, while trolleycars and automobiles
made it difficult to cross the streets safely. Despite the Spanish origins of
La Reyna de Los Angeles, English prevailed in the language of daily
commerce, while more foreign dialects like Chinese and Yiddish mingled
equally with the mother tongue of Mexico to form a cacaphonous ethnic
symphony on the streets of the city. Despite familiar-sounding place-
names, a congenial climate, and an occasional structure which looked like
home, Los Angeles was indeed a strange environment for a person who
had only recently shook the dust of rural Mexico from his or her shoes.

Few Mexican immigrants could have understood that Los Angeles
was as alien for the majority of Anglo American residents as it was for
them. In 1890 and 1900, no more than one-third of the native-born
white population had begun life in California, and this figure decreased
to one-fourth in the censuses of 1910, 1920, and 1930. Compared with
cities in the East, where over 80 percent of the citizenry were native to
their respective states, Los Angeles' Anglo population consisted primarily
of people new to the region. Even when viewed in relation to other
western cities, Angelinos were exceptionally mobile. San Francisco's
American-born population averaged around 65 percent native Californian through these decades, while 30 to 40 percent of the American
residents of Denver, Portland, and Seattle had been born in their respective states. "In a population of a million and a quarter," as one visitor
described Los Angeles in 1930, "every other person you see has been
there less than five years. . . . nine in every ten you see have been there
less than fifteen years."[1]

Yet even Anglo Americans new to the region took it as their mission
to integrate foreigners into southern California. Their own mobility

prompted a concern to define better the new culture in which they found themselves. By stressing conformity to the American industrial order, they could try to impose stability on a society in rapid flux. They supported and sometimes developed Americanization programs established by progressive reformers to transform the values of the Mexican immigrant. Ironically because of the peculiar burgeoning character of migration into Los Angeles, these efforts often amounted to one newcomer trying to change another while neither was particularly familiar with local conditions or customs. More than simply an examination of cultures in conflict, Americanization programs revealed the assumptions made about Mexican culture, and also the version of American culture which Anglo American migrants to California brought with them from points east. The context of Euro-American migration to the area set the stage for this conflict.

No other city in the United States attracted so many people so consistently for so long as Los Angeles. Between 1890 and 1930 the city's population increased from 50,000 to 1.2 million, while the number of county residents jumped from 101,000 to 2.2 million. In this period, Los Angeles had grown from a modest-size regional city to become the nation's fifth largest city and fourth largest metropolitan area. Every part of the nation contributed to the tremendous volume of migration to the region. Out of every 100 American-born inhabitants of Los Angeles in 1930, only 28 originated in the Far West, while 37 were midwesterners, 13 southerners, 13 easterners, and 8 westerners.[2]

Not only did Los Angeles' population boom; the city also grew in physical area from 29 square miles in 1895 to 442 in 1930, making it geographically the largest city in the United States. Los Angeles consistently annexed neighboring communities (see Map 7) in a concerted attempt to supply the necessary water, power, and other utilities for its unchecked suburban sprawl. Moreover, urban growth generated industrial expansion. Initially sustained by the development of manufacturing for local consumption, this growth was bolstered after 1915 by the introduction of the motion picture and oil industries to southern California. By stressing the availability of workers, land, and markets and the weakness of trade unions, the local Chamber of Commerce also successfully wooed national corporations such as the Ford Motor Company and Goodyear Rubber Company to the city. By 1929, Los Angeles surpassed all other western cities in manufacturing, with a total output value of over $1.3 billion.[3]

Anglo migrants had generally left behind relatively stable communities to settle in this city rapidly undergoing bewildering change. Most of the city's newcomers came from America's heartland—former residents of Ohio, Indiana, Illinois, Iowa, Nebraska, Missouri, Kansas, Michigan, and Wisconsin. Many of these midwesterners, fresh from mid-American farms and rural towns, had been lured to the West Coast by the energetic

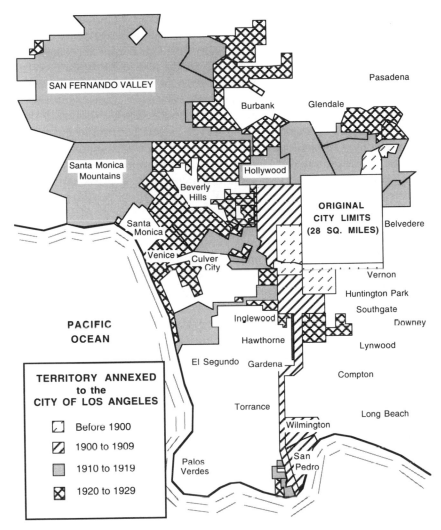

Map 7 Territorial Expansion of the City of Los Angeles
Source: Robert M. Fogelson, *The Fragmented Metropolis: Los Angeles, 1850–1930* (Cambridge, Mass.: Harvard Univ. Press, 1967), pp. 224–25.

publicity efforts of the Chamber of Commerce and the All Year Club of Southern California. Some had caught their first imaginary glimpses of California while strolling down a grocery store aisle in their hometowns, where they encountered vibrantly colored orange crate labels which depicted an idyllic, sun-drenched California lifestyle. These were a product of the Chamber of Commerce's conscious advertising strategy. In 1907, for example, the Southern Pacific Railroad promoted both oranges and migration to California by financing a special Orange Train to Iowa publicizing "Oranges for Health—California for Wealth." That same year the Southern California Fruit Exchange adopted the trademark "Sunkist," a step which marshalled efforts to increase Americans' taste for oranges.[4]

Some migrants first considered moving west after a vacation to southern California. Others came in search of the year-round sunshine and moderate weather which purportedly promoted robust health. The completion of a transcontinental railroad connection by the mid-1880s drew native-born Americans west to Los Angeles. Tourism in the region grew steadily, especially when a rate war between the Southern Pacific and the Santa Fe reduced fares by at least two-thirds in the late nineteenth century. Visits to California often convinced tourists and health-seekers to make a permanent move. When agricultural prices rose across the nation after 1896, particularly in the boom years of 1904–06 and 1910–13, many midwestern farmers and storekeepers marketed their last crops and merchandise, sold their farms and shops, and headed west to start a new life. "We are all 'on the jump' nowadays trying to get in this big corn crop," wrote one Nebraska woman, "so that we can hit the trail for California." Others recalled that enthusiasm in some midwestern towns for a move west "spread through the quiet old street, lined with maple trees, like a panic."[5]

Newcomers to the area overwhelmed native Angelinos, creating a population with several exceptional characteristics. Besides its small proportion of California-born, Los Angeles registered a very low ratio of men to women, an extremely circumscribed percentage of young people, and an exceedingly high number of elderly. None of these characteristics was typical of other western settlements, yet they made sense given the dominant nature of the city's family migration from the Midwest and East. Additionally, Los Angeles settlers tended to be either middle class or well-to-do, a feature characteristic of a successful migrant group looking to replace the drudgery of farm labor with greater comforts and more rewarding livelihoods in the land of the sun. Consequently, the white American-born inhabitants of the region were a relatively homogeneous population.

Yet Los Angeles was not without its diversity. Nonwhites (Mexicans included) composed at least 14 percent of the population in 1930, a proportion exceeded only by Baltimore among the nation's largest cities. European immigrants, dominant in the non-native makeup of eastern and midwestern cities, were joined in Los Angeles by large numbers of Mexicans, Japanese, and black newcomers. The Mexican population of the city, which had numbered around 30,000 in 1920, rose dramatically in the decade to at least 97,000 by 1930. African Americans also migrated to Los Angeles in significant numbers, growing in population from around 15,500 in 1920 to 39,000 in 1930. The Japanese community also expanded from about 11,600 in 1920 to over 21,000 in 1930. The Chinese population of Los Angeles remained relatively small in this period, numbering between 2,000 and 3,000. Diversity in Los Angeles, as compared to that which marked eastern cities, represented a wider range of cultures and peoples.[6]

It was the middle-class midwesterner, however, who dominated the public culture and politics of the city during the early twentieth century. These were the settlers who had been lured by the railroads and the Chamber of Commerce to populate the Los Angeles basin. Real estate agents began carving out former farm and ranch land into suburban plots for single-family residences, complete with front and back lawns. As the newcomers settled into their new homes, they transformed Los Angeles into a sprawling metropolitan area. More important, they reshaped the political and social mindset of southern California, giving it a midwestern flavor.

Midwestern migrants to California brought with them a familiar Protestant world view. This pietistic perspective emphasized faith in a transcendent God, concentrated on the immediacy of personal conversion, and adhered to strict codes of proper behavior to insure individual salvation and prosperity. Adherents strongly supported temperance and opposed parochial schools, and they had little compunction about using politics to advocate their beliefs. Nativism led some to see immigrants as embodiments of evil, repudiating both their Catholicism as well as their occasional pleasure in drink. In describing the social makeup of midwestern small towns, observer Lewis Atherton clearly drew the social spheres which affected the midwestern world view:

> Here, then, was the cultural pattern—a dominant middle-class Protestant group given to religion and stern morality; an upper-class group of "respectable" people who failed to see any necessary connection between pleasure and sin; Catholics; foreigners; and a "lower" class, which ignored the dominant code except perhaps for temporary allegiance following revival meetings.[7]

The dominant middle-class pietists frequently exhibited distinctly anti-urban sympathies, since the city often appeared a corrupting, sinful environment, full of immigrants and liquor. The impersonality of the urban milieu denied pietists the moral leverage that they had exercised in their smaller rural hometowns. No wonder many bolted the Republican party in 1896 to vote for William Jennings Bryan, who canonized the farmer in his famous "Cross of Gold" speech by declaring: "Burn down your cities and leave our farms and your cities will spring up again as if by magic; but destroy our farms and grass will grow in the streets of every city in the country."[8]

But in the 1890s and particularly after the turn of the century, the midwestern suspicion of the city became intermingled with a powerful attraction. Gradually, distrust was overcome by the lure of the metropolis, and it became increasingly difficult to keep the subsequent generation "down on the farm." Upward of 90 percent of California's incoming Americans settled in urban areas, making the state's population the most highly urbanized in the union by 1920.[9] But California's midwestern migrants had opted for a particular kind of environment. They were a special lot, with the resources to make a 1500 mile journey in search of a new home. They had consciously rejected cities like Chicago or New York, areas teeming with impoverished European immigrants and crowded ghettoes. Instead, they sought a land that promoters described as "a warm, dry climate, a varied, even exotic landscape, and a familiar suburban environment . . . an easier, more varied, less complicated, and well-rounded life."[10] Indeed, some argued Los Angeles retained a midwestern flavor. As one promoter observed, contrasting the City of Angels with other regions:

> Witness the difference in the attitude of crowds here and in the East. Those that came from the Atlantic seaboard, the northern half of it, will recall that ill nature of crowds, that sharpness of those that handle crowds. It starts at Ellis Island and gets less and less as you come west. It is the influence directly and indirectly, of a deluge of South and Central Europe peasantry. Who could call those crowds "folks?"[11]

Though rejecting the traditional style of rural life, Los Angeles migrants also brought with them this distinctly anti-urban ethos. They dreamed of a pastoral suburbia, with homes resting picturesquely among valleys and hills. "Ruralize the city; urbanize the country," urged the Reverend Dana W. Bartlett, head of the city's housing commission in 1910. Born in Maine, reared in Iowa, and destined to lead early Americanization efforts in Los Angeles, this minister exemplified the midwestern commitment to social harmony based on the scattering of the populace.[12] He favored a sparse dispersal of residences and encouraged the endless physical expansion of the city to avoid overcrowding and congestion. Ironically, this pattern of urban sprawl would eventually make Los Angeles one of the largest cities in the nation. With over 100,000 people a year moving to southern California during the 1920s, the dream of quiet, peaceful suburban life was doomed.[13]

Midwestern migrants brought with them a desire to recreate a sense of community in their new neighborhoods, hoping to perpetuate the communal familiarity which characterized their former rural and small town lives. Though homeowner associations were organized to maintain racial and class exclusivity, suburbs brought assorted individuals together only superficially. Feelings of isolation and loneliness quickly set in. "I felt so odd among so many and so many miles between me and my home folks," confided a young bride to her diary in 1906. Ironically, the churches lost membership, while secular state societies flourished. Iowans were the first to hold Sunday picnic meetings, and by the late 1920s, 500,000 transplanted midwesterners were members of some state society.[14]

In the city, the newcomers feared lowering standards of personal morality even as they sought a climate and natural environment which

promoted relaxation and fun. Historian Elaine Tyler May has characterized these paradoxical impulses in Los Angeles as representative of a
"Victorian tradition in transition":

> Clearly, these Midwesterners wanted something new, fresh, and exciting.
> But they also wanted a place where they could live, work, and establish
> families in tune with the values of the past. . . . Combining reverence for
> the old with eagerness for the new, they helped set the tone for America's
> first truly twentieth-century city. Desires for newness, however, were con-
> tained within a strong adherence to tradition; the mold in which these set-
> tlers cast their new city was distinctly Victorian.[15]

This paradox was exemplified by various public efforts to enforce morality in the twentieth century. For example, in 1917, Los Angeles voters
approved a local ordinance eliminating saloons and restricting alcoholic
beverages to beer and wine. The city gained a reputation for being the
"dryest" in the state. But this pietist dream eventually turned into an
administrative nightmare. By the mid-1920s, police had their hands full
trying to enforce prohibition on an uncooperative public.[16]

Transplanted midwesterners underwent an acute "search for order"
in California, an extreme version of the nationwide process described by
Robert H. Wiebe.[17] In the new, confusing environment, they sought to
reconcile their small town values with the larger, impersonal bureaucracy
of urbanized Los Angeles. As the dream of a suburban paradise faltered,
many turned to politics for a solution. In 1909, George Alexander, a
former Iowa farmer and merchant, was elected mayor of the city on a
progressive reform platform. The following year, the gubernatorial campaign of reformer Hiram Johnson attracted the attention of countless
disgruntled newcomers to the West Coast.

Johnson directed his appeal to "the people," pitting himself against
the directors of the Southern Pacific Railroad and their oligopolistic control over politicians in the state. His strident message appealed to midwestern moral sensibilities and to the growing national inclination to
oppose large corporate power. Johnson also provided them with an easy
scapegoat for the disappointments they had experienced in their new
home. In addition, Angelinos had concrete reasons to support an attack
on the railway companies. Henry Huntington, nephew of Southern Pacific Railroad president Collis P. Huntington, controlled southern California's electric streetcars, making millions from speculative real estate
investment in the process. Most city residents blamed him for inefficient
transit, congested downtown streets, and costly public services. As a partial consequence, Johnson received the majority of the Republican primary vote in Los Angeles County, carried every southern county in the
general election, and attracted widespread support from voters of native
stock.[18]

It was this progressive coalition that enabled the Johnson administration to enact its most significant legislative reforms. In 1913, still riding

the wave of his electoral victories, Johnson established a permanent Commission of Immigration and Housing, creating the primary body governing the state's immigrant policy for the next decade. The CIH investigated the working and living conditions of immigrants to California and spearheaded efforts to teach English to and Americanize foreigners. Although governmental bodies and private organizations in other states also sought to Americanize Mexicans and other immigrants, California's program was the most successful attempt to combine the efforts of government, business, and private citizens in dealing with the "problem of the immigrant." In a "scientific" and "rational" fashion, the Commission impressively recruited university academics, religious social workers, government bureaucrats, and white middle-class volunteers.[19]

The original impetus for the creation of the CIH came from social reformers in the Progressive party dedicated to the social settlement movement and the "Social Gospel" tradition. At the 1912 Progressive party convention, social workers Jane Addams and Frances Kellor introduced a statement subsequently incorporated into the party platform. The plank squarely committed Progressives to "Governmental action to encourage the distribution of immigrants away from the congested cities, to rigidly supervise all private agencies dealing with them and to promote their assimilation, education and advancement." In Chicago, Hiram Johnson, selected to be Theodore Roosevelt's running mate, was first approached about the CIH by California delegate Simon Lubin, himself a product of social settlement work in Boston's South End and Manhattan's Lower East Side. Lubin, who would eventually head the commission, envisioned an agency which would promote mutual accommodation between the native-born and foreign-born, with a focus on "immigrant gifts"—the cultural strengths that foreigners brought with them to American society.[20]

However, Lubin's initial emphasis on preserving what was positive and vital in various immigrant cultures gradually met with resistance. As World War I heightened anxieties concerning immigrants, nativist sentiment began to affect Americanization efforts through the "100 Per Cent American" movement, a loose collection of interests which sought to insure the loyalty of the immigrant to the United States. Additionally, big business took an interest in Americanization, as leaders sought a method to combat radicalism among foreign-born workers. Employers supported efforts to produce loyal, obedient employees, with at least one ultra-conservative business group in Los Angeles encouraging a "superpatriotism" that included upholding the "open shop."[21]

Moreover, as the decade progressed, Johnson's social agenda lost support among middle-class Angelinos, especially when he took positions in favor of workmen's compensation, minimum wage laws, and restrictions on child labor.[22] Los Angeles County's fickle enthusiasm for Johnson's programs also indicated a more conservative approach to the problems of immigration than that of Lubin and his supporters. While ethnic organizations of Jews and Italians played a prominent role in Americanization efforts to the north, groups such as the Daughters of the American Revolution led efforts in southern California from the start. By 1923, the conservative approach had prevailed statewide, as governor Friend W. Richardson dismantled the Commission on Immigration and Housing and transferred its Americanization duties to the state's Department of Education. Immigrant traditions and customs were now seen as impediments to their rapid, thorough integration into American life.[23] Ellwood P. Cubberley of Stanford University, long an advocate of the application of business methods in the schools, took over as the leading spokesperson for the Americanization work in the state. His philosophy of immigrant education was unequivocal. "Our task," he observed as early as 1909,

> is to break up these groups or settlements, to assimilate and amalgamate these people as part of our American race, and to implant in their children, so far as can be done, the Anglo-Saxon conception of righteousness, law and order, and popular government.[24]

Americanization programs dealing with the Mexican immigrant, largely developed after World War I, were swept up in this change of direction. The originators of California's program envisioned that the opening of the Panama Canal would bring an influx of European immigrants to California's shores; they did not anticipate the curtailment brought about by World War I and the subsequent immigration restriction acts. California progressives also reflected the general distaste for Asian immigration as reflected in their support for the 1913 Alien Land Act aimed at restricting Japanese land ownership or more long-standing calls for Asian exclusion. More apt to see Japanese and Chinese immigrants as "unassimilable," these reformers considered the Mexican immigrant as similar to the European in adaptability. Mexicans might have presented a greater challenge than did Italians or Jews, but California's Americanizers found nothing inherent in the Mexican character to prevent their eventual assimilation into the "American way of life."[25]

This position was clearly at odds with the approach of several groups who also witnessed increasing Mexican immigration in these years. Nativists, including many Anglo American politicians, academics, reporters, and others who believed in Anglo-Saxon racial superiority, waged a long and virulent campaign against unrestricted Mexican immigration. After successfully pushing Congress to limit severely immigration from Asia and southern and eastern Europe in 1920, nativists were dismayed to discover that the law still allowed for the widespread introduction of "foreigners" from south of the border, whom they considered equally, if not more, undesirable. Emphasizing the "Indian" or "Negro" makeup of the Mexican, his or her threat to "American standards of living," and the view that the Mexican represented an unstable element in a democracy, these nativists called for restriction on racial grounds.[26] Kenneth L. Roberts, writing in the *Saturday Evening Post* in 1928, gave voice to nativist sentiment when he observed that in Los Angeles, one sees:

> the endless streets crowded with the shacks of illiterate, diseased, pauperized Mexicans, taking no interest whatever in the community, living constantly on the ragged edge of starvation, bringing countless numbers of American citizens into the world with the reckless prodigality of rabbits . . . .[27]

Nativists were joined by others in their attempt to have Mexican immigration contained or reversed. Organized labor, under the auspices of the American Federation of Labor, viewed Mexican immigrants as cheap competitors with "American" workers. Samuel Gompers urged Congress to include Mexico in quota restrictions, arguing that Mexicans would not be content with farm labor and would soon attempt to enter the trades in the cities. Only months before his death in 1924, Gompers expressed concern that in Los Angeles, "it appeared . . . that every other person met on the streets was a Mexican."[28]

In contrast to these restrictionists, southwestern employers, particularly railroad, agricultural, and mining companies, defended unrestricted Mexican immigration on economic grounds. They were no less racist in their attitudes, but cited the economic advantage of Mexican labor, contending that "white" laborers would not and should not perform certain work. According to these employers, Mexican labor provided the most desirable option for filling labor shortages and was vital for the survival of their industries. To counteract the racial and political arguments of restrictionists, employers stressed that the cultural traits rejected by nativists actually benefited American society. The Mexican worker, they argued, embodied the perfect, docile employee, had no interest in intermixing with Americans, and invariably returned to Mexico once his labor was no longer needed. W. H. Knox of the Arizona Cotton Growers' Association belittled nativists' fears by asking,

> Have you ever heard, in the history of the United States, or in the history of the human race, of the white race being overrun by a class of people of the mentality of the Mexicans? I never have. We took this country from Mexico. Mexico did not take it from us. To assume that there is any danger of any likelihood of the Mexican coming in here and colonizing this country and taking it away from us, to my mind, is absurd.[29]

Whatever their theoretical position on the Mexican immigration issue, however, Anglo residents of Los Angeles had to face the fact that Mexican immigrants were becoming a permanent fixture in the urban environment. From 1910 to 1920, Mexicans had become the leading foreign-born group in the state, forming 12 percent of California's immigrant population. By 1930, that figure increased to almost 19 percent.[30]

While the battle between restrictionists and employers raged in legislatures and in the editorial pages of newspapers, other analysts began to believe that Americanization of these migrants was the only means to insure their cultural allegiance to the United States. Eventually, Mexicans became the primary targets of Americanization programs in California during the decade preceding the Great Depression. Particularly in the southern half of the state, Americanization came to embody the Anglo majority's attitudes toward Mexican immigrants. Borrowing from the arguments of both restrictionists and employers, Mexican culture in the United States was carefully scrutinized in the 1920s and found wanting.[31]

With the appointment of Dr. Emory S. Bogardus at the University of Southern California in 1911, Los Angeles had an academic who would, through his writings and that of his students and colleagues, lead the efforts to provide intellectual justification for Americanization over the next few decades in Los Angeles. Trained in sociology under Robert Park at the University of Chicago, Bogardus saw his mission in Los Angeles to be much like the more famous studies of social conditions of Chicago's ethnic working classes during the 1920s. Bogardus was appointed the first chairman of the Department of Sociology at USC, then organized USC's School of Social Work in the early 1920s, where he served as its first dean. Well into the 1930s, he regularly wrote on the conditions of Mexican immigrants in Los Angeles, and encouraged his many students to take ethnic Los Angeles as a site for many of their studies.[32]

Several of his graduate students rose in prominence as the leading social reformers and activists in the city during the 1920s and 1930s. They included a future director of the Los Angeles Housing Commission, future principals and teachers in the Los Angeles School District, and many social settlement workers in the immediate area. In fact, what distinguished the efforts made by USC's Department of Sociology and School of Social Work from the more famous work of Robert Park at the University of Chicago was its heavily moralistic impetus which had no qualms about mixing Protestant sensibilities, academic research, and public policy. Intimately linked to the growing Protestant social activism of the 1920s, Bogardus and these former students saw their role as awakening the growing Anglo American population of Los Angeles to the social realities and dangers represented by poorer, ethnic newcomers to the region.

Initially, most programs focused on identifying immigrants through their employers, a practice which inevitably made men the targets of Americanization. In the summer of 1917, for example, a railroad company found that converting one of its boxcars into a model house made good business sense. This rolling example of middle-class American culture also served as transportation for the Americanization teachers. "The labor supply had been more steady," reported the Commission of Immigration and Housing, and "the camp had been kept in beautiful condition, and the satisfaction of the workmen was showing itself in the better care of the track."[33]

Soon, however, Americanization instructors discovered that there were many factors which seriously undercut their efforts with male laborers. First, the transient nature of migratory, seasonal labor made sustained language instruction unfeasible:

> Many start [English language night classes] and show rapid progress, and then their work takes them away to another part of the county. When they return they have forgotten much they had learned because they have had little opportunity to practice English. Their class is ahead of them, and they "have shame," and do not want to re-enter, unless there are others in the same situation.[34]

Additionally, many teachers complained that the employers had created a situation in which the average Mexican laborer could not find sufficient opportunity to use the English language. Most stores he frequented had Spanish-speaking clerks to assist him. A few simple English words were all that was needed to negotiate the streetcar system to get to work. Most important, he was employed primarily in a segregated "gang" situation with other Mexican laborers, where the foreman usually spoke Spanish.[35]

These problems, along with the difficulty of securing the aid of enough employers, led Americanization advocates to search for an alternative approach. Since single men seemed more apt to drift from work site to work site and then to return to Mexico, it gradually became apparent that influencing the home life of married men, particularly those with children, might yield the desired results. Beginning with the World War I period, several important employers began to give preference to men with families. In addition, changes in immigration law made it more likely for men to keep their families with them in the city. The professionalization of Americanization work after 1920 also moved the locus of efforts into the schools and community centers, and away from migratory camps and employment sites, where volunteers had previously focused their attention. This shift determined that future Americanization work would center on immigrant women and their children, expressing Americanizers' hope that the influence of the home would extend to the public sphere. Justification for this modification of earlier plans was quick in coming:

> The Americanization of the women is as important a part as that of the men. They are harder to reach but are more easily educated. They can realize in a moment that they are getting the best end of the bargain by the change in relationships between men and women which takes place under the new American order. . . . "Go after the women" should become a slogan among Americanization workers, for after all the greatest good is to be obtained by starting the home off right. The children of these foreigners are the advantages to America, not the naturalized foreigners. These are never 100% Americans, but the second generation may be. "Go after the women" and you may save the second generation for America.[36]

The mechanism by which Mexican immigrant women were to be reached was already in place in the infrastructure of the Commission's activities. In 1915, the California state legislature passed the Home Teacher Act, a law which allowed school districts to employ teachers "to work in the homes of the pupils, instructing children and adults in matters relating to school attendance . . . in sanitation, in the English language, in household duties . . . and in the fundamental principles of the American system of government and the rights and duties of citizenship."[37] After World War I, the home teacher program was expanded, professionalized, and located within the public school system. From 1915 to 1929, the home teacher—usually a single, middle-class, Anglo woman—was the linchpin of Americanization efforts aimed at the Mexican family.

Mexican immigrant women were targeted for a variety of reasons. First, they were assumed to be the individuals primarily responsible for the transmission of values in the home. According to reformer's strategy, if the female adopted American values, the rest of her family would follow suit. Pearl Ellis, who worked with young Mexican women in southern California throughout the 1920s, stressed the important "influence of the home" in creating an employee who is "more dependable and less revolutionary in his tendencies . . . . The homekeeper creates the atmosphere, whether it be one of harmony and cooperation or of dissatisfaction and revolt."[38]

Americanization advocates were interested in the contribution Mexican women could make in transforming their families' habits from those of a rural, pre-industrial lifestyle to a modern American one. Herbert Gutman, in his important essay, "Work, Culture, and Society in Industrializing America," has examined the "recurrent tension" produced when immigrant men and women new to the American industrial order came into contact with the rigorous discipline of the factory system.[39] Because the Southwest lagged behind the rest of the nation in industrialization, local reformers were anxious to introduce Mexican women and men as rapidly as possible to the temperament of industrial society and inculcate Mexican families with the "Protestant work ethic." Targeting mothers was crucial to the overall strategy of Americanization.

Motherhood, in fact, was the Mexican immigrant woman's most highly valued role in Americanization schemes. By focusing on the strategic position of the mother in the Mexican family, Americanizers hoped to have an impact on the second generation, even if the immigrant generation itself turned out to be less malleable than expected. Undeniably, Americanization ideology was infused with the traditional American belief in the exalted role of the mother in shaping the citizenry of the Republic.[40] "As the mother furnishes the stream of life to the babe at her breast," wrote Ellis in a guide for teachers,

> so will she shower dewdrops of knowledge on the plastic mind of her young child. Her ideals and aspirations will be breathed into its spirit, molding its

character for all time. The child, in turn, will pass these rarer characteristics on to its descendants, thus developing the intellectual, physical, and spiritual qualities of the individual, which in mass, are contributions to civilization.[41]

Although Americans had debated for almost three decades the conflicts between women's private family responsibilities and their public roles as workers, Americanization programs demonstrated no such concern when addressing the ideal future of the Mexican American woman. With regard to immigrant women, Americanization advocates were readily capable of blurring the public and private spheres.[42] Teaching the Mexican mother proper American homemaking skills was meant to solve two problems at once: a happy and efficient mother would create an environment suitable for molding workers to the industrial order, and her newfound homemaking skills could be utilized in the cheap labor market outside the home. In 1908, a U.S. Bureau of Labor inspector had regretfully noted that Mexican "immigrant women have so little conception of domestic arrangements in the United States that the task of training them would be too heavy for American housewives."[43] However, black and European immigrant women had not migrated to southern California in large enough numbers to fill the growing demand for domestic labor. Consequently, Americanization teachers targeted Mexican women to help alleviate the shortage of housemaids, seamstresses, laundresses, and service workers in the Southwest.[44] By the 1920s, Americanization programs were busy training Mexican women to perform these tasks.

The most potent weapon used to imbue the foreigner with American values was the English language. All social reformers cited the ability to speak English as a fundamental skill necessary for assimilation. During and after World War I, however, English instruction was intended to provide the immigrant with much more than facility with the spoken language of the United States. In 1917, California's Commission of Immigration and Housing recommended "that employers of immigrants be shown the relation between a unified working force, speaking a common language, and industrial prosperity."[45] In 1918, Mrs. Amanda Matthews Chase, a home teacher with twelve years' experience teaching in Mexico City, was hired by the Daughters of the American Revolution and developed a primer to teach English. Home teachers were instructed to associate their lessons "with the pupils' own lives and affairs."[46] Thus, for example, they used the following song (sung to the tune, "Tramp, Tramp, Tramp, the Boys are Marching") to instruct female pupils about women's work as they learned twenty-seven new English words:

> We are working every day,
> So our boys and girls can play.
>   We are working for our homes and country, too;
> We like to wash, to sew, to cook,
> We like to write, or read a book,

We are working, working, working every day.
Work, work, work,
We are always working,
   Working for our boys and girls,
Working for our boys and girls,
For our homes and country, too—
   We are working, working, working every day.[47]

Yet despite the attention of reformers, Mexican women continued to lag behind men in learning the English language. A study of 1,081 Mexican families in Los Angeles conducted in 1921 found that while 55 percent of the men were unable to speak English, an overwhelming 74 percent of the women could not speak the language. Similar gaps existed in English reading and writing.[48]

Advocates of Americanization blamed the patriarchal nature of the Mexican family for this discrepancy. "The married Mexican laborer does not allow his wife, as a rule, to attend evening classes," reported USC's Emory Bogardus.[49] Americanization teachers consistently criticized as traditional and unprogressive the alleged limitations placed upon the Mexican wife by her husband. According to one Americanization instructor, if left in the home, the Mexican woman's "intellectual ability is stimulated only by her husband and if he be of the average peon type, the stimulation is not very great." The Mexican home, she concluded, "being a sacred institution, is guarded by all the stolid tradition of centuries."[50] If the Mexican home remained such a fortress, Americanization specialists would not be able to accomplish their mission.

Getting the Mexican woman out of her home, therefore, became a priority for Americanization programs because reformers saw this not only as the only avenue available for her intellectual progress, but as the only method by which they could succeed in altering her values. Home teachers visited each individual Mexican family in their district to gain the trust of members and encourage the husband to allow his wife to attend classes. The scheduling of alternative sessions in the afternoon for wives and mothers facilitated this process.[51]

Americanization programs, however, did not mean to undermine entirely the traditional Mexican family structure. Ironically, they counted on the cohesiveness of the Mexican family to achieve their assimilationist goals. Home teachers, even when they did get Mexican women out of the house and into classes, encouraged the acquisition of traditionally feminine household skills. In the ditty, "The Day's Work," for example, home teachers utilized the following sequence of phrases both to teach the English language and to instruct women about the proper organization of the family economy in American society.

In the morning the women get breakfast.
Their husbands go to work.
Their children go to school.

Then the women get their houses in good order.
They give the baby its bath.
They wash, or iron, or cook.
They get the dinner.
After dinner they wash the dishes.
Then they sew, or rest, or visit their friends, or go to school.[52]

Americanization programs sought to maintain the structure of family life while transforming familial habits, especially those concerning diet and health. Reformers encouraged Mexican women to give up their penchant for fried foods, their too frequent consumption of rice and beans, and their custom of serving all members of the family—from infants to grandparents—the same meal. According to proponents of Americanization, the modern Mexican woman should replace tortillas with bread, serve lettuce instead of beans, and broil instead of fry. Malnourishment in Mexican families was not blamed on lack of food or resources, but rather on "not having the right varieties of foods containing constituents favorable to growth and development."[53]

Women in the American reform movement were certainly conversant with the turn-of-the-century domestic science movement—a movement which associated scientific homemaking with moral regeneration.[54] Within the rubric of Americanization efforts, food and diet management became yet another tool in a system of social control intended to construct a well-behaved, productive citizenry. In the eyes of reformers, the typical noon lunch of the Mexican child, thought to consist of "a folded tortilla with no filling," could easily be the first step to a lifetime of crime. With "no milk or fruit to whet the appetite" the child could become lazy as well as hungry and might subsequently "take food from the lunch boxes of more fortunate children. Thus, the initial step in a life of thieving is taken."[55] Teaching immigrant women proper food values became a route to keeping the head of the family out of jail and the rest of the family off charity.

Health and cleanliness represented additional catchwords for Americanization programs. One of the primary functions of home teachers was to impress upon the minds of Mexican mothers and mothers-to-be "that a clean body and clean mind are the attributes of a good citizen." Reformers working with Mexican women were warned, however, that their task would be a difficult one. "Sanitary, hygienic, and dietic measures are not easily learned by the Mexican. His [sic] philosophy of life flows along the lines of least resistance and it requires far less exertion to remain dirty than to clean up." Reformers blamed Mexicans' slovenliness for their poor state of health. Such labeling reinforced the stereotype of the "dirty Mexican" and expanded its usage among Anglo urban dwellers. One eminent sociologist working with Americanization programs noted that Anglo Americans objected to the presence of Mexican children in the public schools for fear that their own children would catch a contagious disease.[56]

Pressing "American" standards of diet, health, and cleanliness upon Mexican women was not the only component essential in creating a healthy home environment. None of the potential gains made by these programs could be considered noteworthy if the Mexican female continued to bear too many children. Americanization advocates worried that unless she learned to limit family size, the Mexican mother would be unable to train adequately each individual member of her household.

Limiting the growth of the immigrant population was a long-standing concern of both Progressives and nativists. Americans first noticed that immigrant groups had a higher birthrate than native-born Americans at the end of the nineteenth century, and fears of "race suicide" had existed in the Anglo American mind ever since. When this fear rose in relation to the Mexican immigrant, both nativists and proponents of Americanization became alarmed: nativists wished to stave off an "invasion," while Americanization advocates viewed all unrestricted population growth as a vestige of Old World ways that must be abandoned in a modern industrial setting.[57]

Americanizers held Mexican women responsible for family planning. They also saw her hampered in these efforts by a number of factors. Traditional early marriage and the "inherent sentimentality" of the Mexican female promoted, they believed, a primitive sexuality and reinforced sexual ignorance. In addition, Catholicism discouraged birth control. Despite these barriers, Americanization teachers reported that Mexican mothers were beginning to exhibit dismay with their large families, and occasionally inquired about birth control measures. Some even warned others to delay marriage on the grounds of "much work, too much children."[58]

Americanists viewed such evidence of changing attitudes as a hopeful sign, because limited reproduction opened up new opportunities for Mexican women within and outside the home. As proper household managers, Mexican women could devote more time to raising fewer and more productive children. But family limitation also created new possibilities for female employment by freeing Mexican women from the demands of continual childrearing. Traditionally, Mexican women's family obligations had barred them from wage labor outside the home. When a Mexican immigrant woman worked, it was almost always in her late adolescent or early adult years before marriage.[59]

As industrialization in the American southwestern economy developed, so too did demands for cheap labor performing tasks that had traditionally been performed by women inside the home. While the garment, laundering, domestic service, and food preparation industries gradually relied more on "women's work" in the marketplace, employers in the region had fewer workers because of the restrictions placed upon Asian and European immigration, and because black migration to the Southwest was still quite low. Moreover, demands of the Anglo middle class for these services increased, exacerbating further the labor supply

problems. Despite all the traditional objections to Mexican women working outside the home, Americanization programs actively promoted
Mexican immigrant women for entrance into these sex-segregated occupations.[60]

This commercialization of traditional female forms of labor made it
easier for Americanizers to advocate instruction in such tasks without
appearing to upset the social order within the Mexican family. For example, skilled needlework was viewed in Americanization programs as a
talent passed down through generations of Mexican women. Americanization teachers fostered such activity so "that we may not lose this valuable contribution to our civilization with the passing of time." Some
reformers argued that needlework instruction should replace academic
courses for Mexican girls as early as the third grade in school. This sort
of activity, supposedly more in line with "traditional abilities," they argued, would keep Mexican girls in school and foster "greater respect for
the school and for our civilization."[61]

Given the dual role reformers envisioned that the Mexican woman
would play within and outside the home, every newly learned skill supposedly benefited American society doubly. When Americanists stressed
the ability to set a table and to serve food properly, they were encouraging Mexican women not only to arrange home meals by American standards but also to learn that "sloppy appearance and uncleanliness of person would not be tolerated in a waitress." In addition, the burden on a
private citizen employing a Mexican woman as a domestic servant would
be considerably lightened if the employee had already been adequately
trained through their programs. As one social worker stated in the late
1920s: "Americans want household help for two or three days a week,
and they can, if they will, take Mexican women and teach them. It requires patience to be sure, but there are large numbers of Mexicans who
can fill the household gap if the proper conditions are made."[62]

Encouraging Mexican women to engage in hard work was also
viewed as an important step toward "curing" the habits of the stereotypical "lazy Mexican." According to one Americanization teacher, " *'Quien
sabe?'* (who knows?) was the philosophy of all of Mexico, and the inability of Mexicans to connect the things that are valued as worthwhile to
the effort necessary to obtain them made Mexican laborers inefficient."
Another felt that "the laziness of Mexicans was due to climate conditions
and inherited tendencies" which only hard work could root out. Consequently, putting Mexican women to work would have the effect of promoting discipline in them, which in turn would encourage them to pass
on a similar level of self-control to their children.[63]

Eventually, as national attention increasingly turned toward restricting future immigration from Mexico, Americanization advocates
found themselves caught in the middle of the controversy, with little
concrete evidence to prove that their efforts had effectively resolved the
"Mexican problem." One of the few quantifiable means by which to mea-

sure success or failure in Americanization was the rate of naturalization, and in this area Mexican immigrants displayed little progress. Statistics from the period simply did not suggest that Americanization had affected the rate of naturalization. In fact, among the Mexican immigrant population in California, which already had the lowest rate of naturalization of any immigrant group in the state in 1920, the ratio of naturalized citizens to the total foreign-born Mexican population declined during the 1920s.[64] Given this trend, and the long-standing ambivalence of reformers toward the immigrant, Americanizers shifted their focus. In 1927, the Commission of Immigration and Housing sided with restrictionists, calling for an end to unlimited immigration from Mexico, and blaming immigrants for "causing an immense social problem in our charities, schools and health departments."[65]

Moreover, the efforts to alter the immigrant generation itself were abandoned in favor of school-based programs which sought to teach American-born children a culture different from that of their immigrant parents. In the schools, socialization in American values and language skills were even more emphatically combined with the goal of social stability. The increased application of I.Q. testing, always administered in English, invariably segregated Mexican children in special classes for the mentally inferior or mentally retarded. Macy Street School, for example, located just east of the Plaza, developed a curriculum at the primary level in which two-thirds of class time was spent on "decorative subjects" such as music, dancing, art, needlework, cooking, and other manual arts. The pedagogical impetus behind this curriculum was that the "Mexican genius expresses itself through activities rather than abstractions." At the secondary level, citizenship classes were integrated into vocational training for laundries, restaurants, garages, household work, and agriculture. By the late 1920s, the promoters of Americanization put their hopes for the future in vocational education and classes in citizenry directed at American-born Mexican children.[66]

The efforts directed at children, like those aimed earlier at their parents, promoted above all the habits of thrift and time discipline. In southern California, business interests ardently favored Americanization programs that advocated promptness and diligence at work. Businessmen learned to cooperate both with the Protestant reformers interested in fostering internal controls over morality and economy and with the social feminists hoping to upgrade women's position within the Mexican family. They understood full well that despite the range of motivations behind Americanization, the price of acceptance for Mexicans into American society via their programs was predicated on the abandonment of a culture they perceived as inherently inferior.

Rather than provide Mexican immigrants with an attainable picture of assimilation, Americanization programs could offer these immigrants only idealized versions of American values. In reality what was presented turned out to be little more than second-class citizenship. The most progressive assumptions behind Americanization programs were never fully shared by the government or business interests involved, and thus they could never be fully implemented. One Americanization teacher who spent the decade working with Mexican immigrants noted with disappointment in 1923 that the newly elected governor of California had eliminated financial provisions for the Americanization program in the public schools from his budget.[67] At least one historian has concluded that the "love affair between the progressive and the businessman" in California inevitably led, in the 1920s, to a blunting of "the cutting edge of progressive social reform."[68]

The halfhearted effort of administrators of Americanization programs limited available personnel and resources and ensured that the programs would never be able to cope with the volume of the Mexican migration. The barrios expanded so quickly in the 1920s that any single Americanization teacher found it impossible to keep abreast of the number of new Mexican families in her district who needed a resumption of her program from scratch. Newer areas of Mexican settlement were usually beyond the reach of established Americanization programs entirely. Furthermore, Mexicans experienced a high degree of geographic mobility in this period that easily wiped out whatever progress had been made by these programs in a given community. According to historian Ricardo Romo, fewer than one-third of Mexicans present in Los Angeles in 1917–18 were present in the city one decade later.[69] Americanization teacher Amanda Chase acknowledged the extent of this problem when dealing with Mexican women: "I have had in my class record book this year the names of about half as many Mexican women as there are Mexican families in the district. But a third of them moved to other districts."[70] Mexican immigrants could not hope to develop allegiances to the United States when the economic condition of their families forced them to migrate consistently in search of an economic livelihood.

In the end, Americanization programs never had the time to develop sufficiently even to approach a solution to the problem of Mexican immigrants in the United States. With the stock market crash of 1929 and the subsequent Great Depression of the 1930s, all attempts to Americanize Mexican immigrants came to an abrupt end. Rather than search for ways to assimilate these newcomers, American society looked for methods to be rid of them altogether. About 500,000 Mexicans left the United States during the 1930s under strong pressure from the government, and up to one-tenth of these individuals had resided in Los Angeles.[71] Americanists joined in these efforts to repatriate Mexican residents; their commitment to improving the conditions of the Mexican had no place in an economically depressed America.

Instead, Americanization programs are an important window for looking at the assumptions made about both Mexican and American culture by progressive Californians during the 1920s. Mexican culture was seen as malleable, but required intense education in "American values" to fit into a modern, industrialized society. These efforts also made clear, however, that Mexicans were intended only to assimilate into the bottom segment of the American work force as low-paid, yet loyal, workers. As we shall see, Mexican immigrants generated their own version of Americanism without abandoning Mexican culture. What they would create would be quite a different product indeed.

CHAPTER 5

The "New Nationalism," Mexican Style

Nineteen twenty-one marked the one hundredth anniversary of the consummation of Mexico's independence from Spain, and Mexicans in Los Angeles were planning their largest patriotic celebration ever to mark the milestone. Organized by the newly formed Mexican Committee of Patriotic Festivities, the month-long program included a parade of citizens of Mexican states, a beauty contest, Mexican music concerts and film exhibits, and a public ceremony culminating in the traditional *grito,* or "yell," for independence on the 16th of September. This was the first year that the Mexican independence day activities were directly sponsored by the Mexican consulate office, working largely through an honorary commission of important members of the expatriate community. Each year for the next decade these events would grow in size and importance, as money for such expensive celebrations was obtained from the consulate office and raised from the immigrant communities themselves.[1]

Yet lurking behind the supposed unity of the Mexican community at these events lay tension and distrust. Another organization, the "Sociedad Hispano Americana," had sponsored their own Mexican independence celebrations for many years previous to 1921 and refused to step aside when the new Committee of Patriotic Festivities was organized by the Mexican consulate. This other organization, also known as the Alianza Hispano Americana, was made up of individuals Mexican in origin but American in nationality and was roundly criticized in the Spanish-language press for their position. This one-day celebration was held in Selig Zoo Park, and included a bloodless bullfight, dancing, and oratories in both English and Spanish. Presided over by the consul of Cuba and Panama, the day's activities also included presentations by Professor Miguel Laris, Attorney Anthony Orfila, and Elena de la Llata, president of the Mexican Blue Cross, or Cruz Azul.[2]

The editor of *El Heraldo de México* called upon the renegade organization to give way to the newly formed committee since it was "genu-

108

inely Mexican," although he acknowledged the fact that as "semi-compatriots, sons of beautiful California," they also had a right to honor Mexican heroes. What particularly irked the editor was that the *Alianza* used the Mexican festivities for financial gain. Since the *Alianza* was at its base a mutualist organization that sold funeral and death insurance to immigrants, one can assume that it used Cinco de Mayo and 16th of September festivities to publicize its product. According to *El Heraldo,* however, the revolutionary "Hidalgo would have forfeited his title as Father of the Mexican Independence if he had taken advantage of the precise moment of the 'grito' to sell his silk products and pottery more easily or at a better price." No doubt another factor in raising *El Heraldo*'s ire was that the official Mexican Committee of Patriotic Festivities had maintained its headquarters at the offices of the newspaper.[3]

This tension revealed the increased presence of Mexican consulate officials in the cultural life of Mexican immigrants in the United States. During the 1920s, organizations affiliated with the Mexican consulate increasingly came to dominate the public life of the Mexican community in Los Angeles. From celebrations of Mexican holidays to the formation of Spanish-language schools, the Mexican government, through its emissary the consulate, played the central role in organizing community life. However, tensions like the one over the 1921 Mexican Independence festivities continued to surface periodically throughout the decade. By the onset of the Depression, it became obvious that the interests of the Mexican government were not always identical to that of large segments of the immigrant and native-born Mexican population.

Chicano historians have yet to explore fully the often contradictory role played by Mexican government officials in immigrant communities in the United States. Indeed, many have treated the development of ethnic nationalism among segments of the Mexican immigrant population as an independent, natural phenomenon which was a reflection of general alienation from U.S. society and a developed sense of peoplehood borne of revolutionary sentiment taken across the border through migration. F. Arturo Rosales, for example, credits the emergence of a "Mexico Lindo" (Beautiful Mexico) mentality among Mexican immigrants in the 1920s to "adverse conditions which they encountered in this country" which "strengthened ethnic bonds forged by a common heritage carried over from Mexico."[4] Rarely have the ideological developments in Mexican immigrant communities been connected to the increased role of Mexican consulates during the 1920s in shaping local institutions and programs to instill patriotism among the "Mexicanos de afuera."[5] Though clearly most working-class Mexican immigrants to Los Angeles took pride in their cultural background and ethnic heritage, they also reflected the ambivalence toward the Mexican state characteristic of other migrants fleeing political or religious persecution and economic destitution.

This chapter will explore these tensions to understand the role the Mexican government played in shaping Chicano community interests and identity during the 1920s. As this government focused on the creation of a national identity south of the border, many facets of the emerging government-sponsored nationalism also surfaced among the immigrant population through the work of the consulate office. In Los Angeles, the consulate office launched Spanish-language schools and libraries as a way to reinforce loyalty to the mother country among immigrants and their children. As in Mexico, these activities and institutions did not always have the interests of the general population in mind, but rather served to promote a conception of Mexican nationalism shaped by a rather elite group of Mexican officials and intellectuals. Like Americanization efforts, these Mexicanization programs helped shape the culture of Mexican immigrants but were unable to control the complex process of cultural creation the newcomers underwent once they arrived in Los Angeles.

The Mexican government had long maintained a rather precarious position with Mexican immigrants in the United States. Although one of the main acknowledged roles of the Mexican consulate was to protect Mexican nationals in the United States, consulate officials since the Porfiriato had generally carried more bark than bite. Repeated incidents of abuse and maltreatment from American employers, particularly in Texas, could rarely be met by more than official protests by the Mexican government, and even these came sparingly in the early twentieth century.[6]

Moreover, segments of the Mexican immigrant community also posed a threat to the stability of the current regime in power in Mexico up until 1920. During the last fifteen years of the reign of Porfirio Díaz, various plots to overthrow the Mexican government were launched from the northern side of the border, most notably by the Partido Liberal Mexicano (PLM) forces of Ricardo Flores Magón, working out of Los Angeles, El Paso, St. Louis, and San Antonio. This situation caused Mexican consulate officials to form intelligence-gathering networks within the Mexican immigrant communities to report back to contacts in Mexico City the latest in subversive activity. In Los Angeles, Mexican American policemen, working for the Mexican consulate, were instrumental in the arrest of Magón himself and others for subversive activity.[7]

This situation did not change with the overthrow of Díaz and the coming of the Mexican Revolution. In fact, given the precariousness of Mexican leadership until the signing of the 1917 constitution, surveillance activities in the United States actually increased. Francisco Madero, who in 1910 launched his successful movement to replace Díaz as president while in exile in the United States, continued overt and covert activities against the PLM, providing consulates with substantial funds to purchase support and intelligence from the expatriate communities. The conservative Victoriano Huerta, who overthrew Madero in 1912, replaced important consuls in the United States and set out to quell the growing Constitutionalist forces organizing in most Mexican communities in the Southwest. With the overthrow of Huerta in 1914 and the resulting anarchy in Mexico, competing forces spied on each other throughout the Southwest. Since different forces were in control of different sections of Mexico, consulates in the United States were in the hands of individuals who were forced to side with a particular faction and use their official position to gather intelligence, suppress dissent, and seek political survival.[8]

In Los Angeles, Consul Adolfo Carrillo sided with Venustiano Carranza and received $80 a month for the support of secret service agents to gather intelligence about conservative supporters of Huerta and Díaz, and leftists backing Pancho Villa or the PLM. *El Eco de México,* a bilingual newspaper that supported the Constitutionalist movement, also received a subsidy of $650 a month from the consulate office. Carrillo consistently asked Carranza for greater support for his Los Angeles office, citing its political importance, while undercutting colleagues at other consulate sites. After numerous complaints about his scandalous behavior and lack of interest in protecting Mexican workers, Carrillo was fired by Mexico City in 1916 but had to be forcibly evicted before he gave up his office. The next consul in Los Angeles left suddenly after publicly brawling with the son of a Cuban general. Taken together, a picture emerges of a rather ineffective agency, its occupants usually more interested in perks, favors, and survival than in the protection of Mexican nationals.[9]

The signing of the Constitution of 1917 brought relative stability to Mexico and increased professionalism to the ranks of the consular service in the United States. It also ushered in a new generation of younger consular and diplomatic officials that during the 1920s came to dominate in the United States. Although most major positions continued to be handed out as political rewards, this generation shared experiences and a collective ideology that made it distinct from before.

Unlike their predecessors, most had actively participated in the Mexican Revolution, although not usually on the military or political front line. They most often had been students who consciously opposed Porfirio Díaz and backed Francisco Madero's middle-class revolution against dictatorial powers. In fact, recent scholarship on the Mexican Revolution suggests that the ideas of the leadership elite which came to power in Mexico after 1917 had much in common with certain tenets of American Progressivism. Writing about Alvaro Obregón, for example, who assumed the Mexican presidency in 1920, historian Ramon Ruíz compares his views on capital-labor relations to Wilsonian Progressives who were "enlightened perhaps but certainly not radical." Obregón saw the role of government as a "neutral arbitrator" between labor and capital, "giving business and industry the opportunity to earn a just profit, and in the process, to provide labor with adequate wages and a decent standard of living." Despite the violent character of the Revolution and his coming to power, Obregón, as early as 1915, expressed the belief that of all reforms needed in Mexico the "most important were moral in character." [10]

With a line of leaders that stretched from Francisco Madero to Venustiano Carranza and to Alvaro Obregón, Mexico's revolution had failed to bring into power leaders such as Emiliano Zapata who espoused radical proposals for reform. Instead, a growing middle class, frustrated for decades by the intransigence of dictator Porfirio Díaz, took control of the country. Grown weary of military action, political maneuvering, and near anarchy that marked the period from 1914 to 1917 following the overthrow of Adolfo de la Huerta, these individuals increasingly backed Carranza's Constitutionalist forces when faced with the unpredictable Francisco Villa in the north and the agrarian revolt of Zapata in the south.

The 1920s would become a period in which the middle classes used nationalism to disarm revolutionary sentiment among the mestizo and Indian populations while legitimating their own rule by emphasizing their claim to the revolutionary heritage. The 1924 commemoration of Emiliano Zapata's death, for example, witnessed Obregón's handpicked successor, Plutarco Elías Calles, claiming that "Zapata's agrarian program is mine." This was an amazing act of cooptation on the part of Calles, who had fought Zapata while alive, was part of the leadership which probably sanctioned his assassination five years earlier, and had hitherto failed to implement even the mildest land reform. When the municipal government of Mexico City formed an Official Committee of Patriotic Commemorations in 1925, it did so to insure ideological uniformity in public ceremonies and to circulate propaganda which closely reflected the policies of the federal government. Many strikes during the 1920s were circumvented by the Mexican government by accusing unruly labor leaders of being unpatriotic and counterrevolutionary. [11] Ironically, Mexico's representatives in the United States would espouse a similar middle-class rhetoric to a largely working-class population through its consulates north of the border.

For the individuals who assumed these positions in Mexico's diplomatic corps in the 1920s, the government that was produced by the 1917 constitutional congress in Querétaro represented stability and order and their own political coming-of-age. Future Los Angeles consul Rafael de la Colina, for example, served as a messenger during the Querétaro Congress, remembering it as an "extraordinary experience." [12] The delegates, as described by historians Michael Meyer and William Sherman, were clearly representative of the emerging Mexican middle class:

> The delegates at Querétaro represented a new breed of Mexican politician and, in a sense, constituted a new social elite. Unlike the Convention of Aguascalientes, miltary men constituted only 30 percent of the delegates. Over half had university educations and professional titles. The large majority were young and middle class; because they had been denied meaningful participation during the Porfiriato, many were politically ambitious. [13]

After 1917, but particularly with Alvaro Obregón's ascendancy to the Mexican presidency in 1920, an ideology of order and control came to be represented in Mexican communities in the United States by the Mexican consuls. Obregón was among the first Mexican presidents actively to encourage consulate offices in American cities to expand their efforts to protect Mexican nationals already working in the United States. In Los Angeles, given the volume of migration from Mexico and the lack of a significant Mexican American middle class, the consul emerged as the central organizer of community leadership.[14]

Although the position of the Mexican government toward emigrants had long been ambivalent, during the 1920s the government began to see them as representing an important source of expertise if they could be convinced to return to Mexico. Therefore, a central goal of all programs initiated by the Mexican consulate was the preservation of the cultural integrity of Mexican emigrants through the establishment of institutions to foster Mexican patriotism, with the long-term goal of encouraging return migration. In order to institutionalize this objective, President Obregón established the Department of Repatriation within the Secretariat of Foreign Relations in May 1921 and allocated funds for those wishing to return during the following year.[15]

At first, the Mexican government had attempted to intervene directly in blocking migration to the north. Venustiano Carranza, president of Mexico from 1917 to 1920, initiated efforts to slow down and control the exodus of Mexican laborers to the United States. Government officials warned their citizens of the rampant prejudice and exploitation awaiting them if they ventured north across the border. But the intensive propaganda campaign had little effect on emigration, since the vast majority of migrants relied on information from friends and relatives who, in spite of unfair treatment, usually painted glowing pictures of better conditions in the United States. When officials in Jalisco refused to issue passports because of complaints of labor shortages from local industrialists and agriculturalists, those wishing to migrate simply went to other states to obtain papers or ventured north without them.[16]

Carranza's successor, Alvaro Obregón, more readily acknowledged the futility of trying to hold workers in Mexico. Instead, he used a different tactic, namely encouraging consulate offices in American cities to expand their efforts to protect Mexican nationals already working in the United States. As Americanization programs intensified, the Mexican government watched with ever-increasing uneasiness, for such programs posed a direct threat to their own aspirations regarding Mexican immigrants in the United States. When Mexican officials gradually realized that a significant proportion of their countrymen crossing into the United States would not automatically return, questions of national loyalty and identity attracted more attention. An important goal of the new programs initiated by the consulates, therefore, was the preservation of the cultural integrity of Mexican emigrants through the establishment of institutions to foster Mexican patriotism.

These efforts were to be carried out largely by the organization of Honorary Commissions made up of members of expatriate communities affiliated with the consulate offices. Though their members received no pay from the Mexican government, consuls repeatedly assured the commissions of their importance to their native country, and members became key spokespersons in their respective communities.[17] In Los Angeles, this Comisión Honorífica, in conjunction with the consulate office, sponsored patriotic celebrations, organized community-wide alliances and educational conferences, and initiated a plan for alternative schools where children could learn the Spanish language and Mexican history.

While members of Mexico's diplomatic corps were quintessential representatives of Mexico City's elite, those gathered to participate on Honorary Commissions were a more diverse group. Although outlying non-urban areas of southern California were often served by respected blue-collar workers, the Honorary Commission of central Los Angeles was almost exclusively limited to middle-class businessmen and professionals. Not a homogeneous group by any means, many were political refugees who had fled the Mexican Revolution and, depending on who was in power in Mexico, could be either in or out of favor with the consulate. Only a relative few were small enterpreneurs who had moved up the social ladder after arriving in the United States. Most had come from Mexico with professional credentials and became the barrios' first doctors, lawyers, and pharmacists.[18]

These individuals shared a conservative world view with which they differentiated themselves from the Mexican masses on the basis of race, class, urban background, and education. In Mexico, they had feared the volatile nature of the mestizo peasant led by a Zapata or a Villa, but in the United States they found themselves cast together with the lower classes through the common thread of nationality and the racism of Anglo American society, which rarely distinguished between rich and poor. Through their affiliation with the Mexican consulate, they were able to cultivate a tenuous leadership role in a mostly working-class population.[19] If the Mexican community had a middle class in the progressive tradition, this was it.

Zeferino Ramírez was one of the few who had risen from humble surroundings in the United States to assume a position of respect and authority among the Mexican community of Los Angeles. On first arrival to the United States during the decade of the Revolution, Ramírez engaged in back-breaking work in mining and railroad maintenance, and was forced to live in an insect-infested room in Los Angeles when unable to find work. He finally managed to save enough money to buy a small home in Belvedere, after working seven years as a highway laborer. In the mid-1920s with the help of Protestant missionaries, he started a business of his own, opening an undertaking establishment, after serving as an apprentice in an Anglo-run mortuary. As one of Belvedere's first Mexican businessmen and a founder of the Mexican Chamber of Commerce, he quickly earned the respect of his neighbors. He spearheaded efforts to establish a Mexican school in Belvedere and served in a variety of capacities on committees organized by Mexico's Los Angeles consulate.[20]

More typical of the sort of individual who sat on the Honorary Commissions was Juan B. Ruíz, a pharmacist and manager of Ruíz Pharmacy, the oldest drug store in Los Angeles. Born in Culiacan, Mexico, in 1896, Ruíz was the son of a large rancher and mine operator in that state who had taken a neutral position in the Revolution. Forced to flee to the mountains during 1913, Ruíz took a government job under Carranza after attending medical school in Mexico City. Although offered a position in the Department of Education under Obregón, Ruíz decided to move to Los Angeles in 1920 and was able to invest $500 in the purchase of a drug store on Main Street. During the 1920s, his establishment became known as one of the most important sites for political hobnobbing; Ruíz played host to Mexican politicians from Alvaro Obregón to Jose Vasconcelos. Ruíz himself took on the job of Vasconcelos's campaign manager in his failed attempt at the Mexican presidency in 1928.[21]

Others in the inner circle of the Mexican consulate had similar, if not as exciting, stories of emigration and ascendancy. Mauricio Calderon was the main businessman responsible for the growing Spanish-language music industry of Los Angeles in the 1920s. As proprietor of Calderon Music Company, he had been active in Mexican social circles since his arrival in Los Angeles in 1915, having fled the revolution in Chihuahua. Physicians R. J. Carreon, Rafael Martin Del Campo, and Camilo Servin all arrived in Los Angeles in the 1920s to attend medical school or to begin practices serving the ever-expanding Mexican immigrant community. These doctors would consistently involve themselves in social welfare work and the establishment of clinics under the auspices of the Mexican consulate, in addition to becoming prominent members of the Honorary Commissions.[22]

Although this group of individuals was busy developing a social and political ideology borne out of its unique leadership status in the Mexican immigrant community, they also shared certain similarities with the wider progressive tradition. Together they saw themselves as operating between the masses of working-class Mexican immigrants, often as "guardians" of their "race," and the larger Anglo American powerbrokers in the city, particularly the capitalists who employed Mexican labor. Yet most important, these leaders developed a progressive ideology in the context of *Mexican,* not American society. While Anglo American reformers increasingly sought either to Americanize foreigners or to exclude them in the 1920s, the Mexican progressives tried to encourage continued loyalty to mother Mexico.

As the decade progressed, more ambitious projects were undertaken. During the mid-1920s, for example, Spanish-language libraries were established in southern California with books donated by both the Mexican government and local Mexican bookdealers. At the end of November 1926, for example, a new Mexican library was opened in the growing barrio of Belvedere by a group known as the "League of Mexican Culture." The Mexican consulate, along with several socially prominent Mexican citizens living in Los Angeles, was invited to inaugurate the building intended for the use of Mexican adults in the Belvedere community. This new library contained almost exclusively Spanish-language books, in addition to copies of Spanish-language newspapers from around the United States, Mexico, and Spain. All material focused on the politics and society of Mexico and the rest of Latin America, and not on events in the United States. Even *La Opinion,* the city's largest circulating Spanish-language newspaper founded that same year, devoted less than 11 percent of its editorials and 12 percent of its news articles to coverage of the Mexican population in the United States.[23]

By the late 1920s directors of Mexicanization efforts, like their Americanizing counterparts, increasingly focused on the children of immigrants by establishing schools for the study of the Spanish language and Mexican history and culture. Although individual Mexican citizens had periodically attempted to establish community-based schools before, La Escuela México established in the Belvedere barrio on May 2, 1926, was the first inaugurated under the direction of the Mexican consulate and Mexican Secretary of Public Education with authorization from the Education Department of the State of California. This arrangement was made possible by the presence of Ms. Margarita Robles, who was certified to teach by both Mexican and California officials. This insured that upon a student's return to Mexico, all courses and examinations would be credited as valid.[24]

Zeferino Ramírez, the aforementioned Belvedere businessman, built the school on his property at 4645 Eugene, then donated it to the Mexican consulate for use by the children of the district. The first desks were also underwritten by another Belvedere businessman, while Ramírez paid for the blackboards. All the needed textbooks, paintings, murals, maps, and other instructional materials were provided by Mexico's Secretary of Education. In a schedule consciously patterned after Japanese and Hebrew schools, Mexican children attended classes from 4 to 6 p.m., after a full day in the American public school system. Admission was open to all children of Mexican parents, provided they observed good conduct and attended regularly. Like the programs at Mexican schools throughout southern California, study included instruction in the Spanish language, Mexican geography, history, and native arts.[25] According to an editorial in *La Opinión,* the intended results of this instruction were straightforward:

> the rehabilitation of the Spanish language in many places where children are losing or corrupting it; a knowledge of our geography and history as a prin-

cipal link which joins our hearts to the nation of origin; a better union between those who live in Los Angeles and who find themselves dispersed and isolated; the gathering of a great number of our children . . . [to] foster among the expatriates a youth solidified and unified in their patriotism.[26]

In a stunning analysis of the textbooks used in Mexican public schools of the 1920s, historian Mary Kay Vaughan examines the educational content which would have been transferred to Mexican schools in the United States. Through textbooks and teacher training, Mexican public education displayed a stubborn persistence of Porfirian ideals of law and order, obedience and discipline. Despite revolutionary rhetoric, textbooks were also full of guidance on the urgency to imitate the European experience and basically portrayed indigenous culture as backward and in desperate need of orientation to "civilization." Written in large part to instill patriotism, Mexican textbooks of the period concentrated on elite heroes and a hierarchy of civilization committed to economic modernization with an underlying admiration of Anglo-Saxon success.[27] Given the perspective presented in the classroom, it should not be a surprise that when the Mexican children enrolled in the Belvedere school were taken on a field trip to see the Indian artifacts in the Southwest Museum, at least one former pupil remembered her feeling that "while we were ethnically linked to these Indians, I didn't relate to them and saw the displays as something out of the forgotten past."[28]

In line with teaching practices of the day, Mrs. Robles and the other Mexican schoolteachers taught by rote. A former pupil remembers that "you did things over and over until you got them right."[29] At times, however, the classroom could transform itself into an arena for exploring the condition of Mexican people in the United States:

> The last 15 minutes of class were so special. Everyone would close his books and settle down and Mrs. Robles would read us a chapter from a book that told the story of a little boy's travels from the old world to the new. Sometimes we would see Mrs. Robles, as she closed the book, brushing a tear from her eyes, carried away with the pathos of the story.[30]

No doubt the children connected with the theme that their own status as immigrants placed them in such a precarious position in a foreign land.

Despite the initial generosity of local citizens, however, the Escuela México of Belvedere, like many of the other schools developed by the Mexican consulate, very quickly ran into financial trouble. Although individual members made donations, the Mexican Chamber of Commerce of Belvedere itself never supported the school, a sore point for many because it often took credit for its success. In fact, in its first year the school suffered when it failed to pay its electricity bill and power was shut off for several days. Ms. Robles, the main teacher, organized a "Mexican Thinker" Society made up mostly of parents of children in the school to help defray the costs of additional teacher aides. Although classes were still free of charge, parents were asked to donate $1 per month to provide modest salaries for the women who were involved in instruction.[31]

In addition to continual financial woes, the school was also affected by disputes between groups of parents over the content of instruction. According to *La Opinión,* one group of parents failed "to understand how important it is to inculcate patriotic spirit among Mexican children who are born here or come to this country when they are very small." According to Ms. Robles, this group was also accused of trying to make the school a lucrative enterprise. Because of this dispute, Zeferino Ramírez withdrew his support for the school and the Mexican consul threatened to remove all teaching materials and equipment, which he considered consulate property. Though this particular school ultimately closed, another group of parents and Mexican consulate officials would try again two years later and inaugurate another Mexican school in Belvedere on December 1, 1929.[32]

Despite all the internal and financial problems of schools like the one in Belvedere, grandiose plans for about fifty Mexican schools were developed in the offices of the Mexican Consul General in Los Angeles in 1927 and 1928. Because of lack of funds, however, no more than ten such schools were ever in operation at the same time. Moreover, the state of California made it difficult to maintain the schools by refusing to accredit additional teachers in 1928 on the basis of their foreign nationality. These rulings pushed the local consulate and Mexico's Education Department to concentrate on plans for one large school with transportation throughout southern California, much like that initiated by the Japanese consulate. This plan, like others before and after, failed to materialize in southern California due to lack of funds, rapid turnover of consulate personnel, and disagreements between member of the expatriate communities. By the end of 1930, only three schools were left in operation in California—reportedly in Pacoima, Van Nuys, and Claremont—serving only 200 of the estimated 80,000 Mexican and Mexican American children in the state.[33]

Many accused the Mexican government of lukewarm support for these efforts. One editorial writer demanded that the government cease to sponsor "useless festivities and banquets, ceremonies in which nothing beneficial results." Noting the schools' continued financial difficulties, he advocated additional funding for them.[34] Mexican officials did not deny their set of priorities. When asked why the Mexican schools in California did not survive, Professor Carmen Ramos of the Mexican Department of Public Education, who worked for two years to establish Mexican schools in California, answered readily:

Because of the lack of money. Our government has only resources to cover the urgent needs of the country [Mexico], and until now could not devote a single portion of the budget to educate Mexican children who live abroad.[35]

In fact, one may ask why the Mexican government was involved in educating Mexican children in the United States at all. To understand this concern, one must consider the place of education in post-revolutionary Mexico. During the 1920s, the Mexican federal government's role in public education, particularly in rural areas of Mexico, expanded greatly. Since Mexican politicians were primarily concerned with containing the revolutionary sentiments of the general population, they adapted Dewey's pedagogy of progressive learning to "consciously try to use the school as a mechanism for legitimizing the new state."[36] Rather than taking a broader structural and historical outlook on Mexican economic backwardness, Obregón's Secretary of Public Education, Jose Vasconcelos, and his successor Manuel Puig Casauranc focused on changing peasant values and behavior to lead Mexico to greater capitalist productivity and nationalist integration. While looking out at the mostly rural nation from the vantage point of Mexico City, political officials displayed an intense "desire to impose urban middle-class standards of propriety on rural communities, which were often viewed as primitive, backward, and savage."[37] Historian Vaughn emphasizes that the urbane Mexican teacher played an ideal role in rural education which mirrored the ideology surrounding American progressive educators:

> The teacher was a civilizer, who introduced proper customs to the community. Thus, the school was preoccupied with such issues as the use of soap and bathing habits, the modification of dress styles, combing and cutting of hair, the use of shoes, beds, and furniture, as well as abstention from the consumption of alcohol.[38]

Not surprisingly, Mexican progressives during the 1920s often shared some of the same ambivalent attitudes toward Mexican emigrants as those Anglo American progressives advocating Americanization. The individuals who promoted a Mexican education for children of immigrants in Los Angeles characteristically reflected a middle-class, Mexico City–focused perspective on the largely working-class immigrant population. Their preoccupation with return migration—and hence their interest in developing schools which stressed Mexican patriotism—was rooted in the belief that the emigrant experience in the United States had "civilized" the part-Indian peasant migrant. According to these officials, the provincialism exhibited by villagers in Mexico had given way to national rather than regional pride. In the United States, workers learned new skills and a work discipline that Mexican leaders believed was desperately needed for Mexico's own development. Mexican nationals who had experienced life in the United States were believed to be potentially more productive and refined than the typical mestizo villager.

This attitude was driven in Mexico by the doctrine of *indigenismo*, a construct thoroughly the product of non-Indians, which sought to exalt the native Indian of Mexico while destroying his culture and land base. Part of the larger effort to institutionalize the Mexican state and legitimize it throughout the population, *indigenismo* reflected the contradiction of an institutionalized revolution dedicated to constructing a sense of unifying nationalism among a diverse and often unwieldy population. With "Indian" being as much a social categorization as a racial one in Mexico, economic progress and unified nationalism required the integration of the Indian and the uplifting of the rural masses to the level of civilization as represented by the middle and upper classes of the Mexican capital. As important a figure as anthropologist Manuel Gamio claimed that two-thirds of the Mexican population was "Indian," putting in the same category the Morelos peasants who followed Emiliano Zapata with the Yaquis of Sonora or the Mayans of Quintana Roo.[39] In this fashion, the largely rural migrants that had made their way to Los Angeles became part of the larger mission of the nationalization of the "Indian" launched by the Mexican government during the 1920s.

The Columbia University–trained anthropologist Manuel Gamio, in fact, becomes a prime vehicle through which to view *indigenismo* in practice. As author of *Forjando Patria* ("Forging the Fatherland") (1916), Gamio was one of the main proponents of *indigenismo* in Mexico, introducing the country to Boasian cultural anthropology through his writings and archaeological undertakings. Serving as director of the Department of Anthropology in the early 1920s, undersecretary of education in 1925, and later as the first director of Mexico City's Museum of Anthropology, Gamio came to dominate the official Mexican view of the Indian past, present, and future for his generation.[40] Between the positions he held in Mexico, Gamio came north of the border to do an analysis of Mexican immigration to the United States sponsored by the Social Science Research Council, thereby displaying the relationship he felt these northern migrants had to Mexico's rural Indian population. Both of the products of this research, *Mexican Immigration to the United States* (1930) and *The Life Story of the Mexican Immigrant* (1931), are considered early classics by Chicano historians, yet the underlying racial and class assumptions of the work have rarely been questioned.[41]

Like other Mexican nationalists of his generation, Gamio certainly affirmed the contribution of native peoples to the history and culture of Mexico, particularly in the creation of folk art and artifacts which he prominently displayed in each of the museums with which he was affiliated. Yet as a prime intellectual of the movement to incorporate and solidify the nation, present-day Indian culture and language stood mostly as an impediment to national unity. As described by historian David Brading, Manuel Gamio's

> *Indigenismo* was thus a means to an end rather than an enduring mission; if incorporation was its aim, then essentially it sought to destroy rather than fortify the peasant culture of native communities. Modernising nationalism of the brand advocated by Gamio certainly found consolation in past glories but its inner vision was based on the liberal resolve to transform a backward country into a modern nation able to defend itself from foreign hegemony.[42]

Like many of his intellectual contemporaries, Gamio's vision of an ideal modern nation was, indeed, the United States. As an anthropologist, however, committed to a large extent to the connection of cultural backwardness with race, Gamio had a curious view of what had made the United States an economically powerful, modern nation. As director of the Anthropological Board of the Mexican Review, Manuel Gamio wrote an article for the committee in 1919 which outlined his perspectives on the nation to the north:

> In our opinion a large part of the great development actually obtained by the United States is originally due to their racial homogeniety and to the unity of their culture, their ideas, habits, customs and language. The North Americans are bound to one another by ethnical affinities founded in their common Caucasian origin. The Indians, insignificant in number, and the colored people which amount to several millions, are fatally condemned to be absorbed by the white population. As to culture—that is, as to ethic, aesthetic and religious ideas, to ambitions, ideals and national institutions, to customs and usages, etc.—a surprising cohesion and uniformity are observed.[43]

After expanding this interpretation to include the importance of a common English language, Gamio concluded that ". . such national unity of so many facets undoubtedly constitutes the main principle, the fundamental basis of the American success."[44] Racial and cultural homogeneity, therefore, was the most fundamental factor in determining the advanced position of the United States.

Moreover, Gamio felt that Mexico's backwardness, along with that of other Latin American nations, was due primarily to a lack of racial homogeneity. He continued his comparison across the border:

> On the other hand, the failure of our Latin-American countries is explained by the heterogeneity of race which implies an ethnical strangeness of the white inhabitants in regard to the natives. . . . May the public welfare be achieved, the nationality be formed and the common fatherland be constituted in countries lacking national unity? Undoubtedly no.[45]

Gamio's solution to this dilemma, which became his lifelong ambition, was to attempt to *"forjar patria,"* or forge a fatherland, by studying Mexico's Indian population in an attempt to incorporate them into the new revolutionary nation.

His view of the United States, of course, made it difficult to conceive of any possibility that Mexican immigrants would ever be considered equals by their Anglo-Saxon neighbors. In fact, since the vast majority of immigrants emerged from those native and mixed cultures he felt to be quite backward, the adjustment to the "highly modern civilization of the United States" became for him "exceedingly abrupt . . . even more intense and contradictory than in present-day Mexico."[46] Although the landing in this new "modern civilization" was often softened by the presence of a go-between Mexican American culture, "the great difference between the purely American and the purely Mexican, together with the factor of race prejudice, makes an intellectual, emotional, and traditional disparity too great to be bridged rapidly and perhaps never completely."[47]

Because of this intractable situation, Gamio advised that although temporary migration by Mexican laborers was desirable for both countries, permanent immigration was undesirable and should be discouraged by all means possible, including establishing plans to repatriate those who had settled permanently in the United States.[48] Repatriation, in fact, became a constant call by Mexican officials and parts of the elite expatriate community in the late 1920s. In 1928, when Jose Vasconcelos considered running for the Mexican presidency from exile in California, he called on Mexican laborers to repatriate themselves so as not to become a loss to the Mexican nation. The president of the pro-Vasconcelos campaign in Los Angeles, Nicolas Rodríguez, announced a special campaign platform which included increased educational programs of patriotism for the children of immigrants, repatriation of immigrants back to Mexico, and new restrictions on emigration from Mexico.[49]

The reason that the "backward semi-Indians" who had ventured north were now valuable to Mexico was that they had experienced life in a "modern civilization," thereby becoming important potential actors in Gamio's plan to create a new integrated nation. Their experience, according to Gamio, had brought them practical skills in agriculture and industry: "they have learned to handle machinery and modern tools; they have discipline and steady habits of work."[50] Besides these practical abilities, Gamio also surmised that the experience in the United States had actually created "Mexicans" out of a people who had little connection to their homeland as a whole before migration:

> We have seen frequently that natives or mestizos in rural districts in Mexico have not much notion of their nationality or their country. They know their town and the region in which it is situated, and this is a "little country" for them. People of this type, as immigrants in the United States, learn immediately what their mother country means, and they think of it and speak with love of it. Indeed, it can be said that there is hardly an immigrant home where the Mexican flag is not found in a place of honor, as well as pictures of national Mexican heroes. Love of country sometimes goes so far that little altars are made for flag or hero, or both, giving patriotism thus a religious quality.[51]

As Gamio's line of thinking makes clear, the role of the Mexican government in barrios in the United States was not simply a benevolent effort to promote an ethnic consciousness—a "Mexico de afuera" mentality—on the part of their expatriates. The efforts to promote Mexican nationalism had ulterior motives that were linked to complex reactions in Mexico concerning emigration to the United States. Mexican government officials were not simply altruistically striving to provide a positive national identity for Mexicans in the United States. Their intent was to convince emigrants to return. The Mexican government realized that the past and present siphoning off of many hardworking citizens hampered its recovery from the ravages of revolution, particularly in the northern states which lacked an adequate labor supply. The presence of so many Mexican citizens in the United States was also an embarrassment to leaders who had fought a nationalistic revolution against the pro-American dictator, Porfirio Díaz. When Americanization programs called for Mexicans to become citizens and abandon their native country and customs, Mexican leaders worried that the United States would steal away its most potentially productive nationals.

Instead the Mexican government stepped in to create institutions which fostered a Mexican patriotism which would bind the immigrant generation and their children to Mexico. As an editorial in *La Opinión* made evident, the success of the Mexican schools in instilling a strong sense of Mexican nationalism in the second generation would make possible

> the ultimate goal which is to one day, when the conditions of our country improve, reincorporate them as factors in real progress; for, they will carry with them the advantage of having two languages and the experience of two social mediums which have marked differences which, once compared and culling from them, could produce a level of superior life.[52]

Thus, when the Great Depression hit Los Angeles with full force in 1930, consul Rafael de la Colina jumped at the chance to help facilitate the repatriation of Mexican nationals and their often American-born children. He worked closely with local American officials to construct plans for county-sponsored trains to return Mexicans living in Los Angeles and on relief back to Mexico. When the Mexican Chamber of Commerce complained that the deportation raids launched by the Immigration Service were disruptive of the local community, they were criticizing the chaotic nature of the raids which made residents afraid to shop or work, not repatriation itself. The Comité de Beneficencia Mexicana, a committee of the Los Angeles Honorary Commission, changed its own focus in 1931 from supporting indigent Mexicans in the city with food, clothes, and medical care to paying for railroad passage back to Mexico for those who could not afford it. The repatriation efforts of the middle-class leadership, therefore, were not simply a reaction to initiatives formulated by racially inspired Anglo officials, but the culmination of efforts began in the early 1920s to keep Mexicans in the United States loyal to their mother country.[53]

Ironically, support for repatriation ended the reign of these Mexican progressives among the leadership of the Chicano community of Los Angeles. The exit of approximately one-third of 150,000 Mexican residents from the city during the repatriation period of the early 1930s ushered in a new period of leadership that would witness the emergence

of American-born Chicano leaders more affiliated with their working-
class communities and organized labor backgrounds. After 1935, the
Mexican consulate would never again play as crucial a role in organizing
local leadership around goals formulated in Mexico City. Increasingly,
the Mexican American community would see its own political future as
wrapped in the context of American civil rights and the fufillment of the
promises of U.S. citizenship.

Despite this eventual turn, during the 1920s a political generation
did emerge in Los Angeles which reflected the nature of both Los
Angeles progressive politics at the time and the institutionalized revolution of the Mexican state. These leaders intended to impose order on the
chaos of the Mexican Revolution and the turbulence of the Mexican situation in the United States, while retaining Mexican national pride and
progress. While trying to change Mexico from afar, they also encouraged
working-class Mexicans to see their future intertwined with a middle-
class leadership of professionals and businessmen. For these leaders, the
bonds of race could provide the social cement which maintained control
in an otherwise hierarchical political and economic order.

Neither Americanization nor Mexicanization programs, however,
were able to control the complex process of cultural adaptation Mexican
immigrants underwent once they arrived in Los Angeles. Mexican immigrants resisted Americanization efforts by retaining much of their culture
on American soil, and most refused to adopt American citizenship. On
the other hand, the Mexican government was unable to keep its citizens
from migrating to the north, and until the onslaught of the Great Depression, virtually all individuals who responded to the call to return to
Mexico were replaced by newcomers from south of the border. The
shared contempt for the Mexican peasant by those advocating Americanization and Mexicanization generated widespread distrust of both American and Mexican government officials among the working-class population. Additionally, neither program had the necessary funding nor
unequivocal political support to complete its task. American officials remained somewhat ambivalent about transforming Mexican workers into
American voters, while Mexico could hardly expect to build schools for
its citizens in the United States when it was unable to educate adequately
the children within its own borders.

Yet both had meaningful, if unforeseen, consequences for the Chicano community in Los Angeles. Americanization efforts did encourage
an American identity in the second generation, even while this American-
born group continued to be denied equal status as citizens. Conversely,
Mexicanization promoted an ethnic pride within the community, a pride
which ironically fostered an identity as an ethnic American, rather than
encourage a return to Mexico. Eventually, a pivotal segment of this immigrant community learned English, acclimated themselves to American
life, and saw their future as residents of the United States, not Mexico.
But as commentator Rodolfo Uranga pointed out in *La Opinión* in 1926,

the "Mexican soul" continued to exist within the immigrant laborer and his or her children. "To all it is clear that although our workers dress, eat, and entertain themselves like Americans, they have well defined and indelible characteristics, and the majority preserve their customs, their native language, their parents' religion, and a deeply rooted love of their native land."[54] The battle for cultural allegiance waged between the American and Mexican governments mirrored a more poignant struggle within each individual immigrant, as he or she learned to balance nationalistic sentiments with a new ethnic identity.

PART THREE

SHIFTING HOMELANDS

It might be well to consider the meaning of that word patriotism. It connotes love of country. But what is meant by country? Is it the ground we walk on, or our fellow citizens, or the government, its institutions and its laws?

—Father Vaughn, popular radio priest in Los Angeles,
discussing the patriotism of Mexican Catholics, 1934[1]

I had always considered returning to Mexico, but the months and years went by, along with the fact that since I earn very little I can't save much.

—Estanislao Gómez, an immigrant from Guadalajara, 1927[2]

Cultural identities come from somewhere, have histories. But, like everything that is historical, they undergo constant transformation. Far from being eternally fixed in some essentialist past, they are subject to the continuous "play" of history, culture and power. Far from being grounded in a mere "recovery" of the past . . . identities are the names we give to the different ways we are positioned by, and position ourselves within, the narratives of the past.

—Stuart Hall, 1990[3]

At last he came to this county and rented an apartment
Without knowing that in this town one dies working in the cement.
When he felt that he had money he began to buy on time;
And when he bought ill-fitting suits he felt himself equal with Carranza.
It is true, it is true, it seems a lie that being a Chicano
So spirited and so healthy that he would come from over there.

He rented a radio and aerial with light bulbs and buttons.
Because his house was very quiet without music or songs.
At the hour that they transmit the concerts to Chicanos
It happens that they advertise pork and the best country gravy.
It is true, it is true, it seems a lie that in place of songs
Those city people would advertise cantaloupes.

After three quarters of an hour they sing us some fox trot,
Then they announce the lady who makes good *tepache* [a mexican beverage made
    of pulque, water, pineapple, and cloves].
Other subjects follow, illustrating the bargains
That they will make to the dead if they buy good coffins.
It is true, it is true, it seems a lie that they would vex us
In these places, those of the city; it seems a lie, that they
    would vex us, those of the city.

—Excerpts from the *corrido,* or folk song, "Radios and
Chicanos," ca. 1930s[4]

CHAPTER 6

Family Life and
the Search for Stability

As Guadalupe Salazar looked out of her train window, her mind was full of images of the past and questions about the future. Heading from Chicago to Los Angeles, she realized that her life in the United States had not turned out as she had hoped. It was the middle of the Great Depression, and Guadalupe had just ended a marriage that had lasted only a few years. Her ex-husband, Arcadio Yñiguez, had crossed the border in 1913 as a teenager from Nochistlán, Zacatecas, fleeing the violence of the Mexican Revolution. Working at a variety of odd jobs, he finally settled in Chicago during the 1920s, and there met and married Guadalupe. When the two split up, Arcadio returned to Nochistlán, while Guadalupe and their five-year-old son left for California. She was determined to start a new life in Los Angeles, where her father resided, although she had not seen him since his impressment into military service during the revolution twenty years earlier. A single female parent in 1931, Guadalupe Salazar saw her immigrant dream fade into a painful reality of insecurity.[1]

The generation of scholars who wrote during the post–World War II decades about European immigrant family life would not have been surprised by Salazar's experience. Their work emphasized the sharp discontinuities between traditional family relations in Old World peasant villages and the life immigrants encountered in modern, industrial cities after migration. Rooted in an unbending model of modernization, their studies found family disintegration to be an unfortunate, but inevitable consequence of the immigrants' undeniable break with their past. Guadalupe's failed marriage might easily have been portrayed by this school of immigration history as the result of a futile attempt to construct an orthodox union in a new and hostile environment. As Oscar Handlin put it: "Roles once thoroughly defined were now altogether confounded."[2]

Yet Guadalupe's story defied such characterization. Reunited with her father, she built a new life in Los Angeles out of which emerged a remarkable family. Her second husband, Tiburcio Rivera, had been a band musician in Mexico. He knew Guadalupe from Chicago, where he briefly owned a pool hall. They did not court until he too moved to California. In addition to Guadalupe's son, the couple had four daughters, all of whom grew up in East Los Angeles. The family endured the Depression, frequent bouts with overt discrimination, and hazardous work conditions. In spite of these hardships Guadalupe and Tiburcio provided their children a stable working-class family life. Fifty-five years after her arrival in the city, Guadalupe Salazar received the "Mother of the Year" award from the senior citizen clubs in East Los Angeles. Mother of five, grandmother of 28, and great-grandmother of 10, she had become the respected elder of an extended family that totaled more than 200. Asked about her success, she responded: "You have to have family unity."[3]

Critics of the "Handlin school" of immigration history have pointed to the stability and resiliency of immigrant families such as Guadalupe Salazar's. Their depiction of immigrant adaptation stresses the retention of traditional values and the durability and adaptability of social relationships, all of which helped to withstand the changes wrought by migration, settlement, and adjustment. In particular, these historians understand the critical role of kinship networks which allowed Salazar to reestablish herself in Los Angeles. Her relationship with her father, though strained because of the separation in Mexico, was rebuilt in the United States. In fact, this family was strengthened by Salazar's decision to call upon kin in time of need. As revisionist historian Virginia Yans-McLaughlin has pointed out, "immigrants put their Old World family ties to novel uses in America," essentially putting "new wine in old bottles."[4]

Just as historians of immigration debated the degree of cultural persistence inherent in immigrant family life, Chicano social scientists were examining the dynamics of the Mexican immigrant family. These scholars depicted *la familia* as warm and nurturing, an environment of support and stability in times of stress. They surmised that since roles and expectations continued to be circumscribed in the traditional manner, conflict within the family was kept to a minimum. From this perspective, *machismo* was not so much a maladaptive response which solidified male dominance, but rather represented an appropriate mechanism to insure the continuation of Mexican family pride and respect. Although noted in the literature, the oppression of women within the family was dismissed as a necessary evil in order to maintain family stability and tradition.[5]

Ironically, this approach had much in common with another, older body of sociological literature that depicted the Mexican family as pathological. These psychoanalytically oriented studies were the product of decades of stereotypical accounts examining "the problem" of the Mexican. They viewed Mexican families as authoritarian and *macho*-dominated, impeding individual achievement and independence while promoting passivity and familial dependence. Thus, the same values that some Chicano scholars characterized as positive were viewed as "a tangle of pathology" by Anglo American social scientists.[6] What both groups shared was a unidimensional view of the Mexican family, a caricature suspended in time and impervious to the social forces acting upon it. Such a perspective found any acculturated family to be atypical.

When placed in historical context, both characterizations of the Mexican immigrant family are problematical. First, and most important, they ignore the great diversity among Mexican immigrant families. Although many Mexicans migrated from rural villages, others came from cities. Many families migrated as entire units, while others were involved in chain migration. Some immigrants settled in largely Mexican communities along the border; others ventured further inland where the Anglo American population dominated. Before 1940, thousands of families and individual family members were in this country only temporarily. Perhaps the majority came as single migrants, and reconstituted their families in the United States. These families were often mixtures of Mexicans born on both sides of the border. They occasionally included a non-Mexican spouse. Moreover, individual families acculturated and adapted to American life in a multitude of ways.

Second, both conflict and consensus existed within each family. Individual members of a family might disagree over a particular family decision. Over time, positions would reverse themselves as other situations arose. Difficult periods of maturation, like a child's adolescence, could prove to be a time of family conflict, while family unity might be invoked during periods of crisis and abrupt change, such as the death of a parent or a new marriage. Moreover, while Mexican family members often gave highest priority to the welfare of the family, specific family decisions could mask the range of compromises made by individuals involved in that resolution.

Finally, every Mexican who came to the United States made adjustments. Though most families did not disintegrate under the weight of changing circumstances, they certainly acclimated. The nature of this acculturation varied, depending on the setting, and different strategies were developed to fit the needs of the historical moment. A new identity was continuously being formed.

To understand the diversity of family experiences among Mexican immigrants in Los Angeles, we must examine critically assumptions regarding family life in turn-of-the-century Mexico in regions that contributed migrants to the United States. Most interpretations characterize Mexican families as hierarchical, rigidly patriarchal, solidified by age-old customs rooted in peasant values and Catholic tradition. Mexicans were characterized as having large, extended family structures in which gender roles were strictly separated, reinforced by stern parental discipline and community pressure. Each individual village usually consisted of a few extended families linked to each other through generations of intermarriage and other kin relationships, including *compadrazgo,* the interlocking bond created by parents and godparents of a child.

Recent studies challenge this interpretation of Mexican family life, depicting much more flexibility within family patterns. As noted in Chapter 1, economic challenges brought about by the penetration of market capitalism into all but the most isolated villages during the Porfiriato forced families to adapt. As land prices were driven up, families were forced to send members, usually adolescent boys and young husbands, into the wage economy. Women were also swept into the cash-based economy. Some marketed surplus food raised on family plots, while others sewed for profit utilizing Singer technology. Central markets in most villages became more active points of economic exchange. A family's own land was increasingly attended to by those outside this cash nexus, usually by women and children closer to home or those who returned from various forms of wage labor in time to complete a harvest.

Rigid gender roles could hardly be maintained under these circumstances. The Mexican family showed that it was capable of flexibility and adaptability, even under the most distressing circumstances. In addition to migration brought about by economic conditions, most villages contained families that had experienced the death of their male heads of household. Widows were often able to maintain a family's well-being, aided by older adolescents or nearby relatives. Female-headed households, a result of either death or desertion, were not uncommon at the turn of the century, although marriage continued to be the preferred societal norm for all adult women.

Most families participated in economic migration in order to maintain a life that they identified as rooted in traditional values. Working for the railroad or in the mines was intended as a short-term solution to an emergency. Yet the Mexican government's economic and social policies around the turn of the century transformed these strategies into a way of life. Porfirian economics demanded a large, growing wage labor pool, as did economic developments in the United States. Families found themselves caught in a cycle of economic uncertainty, necessitating the flexibility of "traditional" roles and norms for survival.

After 1910, the Mexican Revolution only intensified these patterns. Geographic mobility increased, often forcing entire families to flee their native villages to avoid the danger of incoming troops. More often, male family members were sent scurrying, either to avoid conscription or to join one of the military factions. It was uncommon, in the absence of men, for women to perform most day-to-day economic functions related to a family's property and sustenance. If not touched directly by the fighting, families found that destruction of neighboring fields, markets, or transportation could force them to engage in more extensive migration to market their goods or earn wages for their labor.

Mexicans who migrated to the United States generally came from families engaged in years of creative adaptation to adversity. Unlike European immigrant families, whose movement into American society could best be described as chain migration, Mexican families were much more likely to be involved in a pattern of circular migration. Although most European immigrant groups also had high rates of return migration, ranging from 25 to 60 percent, only Mexicans exhibited a pattern of back-and-forth movement that would continue for years.[7] Men ventured north across the border to engage in seasonal labor, then returned south for a period of a few months or a couple of years. If economic circumstances once again necessitated extra cash, the circular pattern began anew. During World War I and up until 1921, the United States government contributed to this pattern by giving entrance visas to temporary workers in order to regulate their movement back into Mexico at the end of a season.

Changes in U.S. immigration policy, however, made it more difficult to engage in this practice after 1921. An enlarged border patrol, enforced literacy tests, and higher visa fees made back-and-forth migration more risky and more expensive during the 1920s. Workers who had grown accustomed to legal, relatively easy passage across the border were now faced with the prospect of venturing north illegally or being held up indefinitely in border cities. Increasingly, Mexicans were forced to decide where they wanted to reside permanently. While many returned to Mexico, the large increase in the Mexican population of Los Angeles during the 1920s suggests that a significant proportion determined to make their homes in the north. For single, independent migrants, the decision meant a reorientation to the experience of working and living in the United States. Heads of households were required to move whole families across the border.

The process of family migration was often tortuous. It was likely to involve careful decision-making concerning which family members should be on which side of the border, taking place over several years. Economic opportunities and emotional attachments had to be weighed. Individual preferences could not always be ignored for the sake of the family good. Others besides immediate family members were often involved in the move; some provided resources while others provided short- and long-term care of minors.

The experience of one family, accessible to us through archived transcripts of the Board of Special Inquiry of the Immigration Service, may serve as an example of the complex process of family migration to Los Angeles.[8] On August 25, 1917, three individuals—María López de Astengo, her twelve-year-old son, José Jr., and Mrs. María Salido de Villa—presented themselves to American immigration authorities in Nogales, Arizona. Mrs. Astengo and her son had ventured north from Rosario, Sinaloa, on the western coast of Mexico. María's husband, José Sr., had fled their ranch two years earlier to avoid the danger associated with the revolution and to earn income for the family. A bookkeeper in Mexico, he used his experience to gain employment as an office clerk in Los

Angeles, earning $2.50 a day. Mrs. Astengo did not intend to cross into the United States herself; rather, she was sending her son with a friend of her family, Mrs. Villa, who was going north to visit her own two children who lived in Los Angeles. Mrs. Villa's son had married in the United States and had lived in Los Angeles for the last five years. Her daughter arrived in Los Angeles in 1915. Mrs. Villa intended to return to Mexico in November or December when "the weather gets cold."

The following spring Mrs. Astengo sent Enrique, her next eldest son, to live with his father. He traveled with three other young Mexicans from Rosario, none family members. María Valdez, age twenty-seven, headed the group, guarding everyone's money during the passage. María, accompanied by her fourteen-year-old brother Jesús, came north to see her younger sister Josefina, who had been in the United States for about a year. Josefina was single and supported herself by working as a laundress. She lived in Los Angeles with a widowed second cousin. Josefina had been instructed by her mother in Rosario to put young Jesús in an American public school. María herself intended to stay only for about six months before returning to Rosario.

The fourth member of the group was Jesús Cambreros, a seventeen-year-old girlfriend from Rosario, who came to Los Angeles to live with her married sister Elisa, also a laundress. Elisa's husband, Luis Martínez, worked for Wells Fargo Express, earning seven or eight dollars a week. Since coming to the United States as a boy from Chihuahua, Martínez had also been a baker and a foundry worker. The couple had a baby and lived in a six-room house in the downtown area, renting out space to two other adults.

A few months later, Mrs. Astengo and the rest of her family joined her husband and sons in Los Angeles. But the migration of relatives did not end there. That summer, José Astengo urged his sister in Rosario to send her son to Los Angeles to attend school, rather naïvely noting that the city was "very clean . . . perfectly safe and pleasant" with "no saloons, gambling houses, or houses of prostitution." At the beginning of July, Carlos Osuna made the trip through Nogales, accompanied by José's brother. Both planned on living with José's family while attending school. Another brother, Jesús, was also reportedly working in Los Angeles.

These reports of three distinct border crossings suggest the intricate nature of Mexican family migration to the United States. The Astengos first sent their husband north as a temporary measure. Younger male sons followed, once José had established himself in Los Angeles. María Astengo and the youngest children were the last to leave the homeland. Complicating the picture, brothers of José and a nephew also ventured northward when opportunities presented themselves for work or education. The Astengos sent family members north via the train, but each trip was facilitated by other relatives or hometown friends who accompanied the travelers. Regular communication between family members on both sides of the border, including periodic visits and oral messages sent through family friends, enabled José to monitor the migration process. It is more difficult to assess the decision-making power of María, since we are not privy to their personal correspondence. One son reported that María maintained a family store in Rosario while José was in the United States, a fact which indicates some level of economic autonomy. Although the Astengos were better off than the average Mexican family, their experiences with immigration characterize many of the ways Mexicans took advantage of economic opportunity.

Other families who emigrated illustrate the many dimensions of familial migration. Older adolescents and young adults formed the bulk of the permanent emigrants. In particular, single men ventured north to find work, often aided by relatives or friends when they arrived. Young women also moved north, but were invariably accompanied by other family members and had relatives waiting for them in Los Angeles. The migration of these young adults' parents was often more problematic, but many visited their children, at least until the tightening of restrictions during the years from 1921 to 1924.

Single male migrants served as initiators of most Mexican migration. Although many European immigrant groups displayed high levels of family migration, the Mexican pattern seems to be similar to that of Italians, whose single migrant rate was around 75 percent. Among male Mexican immigrants who chose to naturalize in Los Angeles, in fact, only 10 percent had first ventured to the United States as married men. As in the Italian case, single Mexican migrants were also more likely to return to their homeland than those who were married and accompanied by their spouses.[9] Single migrants, like those married but traveling alone, generally remained in touch with their families in Mexico. As long as those ties remained strong, a high proportion of single males returned.

For single male migrants through the mid-1920s, the central Plaza area of Los Angeles remained the most important area of introduction to the city. Although this area also contained recently arrived families, single men dominated community life. Theatres, restaurants, bars, dancing clubs, and pool halls nearby catered to this male clientele. The Plaza itself was often used as a employment recruitment site, and on the weekends served as a locus for political discussions. Rental housing, including boarding houses for single men, was the norm in the barrio around the Plaza. Upon arriving in Los Angeles with eight other single men, Arturo Morales, a twenty-eight-year old from Acatlán, Jalisco, remembered being directed to a rooming house run by a woman from his home state. Within a week, all had obtained work, sharing two rooms in the boarding house between the eight of them.[10] Although other ethnic newcomers to Los Angeles increasingly flocked to the Plaza in the 1920s, most notably Italians and Chinese, Mexicans remained the largest group in the historic Mexican pueblo plaza area.

Many, if not most, of these single Mexican men stayed in Los

Angeles only temporarily. Often they entered the city with the idea of earning money quickly, then returning to their families in Mexico. Living in the central Plaza area made this plan more possible. A male worker traveling alone could find employment through the various employment agencies with offices near the Plaza, or simply stand around in the early morning and wait for a prospective employer's call. Housing, though crowded and often unsanitary, was relatively cheap in the district and was tolerated by laborers hoping to stay only briefly in the city. With images of loved ones waiting across the border in need, many single men found Los Angeles to be a relatively easy stop to find a job and earn extra cash before returning home.

On the other hand, the loosening of ties with the Mexican family of origin was crucial in generating a permanent immigrant population in Los Angeles. Although exact figures are not available, a significant number of single male migrants, who formed the vast majority of the transient Mexican population in the city, reoriented themselves toward permanent residency in the United States. While family considerations were fundamental to Mexicans who contemplated leaving their homeland, breaking those connections was crucial if a migrant was to stay in the United States. This process was aided by the restrictive immigration requirements which originated in 1917. But other factors were also important in solidifying this pattern.

The regional and state origins of immigrants were important factors which determined whether a newcomer planted roots in Los Angeles or not. According to Manuel Gamio's pathbreaking study, migrants from Mexico's agricultural central plateau were much more likely to send money back to their families. Although Los Angeles's Mexican population contained a considerable portion of members from this region, equally significant were migrants from urban areas and northern Mexico. These individuals were less likely to be supporting family members in Mexico. Familiarity with the United States and U.S.-Mexico border communities made it much more likely that single men migrating from northern border areas settled in Los Angeles permanently. Urban migrants were less likely to be involved in the supplemental cash economy which allowed many migrants to retain their agricultural land in Mexico.

The passage of time itself, of course, loosened ties to Mexico. Although many migrants, no doubt, originally intended a short visit to Los Angeles, thousands never achieved their goals. More often than not, Mexicans could not save much from the meager wages they received. It was easy to postpone a return to Mexico until the ever-elusive extra dollar was earned. As Estanislao Gómez, an immigrant from Guadalajara, put it: "I had always considered returning to Mexico, but the months and years went by, along with the fact that since I earn very little I can't save much."[11] Furthermore, Los Angeles was not a border community, and conditions there made regular contact difficult. Urban jobs, unlike agricultural employment, were less likely to be seasonal and were inflexible in providing time to visit relatives in Mexico. Periodic visits also required surplus cash which many migrants were never able to accrue.

Ironically, it was often the establishment of new family ties which broke a single male's connection to his family of origin. When marriage occurred in the United States, ties to families of origin immediately became secondary. As stays in the United States were lengthened, the likelihood increased dramatically that a young single man would encounter a woman to marry in this country. This turn of events changed the orientation of Mexican men living in the United States from that of expatriates temporarily working here to heads of households formed in the United States.

Unlike men, Mexican migrant women in this period rarely ventured to Los Angeles unattached to their families or unaccompanied by relatives. Even if their family of origin remained in Mexico, they lived in Los Angeles with extended family—siblings, cousins, uncles, or aunts. Most came to the city with their family unit, either as wives or children, directly from Mexico or from another part of the American Southwest. Those single adult women who came north migrated only after some personal or family tragedy. Juana Martínez, for example, migrated from Mazatlán, Sinaloa, with her mother and two sisters only after her divorce and the death of her father. Leova González de López also left Mazatlán, but only to escape the slanderous talk that surrounded her decision to raise her brother's son as a single parent. Tellingly, González was an orphan herself, who migrated to Los Angeles under the guidance of her aunt.[12]

From the start, women's orientation toward the United States was formed in the confines of a Mexican family, not as single, independent migrants living alone. Eventually many of the Chicanas who migrated to Los Angeles as children, whether Mexican or American-born, found employment as young adults to help support their families and often to provide themselves with independent income. A small minority tried to live alone or with girlfriends, away from the watchful eye of intruding relatives.

Perhaps because the largest single concentration of unmarried men lived in the crowded housing around the Plaza, this area was strictly off-limits to most women living alone. Instead, the majority lived in the adjacent metropolitan areas to the south and west of the Plaza. Unlike the barrios developing east of the river, housing alternatives to the single-family home emerged. Small apartments, a few boarding houses for women, and households willing to take in a non-related young female were much more common in this part of the city than in other areas populated by Mexican immigrants. Close to downtown, these households provided easy access to both the industrial labor and white-collar employment available to young Mexican women.[13]

The areas west of the river were also the communities most integrated with other working-class ethnic groups and, with the exception of the Plaza, least solidly Mexican/Chicano in their cultural orientation. Women and men who lived here were exposed to the cultural practices of myriad ethnic groups, while enjoying the anonymity of living in a big city. "Here no one pays any attention to how one goes about, how one lives," declared Elenita Arce, pleased at the greater freedoms allowed unmarried women.[14] These areas also seemed to provide a haven for immigrants who went against traditional Mexican family practice. Knowledge and use of effective birth control, for example, seemed concentrated in a small group of Mexican women living in these downtown communities.[15] Also, most single Mexican immigrant men and women lived west of the Los Angeles River, while Chicano family life was increasingly centered east of the river during the 1920s. Over three-quarters of all the single migrants sampled lived in the barrios west of the river. Of the single migrants over age 29—and therefore much less likely to ever marry—most also lived west of the river.[16]

Marriage, however, continued to be part of the expected practice for both Mexican women and men. In Los Angeles, a wide range of possible marriage partners was available. Not only did immigrants from a variety of different Mexican locales reside in the city, but a rapidly growing group of American-born Chicanos provided other potential partners. Non-Mexicans were also potential marriage partners, although prejudice and limited contact kept their numbers relatively small. Still, both native-born Anglo Americans and foreign-born whites were listed among the husbands and wives of Mexican immigrants who applied for naturalization before 1940.

An examination of marriage patterns between Mexican immigrants and other groups reveals figures similar to those offered by earlier historians and social scientists.[17] Almost 83 percent of the marriages involving Mexican immigrants in a sample of 1,214 marriages took place within the Mexican/Chicano community. Some 209 marriages, or 17.2 percent, were between Mexican immigrants and non-Chicanos. Not surprisingly, intermarriage was significantly more prevalent among Mexican immigrant women who chose to naturalize, involving one-third of those in the sample.[18] Marriage to non-Chicanas born outside the United States accounted for only 1.9 percent of the marriages of Mexican men, yet Mexican immigrant women married foreign-born Anglo American men more often than American-born (see Table 9).

A profile of the Mexican immigrant men who married Anglos uncovers some revealing patterns. Mexican men who married non-Chicanas were more likely to have migrated to the United States before age twenty and to have come from larger urban areas in Mexico. Four-fifths of the Mexican immigrant men who intermarried arrived in this country before age twenty, while men in the sample who married Mexican immigrant women were more likely to have come as adults.[19] Most of the future spouses in intermarried couples came as children to the United States and therefore grew up in similar conditions as American-born Chicanos.

Table 9. Marriage Patterns of Mexican Immigrants in Los Angeles

Background of spouse	Men		Women		Total	
	Number	*Percent*	*Number*	*Percent*	*Number*	*Percent*
Mexican immigrant	670	60.7	47	42.4	717	59.1
Mexican American	261	23.7	27	24.3	288	23.7
Total Chicano	931	84.4	74	66.7	1,005	82.8
Anglo American	151	13.7	18	16.2	169	13.9
Foreign-born Anglo	21	1.9	19	17.1	40	3.3
Total Anglo	172	15.6	37	33.3	209	17.2
Total	1,103	100.0	111	100.0	1,214	100.0

Source: Analysis of naturalization documents, National Archives, Laguna Niguel, California.

Additionally, urban areas, and to a lesser extent coastal areas, were more likely to produce immigrants who intermarried, largely because they were more familiar with American culture and urban life. Señora María Rovitz Ramos, for example, married a young, bilingual Anglo American. She had grown up in Mazatlán, where her father was owner of a hotel catering to European and American tourists.[20] Immigrants born in Mexico City were particularly likely to intermarry. In fact, the sample revealed that more immigrants from the Mexican capital married Anglo Americans in Los Angeles (38%) than married other Mexican immigrants (24%).

Non-Chicanos who married Mexican immigrants also shared certain characteristics. Typically, they were also migrants to Los Angeles, often coming as adults. Mexican immigrant women were just as likely to marry a foreigner as someone born in the United States. For Mexican immigrant men, intermarriages most often were made with newcomers from the Midwest or East, although many of these spouses were American-born offspring of Italian or Irish Catholic immigrants.[21] Like Mexican women, these Anglo spouses also tended to marry young—age twenty-two on the average. As recent arrivals to Los Angeles, they shared with their Mexican spouses the disruption of family ties and the need to acclimate oneself to life in Los Angeles.

Not surprisingly, intermarried couples were more likely not to live in the barrios around the Plaza area and in East Los Angeles. In fact, well over half of all intermarried couples lived in the larger metropolitan area to the south and west of the Plaza, compared to only one-third of the all-Mexican couples in my sample. Many reasons account for this distribution. According to contemporary observers, Mexicans who intermarried were generally lighter-skinned, and thus more easily able to move into areas restricted from dark-skinned Mexicans.[22] Entry was usually eased by a non-Chicano spouse.

Second, many intermarried couples were better off financially and

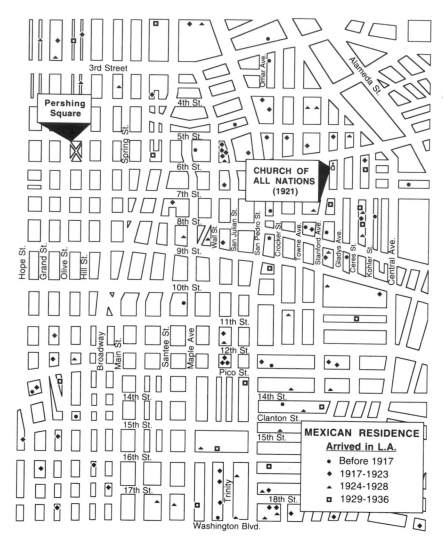

Map 8 Mexican Residences in Downtown Los Angeles
Source: Naturalization Documents, National Archives, Laguna Niguel, California.

could afford to live outside the barrio. One well-to-do immigrant couple saw three of their Mexican-born children marry Anglo Americans, even though the father felt the spouses "didn't belong to [our] society." The rest of the family, including four other unmarried children, lived in Hollywood, where most of their relations were with Americans. A successful real estate broker, a light-skinned intermarried Mexican woman, admitted: "Although I like my people very much I don't want to live with them, especially on the East Side, because they are very dirty there, there are many robberies and one can't live at ease."[23]

The area west of the river and south of the Plaza (see Map 8) providcd shelter for Mexican immigrants who were searching for greater cultural freedom, independence from their families, or interaction with other ethnic groups. Still solidly working-class, this community was a secondary one for many different European and Asian groups. It also housed, along Central Avenue, the largest community of blacks in Los Angeles. A substantial number of Mexican immigrants lived here, in more integrated surroundings than in other parts of the city. This integration occurred, however, largely separate from the city's Anglo American middle class. Here was a neighborhood of ethnic mixture, a polyglot zone of working-class people from around the globe.

For Mexicans, this community was more than a haven for intermarried couples and single women living alone. It represented the social freedom found in the United States, especially for women who were caught between the restrictive practices of Mexican families and the more liberal views of Anglo Americans. One representation of this battle within the family was over issues of dress and appearance of young daughters. Angelita V., for example, asserted her independence from her family of origin upon getting married at age nineteen:

> The first thing I did was to bob my hair. My father would not permit it and I have wanted to do [it] for a long time. I will show my husband that he will not boss me the way my father has done all of us.[24]

Other families exhibited tensions over a daughter's refusal to wear a rebozo as head covering or whether makeup would be permitted.[25]

Another aspect of that independence was less supervision over young single women, a situation that provided greater opportunities for young men and women of all nationalities to meet. This greater liberty allowed for more widespread sexual experimentation and subtle changes in sexual mores. Tellingly, almost one-third of the women involved in cross-cultural marriages had conceived a child before marriage, compared with one-fifth of first births among all-Mexican couples. When both partners had been born in Mexico, strict cultural prescriptions against sex before marriage seemed to prevail.[26] On the other hand, more than half the Mexican immigrant women sampled who married American citizens, Chicano or Anglo, had already given birth or were pregnant at marriage (see Table 10).

Despite the increase in premarital sex among women who did not marry Mexican men, widespread cultural values shared in both the Chicano and Anglo communities encouraged men and women to marry if pregnancy occurred. In every group, less than 20 percent of births occurred outside of marriage, and only 2 single women in the sample had children. Many of the Chicano couples who contributed to the 20 percent were probably common-law marriages that were legalized in preparation for naturalization. While American officials and social workers often saw common-law marriages as evidence of moral decay, the Mexican immigrant community viewed these unions as legitimate.[27]

Table 10. Marriage and Conception Among Mexican Immigrants

Type of marriage (husband/wife)	First Birth				
	Outside of marriage	Conceived before marriage	In first 3 years of marriage	After 3 years of marriage	Total
All-Mexican Immigrant	51 (14.5%)	20 (5.7%)	219 (62.2%)	62 (17.6%)	352 (100%)
Mexican/ Mex. Am.	30 (14.9%)	33 (16.3%)	112 (55.4%)	27 (13.4%)	202 (100%)
Mexican/ Anglo	13 (14.9%)	11 (12.7%)	44 (50.6%)	19 (21.8%)	87 (100%)
Mex. Am./ Mexican	4 (20.0%)	7 (35.0%)	9 (45.0%)	0 (0.0%)	20 (100%)
Anglo/ Mexican	3 (20.0%)	6 (40.0%)	6 (40.0%)	0 (0.0%)	15 (100%)
Total	101 (14.9%)	77 (11.4%)	390 (57.7%)	108 (16.0%)	676 (100%)

Source: Analysis of marriage dates and birthdates of eldest children from naturalization documents, National Archives, Laguna Niguel, California.

Experimentation among Mexican immigrants living in Los Angeles also led to an increase in married women who worked outside the home. Both Mexican and American cultures designated men as the principal family wage earners. Whether or not a newly married woman worked for wages was often a source of discussion and consternation, although many families found the income generated by wives essential. Among the married women sampled where information concerning employment status was known, about 40 percent were engaged in wage labor outside the home.[28] While this figure is similar to the proportion of married Chicanas found working in other studies of southern California communities in this period, it seems to be a higher rate than that found among married women along the border.[29] This proportion of working Chicano married women is much higher than that of other married women, including most immigrant women. In 1920 nationwide, only 6.3 percent of married native white women worked for wages outside the home, while 7.2 percent of foreign-born wives were in the work force. Only the proportion of black wives who were paid laborers, 32.5 percent, was similar to that found for Chicanas in Los Angeles.[30]

In contrast to earlier historical arguments, Mexican-born women were more likely to be employed than American-born Chicanas. Almost half of those women were working for wages, as compared with only 20 percent of the American-born Chicana population. These figures call into question Richard Griswold del Castillo's argument for the nineteenth century that Mexican women "had a more traditional frame of reference" and therefore were less likely than native-born women to enter the job market.[31] One reason for this discrepancy may be that different cultural prescriptions were at work during the early twentieth century that made immigrant women more likely to engage in wage labor after marriage. The flexibility demanded of the Mexican immigrant family for survival in a rapidly changing economy overrode "traditional" frames of reference. The shift had begun in Mexico, where women in migrant families were called upon to head households temporarily while men looked for work elsewhere. In Los Angeles, many Mexican immigrant women entered the labor force when their husbands were unable to find employment, were temporarily laid off, or when family expenses became burdensome.[32]

Although much of the changes in women's roles occurred to the west of the Los Angeles River, repercussions were felt throughout Chicano Los Angeles. Young Chicanas living in East Los Angeles with their parents increasingly challenged the elders' notions of dating and courtship, even while they maintained a deferential attitude toward them in other areas. Moreover, many of the skilled workers who bought homes in the east-side neighborhoods of Lincoln Heights and Brooklyn Heights during the 1920s were able to do so because their wives continued to work after marriage, thereby increasing family income.

Family life in the barrios of Los Angeles ranged from conventional to experimental, and often these families lived in close proximity to one another. Even within a family, certain members could exhibit behavior that others might consider inappropriate or "un-Mexican."[33] Freedom could be positive or negative depending on one's position in the family. One Mexican mother, living with her unmarried children west of the Plaza, enjoyed the freedom to go wherever she wanted without restriction. In Mexico, she had felt oppressed by prescriptive social customs. Nevertheless, she did not like the behavior of young women in this country. "Liberty," she stated, had been "contagious" to her daughters, and this bothered her a great deal.[34]

The creation during the 1920s of a more concentrated Mexican community east of the river, however, offered an opportunity to reassert certain family practices deemed traditional in a wholly different setting. The settlement of Mexican families in East Los Angeles implied a permanency which was not characteristic of Mexican communities west of the river. The stability of permanent settlement in the United States, for example, allowed opposition to married women working to regain ascendancy. Married women living east of the river in Belvedere and Boyle Heights were less likely to be employed than those elsewhere. Having achieved a sense of stability through the extra earnings of female employment, many married women left their jobs after moving to an east-side neighborhood.[35]

This process of claiming certain family practices as traditional in a new setting is crucial to understanding Chicano culture. Migration itself inevitably disrupted the family, often forcing members on both sides of the border to adjust to a new constellation of individuals. As migrants reached important life stages, however, they had the opportunity to influence their own "culture"—shaped, of course, by their conception of tradition. This process was influenced by the fact that widespread segregation of Mexicans in the American Southwest kept many cultural practices insulated from those of the Anglo American majority.[36]

Marriage, and the related practice of courtship, was one life stage in which Chicanos were able to alter cultural practices. The age at which men and women married is one indication of this transformation. Migration itself had tended to delay marriage, particularly for women. The average age at marriage for men sampled was approximately 26.3 years. Women married at a substantially younger age, 23. Men who married in Mexico averaged just over 25.7 years of age and women's average age was barely over 21 years. Even more interesting, however, is the fact that those who migrated from Mexico as children married younger than all other groups. Men who had migrated under the age of 15 married at approximately 24.5 years of age, while women who were child migrants married, on the average, under age 20.[37]

These figures suggest that Mexican immigrants did not delay marriage even after being exposed to the American custom of later marriage. The instability of the migratory process itself caused many young adult migrants to postpone marriage until settled. But once established in the new environment, Mexicans who grew up in American society were likely to marry younger than their counterparts in Mexico. Perhaps the erroneous assumption by Mexican parents that Mexicans in the homeland married very young encouraged them to urge their children—particularly the girls—to marry early.[38]

For children seeking greater independence, young marriages provided an escape from strict immigrant parents. Henrietta from Belvedere, age eighteen, expressed anger that "as soon as I was sixteen my father began to watch me and would not let me go anywhere or have my friends come home. He was born in old Mexico but he has been here long enough to know how people do things."[39] This strict discipline could backfire, ironically leading some young women to flee to their own marriage in order to be free of their parents. Concha, also from Belvedere, used her knowledge of Mexican mores to make her own marital decision:

> My father would not let Joe come to the house. He said when it was time for me to get married, he would have something to say about who my husband would be. So Joe and I fixed that. I ran off with him and stayed with his family. We knew that my father would make us get married then.[40]

Single migrants who lived west of the river often moved to East Los Angeles once they were married. This act usually involved a conscious decision to live in the barrios of the east side, among the growing community of Chicanos. It symbolized the reassertion of community life, this time in the context of an American barrio. It also signaled the passing from a migrant to a more settled mode of existence. For some, particularly women, it could also mean the surrender of freedoms gained through work outside the home and living beyond the cultural dictates of family.

If the communities to the south and west of the Plaza were more conducive to ethnically intermarried couples, Mexican immigrants who married American-born Chicanos congregated in East Los Angeles. These mixed nativity couples found the barrios east of the river, particularly Belvedere and Boyle Heights, particularly appealing. Over one-third of the families in these two neighborhoods displayed this foreign-born/ native-born marriage pattern, as compared with one-quarter of the total sample. As these communities grew during the 1920s, they gradually became the locus of Chicano cultural development. Since integration of the Mexican immigrant population with American-born Chicanos contributed to the creation of a distinctive barrio culture, both Belvedere and Boyle Heights became important settings for the definition of Chicano life in California during the twentieth century.

Like those who intermarried, almost all Mexican immigrants who married American-born Chicanos arrived in the United States before the age of twenty. Unlike the intermarried, however, their places of origin in Mexico were more broadly representative of the entire immigrant group. Border states consistently produced immigrants who married second-generation Chicanos, with 40 percent of immigrants from Sonora, Chihuahua, and Nuevo León engaging in this marriage pattern. Obviously, the interaction of the Mexican population living along the American-Mexican border gave immigrants from this area a sense of common purpose and tradition that fostered intergenerational marriage.

Other immigrants, however, refused to consider marriage to Mexican Americans. Juana Martínez, who had migrated to Los Angeles with her sisters and mother after a failed marriage and the death of her father, worked as a dance-hall employee in the Plaza area. She felt strongly that if she remarried, it would be with a fellow immigrant. "The Americans are very dull and very stupid. They let the women boss them. I would rather marry an American than a *pocho*, however." ("Pocho" refers to the American offspring of Mexican immigrant parents.) A fellow coworker, Gloria Navas, agreed, saying that she preferred immigrant men because "they know how to behave, they are not as 'rough-neck' as the *pochos*."[41]

Many Mexican immigrant men refused to consider marrying American citizens because American-born Chicanas appeared to exercise greater independence from their husbands. "Here the old women want to run things and the poor man has to wash the dishes while the wife goes to the show," exclaimed thirty-year-old Ignacio Sandoval from Fresnillo, Zacatecas. Another man who had lived in the United States for twenty-five years had remained single because he felt that women in this country were "very unrestrained." He surmised that "they are the ones who con-

trol their husband and I nor any other Mexican won't stand for that."
He argued that even Mexican women who migrated to the United States
took advantage of laws protecting women and became like American
women.[42]

Unions between Mexican immigrants and American-born Chicanos,
however, did occur often, but could result in continued tensions. One
Mexican-born husband expressed resentment that his American-born
"wife does not want to stay home and take care of the baby. She learned
how to work in a beauty parlor and now she wants to start a beauty
parlor and make money."[43] In another cross-generational couple, it was
the American-born wife who had complaints:

> My husband is a good man but—too many kids. I am twenty-three years
> old and I have five. American women do not look old and tired when they
> are twenty-three. They are still girls. Look at me. My father picked out this
> husband for me, but he should have sent to Mexico for a girl for him if he
> wanted to have one.[44]

Although most Mexican immigrants were married to other Mexican
immigrants in Los Angeles, only about one-quarter of these marriages
involved individuals from the same Mexican state of origin. In light of
all the various possible unions in the city, then, no more than 15 percent
of Mexican marriages in Los Angeles possibly involved immigrants from
the same state in Mexico. Compared with Italians in San Francisco, for
example, 65 percent of whom married immigrants from the same com-
mune, the figure for Mexicans is exceedingly low.[45] One possible explanation is that in Los Angeles racism set all Mexicans apart from American
society and obfuscated cultural divisions that had existed in Mexico. Sustaining allegiance to a certain area in Mexico became much less important
than beginning a new life as an ethnic family in the United States. A
more generic form of Chicano identity—different from other ethnic
groups in America—began to dominate Mexican American cultural life
in Los Angeles.[46]

The act of marriage, of course, only began the process of redefining
cultural values within the family. The actual nature of the union between
husbands and wives varied tremendously, depending on the individuals'
perspectives. Recent attempts to describe the Chicano family have portrayed it as an institution closely paralleling that of other immigrant families, something akin to a "father-dominated but mother-centered" family
life.[47] Countering the image of the traditional Mexican family as a rigid
patriarchy, these interpretations have stressed the flexibility of roles in
given social and economic circumstances. Some have begun to place emphasis on the mother-centeredness of the Chicano family, while others
have continued to examine the implications of male domination.[48]

One aspect of family life which had profound impact on the relationship between husband and wife was the pattern of childbearing. The
number of children a couple had often reflected cultural values regarding

Table 11. Number of Children for Various Types of Marriages

Type of marriage (husband/wife)	No children	1–2 children	3–5 children	6–8 children	Over 8 children	Number (Average)
All-Mexican immigrant	66 (15.8%)	129 (30.9%)	140 (33.5%)	72 (17.2%)	11 (2.6%)	418 (3.17)
Mexican/ Mex. Am.	34 (14.4%)	100 (42.2%)	77 (32.5%)	20 (8.4%)	6 (2.5%)	237 (2.71)
Mexican/ Anglo	44 (33.9%)	59 (45.4%)	25 (19.2%)	2 (1.5%)	0 (0.0%)	130 (1.36)
Mex. Am./ Mexican	7 (25.9%)	15 (55.6%)	5 (18.5%)	0 (0.0%)	0 (0.0%)	27 (1.52)
Anglo/ Mexican	18 (54.5%)	11 (33.3%)	3 (9.1%)	1 (3.0%)	0 (0.0%)	33 (1.00)
Total	169 (20.0%)	314 (37.2%)	250 (29.6%)	95 (11.2%)	17 (2.0%)	845[a] (2.63)

Source: Analysis of Naturalization documents, National Archives, Laguna Niguel, California.

[a]403 marriages in the sample contained no information on children.

family life. However, some immigrant historians have noted substantial variation among immigrant groups from agricultural backgrounds who settled in America's urban centers. For example, Dino Cinel has argued for Italians in San Francisco that "the crucial point is not the transition from rural to urban life, but the way people perceive the transition."[49] Mexican immigrants to Los Angeles exhibited an assortment of birthrate patterns which corroborate Cinel's assertion.

Mexican immigrant families sampled had an average of 2.62 children, with the largest families containing eleven children. The relatively low average number of children—compared with popular notions of Mexican family size—is undoubtedly a result of the youth of the group which applied for naturalization. Many were couples who had only recently married. Stark differences can be noted in the average number of children, however, when one compares all-Mexican marriages with those involving one non-Mexican immigrant (see Table 11).

Intermarriage with an American-born or foreign-born Anglo resulted in an average of only 1.29 children, as compared with all-Mexican marriages which averaged 3.17 children per family. Marriages involving one Mexican immigrant and an American-born Chicano fell between these two extremes, with 2.59 children per family. Mexican immigrant women who married a man born in the United States, whether a Chicano or an Anglo, were likely to bear substantially fewer children than if they married a man born in Mexico. Mexican immigrant men, on the other hand, were likely to have large families as long as they married within the Chicano community. Only marriages to Anglo women substantially reduced the size of families of Mexican immigrant men.

Not surprisingly, large families were more readily found in East Los Angeles than west of the Los Angeles River. Every barrio east of the river averaged at least 2.8 children per family, while the average west of the river was under 2.5. Given the prevalence of single migrants to the south and west of the Plaza, along with smaller families in these communities, children were a more dominant presence in the barrios on the east side than they had been in the more integrated, working-class neighborhoods around downtown.

Despite variance in the eventual sizes of families, marriage invariably led to childbirth for women in all possible unions in the sample. Between two-thirds and three-quarters of all women had given birth within eighteen months of marriage. Childrearing continued to be the main expectation for married women of this period, even if they continued to work after marriage. The differences that did exist in childbearing practices between families reflect the spacing between births and the curtailment of childbearing among mature unions. The age of marriage and the interval between marriage and first birth were the same for all types of married couples.

As east-side barrios began to grow, the construction of family ties proved a strong basis upon which to promote a sense of community. Powerful religious sanctions against marital breakup kept the numbers of female-headed households low. Despite widespread male migration and the cultural breakdown historians have attributed to both European and black newcomers to the cities, no more than 55 marriages out of 1,249 unions in this sample were affected by divorce or separation. This 4.4 percent rate of divorce is low by American standards in the period, since a 1916 study found one divorce for every five marriages in Los Angeles. More than one-third of these separations occurred between Mexican immigrants who had married Anglos, even though less than 20 percent of the unions in the total sample were intermarriages.[50]

Moreover, three-fourths of these breakups occurred during the 1930s and the instability of the Great Depression. Although it is impossible to know for certain, it does appear that the economic crisis, and the accompanying stress it placed on families, was a direct cause of many of these divorces and desertions. Unemployed men sometimes found it easier to abandon their families than to watch helplessly as their loved ones struggled. Many relief organizations concentrated aid to families with no male head, and perhaps some fathers may have discerned leaving as the best option.

The liberal divorce practices in the United States did provide an alternative generally unavailable in Mexico for women caught in bad marriages. Exercising this option, however, often forced a confrontation with deeply held beliefs concerning proper family relations. Minnie Ortiz, who spent years tolerating her husband's philandering and lack of economic ambition, finally had enough after her husband struck her. A lawyer advised her to apply for a divorce. She did so promptly, but remembered "crying my eyes out, thinking of my shattered home life and of my fatherless girls."[51]

Tellingly, more Chicano families were broken up by death than by divorce or desertion. Sixty-six spouses in my sample were widowed while living in Los Angeles, a rate of 5.3 per 100. Dangerous conditions at work made men more at risk. Some women who lost their husbands moved in with relatives, but most were able to continue as heads of their households. Men who lost their wives often asked relatives to raise their children, since most Chicanos believed that female nurturing was crucial to childrearing. Whatever the situation, family and community networks were called upon in time of family tragedy. As one local Anglo American official acknowledged: "The Mexicans respond to appeals on the basis of their responsibility toward children, on the duties of sisters, aunts, uncles, etc. This is not so in the case of Americans, who are more individualistic."[52]

Many observers who disagree on the strengths and weaknesses of the Chicano family agree that Mexicans are familistic in orientation.[53] Critics of this family orientation accuse the Mexican American family of retarding individual development. These observers blame economic reverses on a family life which encourages members to seek semiskilled jobs with immediate, though circumscribed, rewards. Moreover, familism is often blamed for the lack of a strong sense of public duty, particularly the tendency of Chicanos to dissociate themselves from American politics and public organizations. When emphasis is placed so strongly on the family, they allege there is little time for contemplating the needs of society.

A strong sense of family, however, enabled Mexican immigrants to survive in a hostile American environment, and contributed to a strengthening of community sentiment inside the barrio. Lack of economic opportunity and outright racial discrimination were at the root of limited mobility, and strong family networks allowed Mexicans to persevere in difficult economic times. As following chapters will make clear, even those who eschewed family solidarity rarely moved up the economic ladder. If anything, familism contributed to slow, but steady, economic advancement, and it was often a family tragedy or widespread economic misfortune which sidetracked Mexican immigrants and Mexican Americans in their quest for greater economic security.

In the period directly following migration, Mexicans were unable to settle down because they could rarely count on the extended family networks available in their native villages. As individuals, they had only their own limited economic resources. Unemployment often led to more migration. Even so, many called upon cousins, distant relatives, and friends from hometowns to aid in this difficult period of transition.[54] As immigrants married, particularly if they married American-born Chicanos, they established roots in a new community which they hoped would bring greater stability to their lives. In time, these barrios came to serve as places which made other newcomers to the city feel welcome.

Creative, adaptive strategies predominated among Mexican immigrants who settled in Los Angeles. Only strong, flexible family ties insured the survival of all members. Certain individuals chose to go it alone, and others left the barrio altogether. Yet, for most immigrants, family and community came together in the emerging neighborhoods east of the river. At times, the barrio was for some a stifling, restrictive environment. Strong cultural norms were enforced which kept the community at least outwardly familiar to most newcomers from Mexico. More often than not, however, the barrio provided a haven for Mexican immigrants and American-born Chicanos. There they could adapt to American society while still retaining in their daily lives much of the flavor of Mexico.

CHAPTER 7

The Sacred and the Profane: Religious Adaptations

It is becoming increasingly evident that the Latin races within our gates will retain their hold on the faith of their fathers only in so far as we help them to cling to it. We can no longer wait for them to come to us, we must go out to them and "Compel them to come in."

—From the 1922 Annual Report of the Associated Catholic Charities of Los Angeles[1]

On Sunday mornings during the 1920s, Los Angeles' central Plaza came alive with the sights and sounds of Mexico. La Placita, as it was affectionately called, was home to the city's oldest Catholic church, where Mexicans from throughout Los Angeles came to worship. The Plaza, however, also contained contradictions reflecting the complicated character of Mexicans' religious life in the United States. Just across from the Catholic church Nuestra Señora la Reina de los Angeles, Methodists had erected their own house of worship, the Plaza Methodist-Episcopal Church, located on the Plaza specifically to entice Mexican immigrants to join the Protestant fold. With its doors open to Olvera Street and a Spanish-speaking minister in charge, the Methodists were able to lure a number of Mexicans away from the faith of their fathers and mothers.[2]

Although religion is only one of the many outward manifestations of the transformation of immigrant culture in Los Angeles, it can serve as a significant indicator of larger forces affecting Chicano life during this period. Religious identity, however, is a particularly elusive and complex problem to investigate historically because of its private and individual nature. While it is clear that life in the United States was dominated by secular concerns, the role religion played in the minds and hearts of Mexican immigrants is less discernible. But by exploring some of the more outward manifestations of this identity, one can better understand the impact of the institutional practices of the Catholic Church and the various Protestant denominations and attempt to reconstruct belief systems and traditions characteristic of Mexicans in Los Angeles.

The relationship of ethnicity and religion is still a relatively unex-

151

plored area for historical research, particularly in Chicano history.[3] Unlike other largely Catholic immigrants to the United States, Mexicans did not bring significant numbers of migrating clergy with them. Indeed, they were often served (or not served) by priests from other ethnicities, particularly the Irish. While this fact muted possibilities for Mexican American leadership to develop within the Catholic hierarchy, it created new opportunities for various Protestant denominations to encourage Mexican branches of their churches. Moreover, even among the large majority of Mexicans who remained at least nominally Catholic, religious reference points were often intertwined with an ethnic identity that existed outside of the official framework of the Catholic Church. To understand these developments, one must recall the religious context of society in Mexico.

More than almost any other aspect of life, religion threaded itself into the fabric of village life in Mexico. The local Catholic church was a cornerstone of community stability, invariably located in the central plaza. Men, women, and children together celebrated yearly religious rituals. The priest presided over these festivities, and his role extended far beyond purely religious matters. If one wanted the support of the community for any particular endeavor, the local priest's blessing was the first to be secured.[4]

A severe shortage of clerics at the turn of the century, however, meant that many parishioners were without their own village priests. Often clergymen were forced to go from village to village, serving a population spread out over many miles. This situation was particularly prevalent in northern Mexico, where the number of clerics could not keep up with the rapid population growth. The Church recruited priests from Europe, particularly Spain and France, but most chose to stay near the more populous urban centers in central Mexico. In addition, few priests were homegrown because of limited educational opportunities in the villages.

Strained relations between the Mexican federal government and the Catholic Church also played an important role in framing religious practice in the countryside. Since Mexican independence in 1821, government officials had tried to minimize the authority of Catholic bishops by launching a series of reforms officially separating church and state. Eventually church property was nationalized and marriage was declared a civil contract. This church/state split not only was a question of power and authority but also represented a major conflict between two competing ideologies. Catholic theologians of the nineteenth century rejected the basic tenets of liberalism, especially the belief in progress, individualism, democracy, and the perfectibility of man. They viewed unbridled economic competition as the cause of social disorder.[5]

In contrast, the liberal thinkers who dominated Mexico's educational and financial systems during the thirty-five-year reign of dictator Porfirio Díaz (1876–1911) promoted science, progress, and social Darwinism. They admired Anglo-Saxon nations and identified with the growing middle class of professionals and property owners, rejecting what they perceived as the Indian "backwardness" of their culture, even if it meant discounting 90 percent of Mexico's population.[6] Porfirio Díaz kept tensions between church and state in check during his reign, but within this environment of toleration, subtle changes occurred.

In order to foster a "Protestant work ethic" among the working class and peasantry, the Díaz administration encouraged American Protestant sects to penetrate Mexican society. Methodists and Presbyterians in particular made inroads among workers caught in the profound economic changes of Porfirian society. Even the Mormon Church was able to establish settlements, albeit relatively small and unstable, in the northern and central states of Mexico during the Porfiriato. The peasants dispossessed by the Porfirian land policy usually ended up in railroad construction gangs or mining encampments in central and northern Mexico. It was among these groups of single young men that Protestant proselytizing often proved most effective.[7]

One young man who later emigrated to the United States from Durango described his conversion to the Baptist faith around the turn of the century: "Out of curiosity, I began with a friend to visit the evangelical centers that there were at that time in Durango. As they gave us texts, little tracts and papers, I read them with interest but with suspicion, for I thought they had something to do with the work of the Devil." After staying away from these centers for some time, he returned one evening when a friend persuaded him that they should go listen to a much-talked-about visiting preacher. "We went, and this time I was convinced. That preacher said things that were true, which opened my eyes." Rejected because of his conversion by his family, friends, and girlfriend, he enrolled in the Baptist Seminary, eventually made his way north across the border, and became a Baptist minister among Mexicans in the United States.[8]

Others did not become attracted to Protestantism until arriving in the United States. One young man who converted to Methodism related his first encounter with a Protestant church in Pasadena, California:

> As soon as I was in the church building, I doubted that that place was a church. I did not see there what I had seen in the church. There were no images and no ornaments such as there had been in the churches I had known. A minister appeared who was dressed in a very humble manner.[9]

The sparseness of this Methodist Episcopal church, along with the lack of clerical lavishness, struck this twenty-one-year old and framed its appeal to him as a more personal, less distant religion than Catholicism.

Protestantism flourished among the displaced and served as a stabilizing force in a world rapidly being transformed by technology and transportation. Catholicism appeared to be an ancient religion rooted in a disappearing village life. The new order encouraged the adoption of new values, such as punctuality and the prohibition of alcohol—values which Methodism and other Protestant churches emphasized. The rhythm of work life, constantly interrupted by Catholic religious festivals, no longer appeared functional. Significantly, a large proportion of the earliest Mexican immigrants to the United States came from migrant communities that had been centers for Protestant conversions.[10]

Of course, Mexican immigrants brought a wide range of religious attitudes with them. Despite Protestant encroachment, relatively few became practicing Protestants, though many had been exposed to other denominations besides Catholicism in Mexico. Some came with strong anticlerical views, views which were often reinforced in the United States. Others came as deeply practicing Catholics. After 1910, some members of the middle class, for example, sent their children north to attend American Catholic schools and avoid the violence of the Revolution, while others harbored strong feelings against the institutionalized Mexican Catholic Church. Almost all the peasant migrants considered themselves Catholic, and practiced a "folk Catholicism" full of the rituals of rural village culture. Many working-class migrants had grown up in areas of Mexico where migrant communities had been served by no Catholic clergy. Among most immigrants, skepticism toward the institutional Church ran high. As one grandmother reportedly told her grandson in the early 1900s: "My son, there are three things that pertain to our religion: the Lord, Our Lady of Guadalupe, and the Church. You can trust in the first two, but not in the third."[11]

Religious attitudes and practices in the United States were invariably shaped by the different settings in which Mexican immigrants found themselves. In south Texas and New Mexico, for example, newcomers were located in communities where Mexican Catholics were the numerical majority. But in Los Angeles during the early twentieth century, the city was dominated by a Protestant population new to California. This fact made the religious dimension of life for the Mexican immigrant population in Los Angeles unique, although similar circumstances affected Chicanos elsewhere. As one immigrant from Zapotlán, Jalisco, made clear, Los Angeles presented Mexicans with a diversity of faiths not found anywhere in their native country. "In Los Angeles there are all the religions which one might wish and no one cares whether one is of this or that religion," explained Pedro Nazas. "It makes no difference whether one belongs to one religion or another; one can even be an atheist, no one will say anything."[12] Though the wide variety of creeds gave Los Angeles an aura of tolerance, the city actually provided an arena of stiff competition between Catholics and Protestants.

Protestant domination in Los Angeles had prevailed only at the end of the nineteenth century. As late as the mid-1860s, the Los Angeles *Star* warned that "Protestants who die here . . . have to take their chances of having any religious ceremonies at their graves and so far as getting married, their only show is to employ a Judge or J.P." But by 1880, with the completion of the transcontinental railroad and the decline of the old ranchero economy, there were eleven Protestant churches in Los Angeles, and Protestants controlled the larger religious and social life in the city.[13]

Yet none of the denominations initially showed any inclination to proselytize among the Mexican Catholics. In fact, the first Presbyterian minister in Los Angeles left the city in disgust after less than a year, warning others that "the name of this city is in Spanish the city of the Angels, but with much truth might it be called the city of demons." Rather, each denomination waited patiently for Anglo Americans already allied with their respective faiths to arrive from the east. After the turn of the century, Protestant congregations, having secured substantial membership and large church buildings, began to investigate potential areas of social concern. The result was the creation of several interdenominational groups devised to turn their attention to Los Angeles' Mexican population.[14]

A portion of this early activity centered around providing much-needed services in the barrio and dispensing charity to the poor. The Methodists, for example, instituted programs for training Mexican youth for industrial occupations. The Methodists believed that "if they [Mexicans] are removed from the bean fields, and placed in more modern occupations, they may be more able to adopt the Protestant faith."[15]

Protestantism in Los Angeles was closely associated with economic progress, and residents often linked a strong "work ethic" to religious conversion. Virtually all of the city's industrialists were active members of Protestant congregations. In fact, when the Los Angeles Chamber of Commerce began actively to recruit young industrialists from the East beginning in 1897, only one of the 34 men recruited did not immediately join one of the city's influential Protestant congregations.[16] To these individuals, it served both God and the Chamber of Commerce to train Mexican laborers to be punctual and observant, docile and politically impotent.

Efforts at conversion were substantially accelerated by the development of the Americanization movement in the second decade of the century. Since Protestants dominated government posts in the city, most programs developed by civic authorities had a strong religious undertone. In 1915 the school board, for example, instituted Americanization and citizenship classes in the predominantly Mexican schools. The materials used to conduct these classes came exclusively from Protestant denominations that had devised pamphlets and other instructional material to be used for this purpose.[17]

Although in the minds of most Protestant worshippers Americanization meant adopting a Protestant faith, Protestants defined conversion rather loosely, at least in the beginning. In 1936, for example, one Protestant minister was asked by a visitor to his congregation why Mexican women were allowed to sit in the back of his church saying the rosary. "We can't take everything away from them at once," he replied.[18] Other

Protestant ministers were also hard pressed to alter fully the traditional religious practices of their flock. They often had to make compromises.

In addition, most Anglo Protestant congregations were proud of their homogeniety and did not look favorably on the introduction of hundreds of "reformed" Catholic Mexicans into their churches. Almost all denominations, therefore, approached the Mexican community via missionary activity which allowed members to contribute financially to the Mexican efforts without having to accept converts into their own social circles. These denominations encouraged the development of separate churches in the barrio which served an exclusively Mexican clientele.

Typical was the Presbyterian Church, which appointed Robert McLean and his son, Robert Jr., to serve as Superintendents of Mexican Work from 1913 to 1932. In 1914, McLean Sr. appointed the Reverend José Falcón as a full-time missionary, and they both immediately set out to organize the Church of Divine Savior. Special funds were made available by the Home Missions Board for the construction of a spacious structure just east of the Los Angeles River in Boyle Heights. By 1928, the church had a membership of 371 and a Sunday school attendance of more than 400, making it one of the largest Protestant Mexican churches in the country.[19]

Robert McLean, Jr., took over from his father in 1918 and quickly became known as a leading expert on Mexican immigrants in the United States. He wrote several books on the subject during the 1920s. In particular, he claimed that "to win the good will and the friendliness of the people is a prerequisite to success along evangelical lines."[20] Thus, he promoted combining proselytization with social service. His efforts included free night classes in English and domestic science, the establishment of an employment agency, a free first-aid clinic, kindergarten care for children of working mothers, and summer recreational activities for Mexican youth. Religious instruction was a basic element in each of the programs, even if it amounted to no more than handing out a tract attached to medical forms.[21]

All of this activity, of course, did not sit well with officials of the Roman Catholic Church. They viewed Protestant proselytizing as a direct attack by outsiders on their inherited fold. Archbishop John J. Cantwell, appointed bishop of the Los Angeles–Monterey area in 1917, was particularly troubled by Protestant efforts. Cantwell condemned "the molesting hands of proselytizers who seek to tear out of the heart of the foreigner the religion which he has and which alone will save him from becoming an anarchist . . . . "[22] The contempt of Catholic officials for the work of Protestant denominations with Mexican immigrants was, at times, so virulent that it appeared as if nothing less than a holy war for the allegiance of immigrants was at play in the churches of Los Angeles:

Various Protestant Churches have divided Catholic immigrants into territorial districts to "Christianize them" and "bring them into the Kingdom."

When our Divine Savior hung upon the Cross His executioners divided His garments among them, but upon the seamless robe they cast their lots. Today even the seamless robe has been divided in the name of God and of religion. Scores of non-Catholic workers paid by their respective churches, are attempting to lead the Catholic immigrant to their particular belief—or unbelief.[23]

The Church's new concern came in sharp contrast to its traditional nineteeth-century approach to its Mexican parishioners. After the American conquest, the Irish-dominated North American Catholic Church had quickly asserted itself over Mexican priests in the Southwest through a campaign of "Americanization." In California, the Church made every attempt to stamp out local practices in favor of ritual more in line with American customs. For example, a long struggle took place over the introduction of the practice of tithing, not practiced among the Californios. Bishop Joseph S. Alemany went so far in the 1850s as to disrupt services to the population when they protested paying the financial subsidy to their local priest.[24] Indeed, the Catholic Church in California quickly established itself to serve the needs of Anglo migrants to the state, especially European immigrants. This fact is evident judging from the 322 priests who served in the state from 1850 to 1910. Some 83 percent had Irish surnames, 53 percent of whom had been trained in Ireland.[25] Anti-Catholic nativism in the United States also encouraged the Church in the Southwest to attempt to minimize its overt foreignness. This task was much more easily accomplished with English-speaking clerics than with Mexicans.

The insecurity of the Roman Catholic Church in the United States is at least partially to blame for its slow response to the huge influx of Mexican immigrants in the twentieth century. Only fear of Protestant proselytizing prompted the Church to reach out to these newcomers. As early as 1905, two settlement houses were established in the emerging barrio to serve the Mexican population. At least one of these facilities was established by Catholic women who were "deeply impressed with the knowledge that a larger number of the poor Mexican children of Los Angeles were being proselytized and weaned away from the faith of their fathers." This children's home—El Hogar Feliz—was a direct attempt to devise "ways and means of saving these little wanderers from the fate that threatened them."[26]

It took another twelve years and the appointment of Bishop Cantwell for the Catholic Church to go beyond these early efforts. Almost immediately after his installation, however, Cantwell organized an Immigrant Welfare Department within the Associated Catholic Charities to coordinate activities among the foreign-born population of Los Angeles. In November 1918, he secured $50,000 from the National Catholic War Council for "Americanization work" among the immigrant population.[27] Cantwell was fortunate to have Father Robert E. Lucey among his clergy, a native Angelino who later became Archbishop of San Antonio.

In 1920, Bishop Cantwell named Father Lucey, then assistant pastor at Immaculate Heart of Mary parish, to head the diocesan Bureau of Catholic Charities. During the next five years, Lucey saw the budget of his bureau quadruple, and, by 1929, five diocesan community centers with year-round programs for Mexican Americans had been established.[28]

From the start, this social service thrust concentrated on Mexican immigrants. In 1920, the Bureau of Catholic Charities reported that 32 percent of the individuals serviced by its programs were either Mexican or Mexican American. This compared with only a 7 percent Italian service rate and a 4 percent service rate for the Irish. The only group serviced more in Los Angeles was "Americans" at 34 percent. By the late 1920s, the extent of efforts directed at the Mexican population had risen to a reported 52 percent of the total funds used by the Catholic Welfare Bureau.[29]

Largely playing catch-up with the longer established Protestant programs, Cantwell learned from the work of his competitors and attempted to imitate many of their services. In 1920, he established the first medical clinic in the diocese within the Santa Rita Settlement House, in a neighborhood in which 89 percent of the inhabitants were Mexican American. Santa Rita social workers had serviced a total of 770 families in the previous year, including home visits by nurses and volunteers. In addition to Santa Rita, two of the other four centers for religious instruction and Americanization in 1920 were located in predominantly Mexican neighborhoods. A small religious chapel on Clarence Street was run by a Father Ramirez and included efforts directed at women and children in Boyle Heights by the "Mexican Ladies of Charity," a group of sixty young women of prominent Mexican families in the city. According to reports, these "ladies" visited homes in Boyle Heights to give religious instruction to children and instruct mothers "in sewing and other domestic arts . . . dispensing charity and giving instruction wherever needed."[30]

The Catholic Church also experimented with the creation of a recreation center specifically for Mexican men located just off the Plaza at 401 North Main Street. In addition to housing a reading room, this center functioned as an employment bureau, placing up to 2000 men in 1920. The explanation for this effort was that the "Mexican problem" in Los Angeles was, in large part, an economic one. According to the Bureau of Catholic Charities,

> Hundreds of men can be found loafing lazily around the Plaza district, sitting idly in the park, watching their fellows shoot pool or billiards in the myriad of these dens scattered along Main Street, or strolling leisurely from window to window of the cheap stores. Many of these men are being constantly imposed upon by the greedy and unscrupulous employment agents, often in league with employers just as greedy as they are themselves, and ever ready to take advantage of the Mexican's ignorance of the English language and of our country's laws and customs.[31]

The report went on to concede that if "greater love of country and religion" was to be promoted among the Mexican population, it required that Catholics show "considerate kindness and charitable helpfulness."[32]

The depression of 1920–21 made many of these efforts short-lived, however. By 1922, only two of the Catholic social settlement centers, the Santa Rita center and the downtown Brownson center which served a multiethnic population, were still in operation. Both the North Main Center and the Boyle Heights chapel on Clarence Street, along with an Italian-focused center on Alpine Street, ceased to be supported by the Immigrant Welfare Bureau of the Associated Catholic Charities.[33] Increasingly, the Catholic Church decided not to compete directly with Protestant denominations on providing social services, but rather to emphasize the teaching of Catholic doctrine and tradition. Rather than represent a wholesale abandonment of the social settlement approach, the Catholic Church refocused in difficult economic times on a strategy that more readily took advantage of Catholic background of the Mexican newcomers.

This new effort to emphasize Catholic tradition, however, was as firmly grounded in the concept of Americanization as any effort made by Protestant churches. In fact, the emphasis on religious instruction could easily be mixed with earlier efforts to root out the "foreignness" of the Church in southern California. By reinforcing ties to Catholic Church doctrine in the United States, Catholic officials hoped to promote Americanism. In this fashion, a call to tradition would hopefully serve to change national loyalties and ethnic sensibilities. As made clear as early as Cantwell's 1919 call to action on behalf of Mexican immigrants, this effort hoped to serve the spirit of both the Catholicism and Americanization:

> When it is realized that there are about seventy-five thousand Mexicans in Los Angeles, it can readily be seen what a field there is for Catholic workers. Catholics alone can serve their needs. No one can accuse a Catholic worker of proselytizing. These people are already Catholic by nature and inheritance and need only a friendly and guiding hand to make them good citizens and better Catholics. . . . We believe that in making Catholics better Catholics we shall make them better citizens.[34]

In 1919, a young Catholic woman named Veronica Spellmire recruited ten other women to lead a weekly program of religious instruction in Simon's Brickyard, a Mexican community a few miles east of downtown Los Angeles. As a teacher at the Utah Street School, Spellmire had been exposed not only to the educational difficulties of Mexican children in the early grades but to various school-based efforts to Americanize the children under Protestant influence. She became a volunteer worker at the Brownson House, located in the multicultural downtown area, where she learned the potential of immigrant work under Catholic auspices. Later she worked as an Americanization teacher at the Bridge

Street School in Boyle Heights. Spellmire eventually asked Father Robert
Lucey to help expand the effort she initiated at Simon's directed at Mexican children, since the archdiocese had established few churches and no
schools east of the river.[35]

In 1921, this effort began in earnest as Lucey contracted with a firm
in El Paso, Texas, to provide for Los Angeles Catholic settlements and
clinics one hundred copies of a weekly publication in Spanish, along with
distributing "booklets in the Spanish tongue treating in the simplest language subjects of doctrine and devotion."[36] Finally, Lucey received
Bishop Cantwell's permission to establish the Confraternity of Christian
Doctrine (CCD) in March of 1922, a direct outgrowth of the welfare
work conducted among Mexican immigrants. CCD was intended as "an
organization of lay volunteers organized for the purpose of instructing
the Catholic children attending public schools," and specifically concerned itself with the "religious welfare of foreign nationalities."[37] Spellmire was selected as the first coordinator of volunteers, working directly
under the auspices of the immigrant welfare department of the Associated Catholic Charities.

The first center explicitly devised for CCD instruction after the formal organization in 1922 was located in a movie theatre in Belvedere
Park, on the corner of Riggin and Marianna streets. Some 200 Mexican
children were being taught in the theater by the Holy Family Sisters and
lay female volunteers within four months of its establishment.[38] Within
thirteen years of its founding, the CCD offered religious instruction to
28,500 youngsters and operated 211 centers with 1,279 teachers.[39]
Eventually this effort led Cantwell to initiate a massive school building
campaign in East Los Angeles after World War II, an effort directed
explicitly at educating Mexican American youth in the Catholic tradition.

Part of the preoccupation of the Church with religious instruction
of Mexican children was premised on the idea that Mexicans arrived in
the United States without proper religious training. Indeed, many immigrants had come from areas in Mexico barely serviced by the institutional
Church. Catholic authorities also felt that through religious instruction
Mexicans would reject Protestant attempts to lure them away from Catholicism. As Father Leroy Callahan, director of the Los Angeles CCD
from 1927 to 1937, put it, working among Mexican Americans was "as
truly missionary as the evangelization of the heathen, with the sole difference that we are laboring amongst those who are Catholic by baptism."[40]

The volunteers which worked in early CCD programs were divided
into "Fishers" and "Teachers." "Fishers" were given the task of going
into the homes of families to encourage youngsters to visit their local
catechism center, while "Teachers" offered instruction in the doctrines
and practices of the Catholic Church in more formal classroom settings.
This effort was often designated to women, with various orders of nuns
spearheading activities in specific communities. The Sisters of the Holy
Name, for example, were given the task of teaching catechism to Mexican

children in Maravilla, a poor barrio in the eastern section of Belvedere. In addition to the nuns, 37 paid workers supervised activities at 28 different institutions located throughout the diocese. A legion of volunteers made up the base of the crew designated to lead children out of the morass of nonbelief and faulty religious instruction.[41]

The establishment of CCD instruction under the immigrant welfare division of Catholic social services, therefore, was rooted in the framework of Americanization and within an attitude which viewed Mexican folk Catholicism as deficient and unprogressive. The activist stance which Bishop Cantwell and Father Lucey initiated sought to keep Mexicans in the fold by stressing their historical allegiance to the Catholic Church, while at the same time trying to shift their beliefs toward allegiance with the formal doctrine and hierarchy of the Church and away from practices and beliefs nurtured in Mexico and in American barrios. Although CCD programs were aimed at every Catholic immigrant community in Los Angeles, there was a qualitatively different approach in reaching Mexicans when compared with European immigrant children. Italians, for example, were viewed as having only temporarily put aside the traits of a civilization that had deeply enriched Catholic tradition:

> The church in Italy has gathered from the ranks of these people, whether of noble blood or humble birth, the greatest saints and scholars that the world has known; she has given to civilization the loftiest flight of human thought and from these gentle, simple hearted people she has produced the greatest masterpieces of human genius in painting, sculpturing, architecture, and in music the noblest of the arts. Religion is the inspiration of the Italian people.[42]

Their problem in Los Angeles, therefore, was viewed as simply reflecting the lack of sufficiently trained instructors to bring out the deeply ingrained Catholic values of the Italian population. Male religious orders, such as the Salesian and Don Bosco contingents, were tellingly brought to Los Angeles to specifically serve the religious instructional needs of the Italian community.[43]

On the other hand, religious instruction for Mexican children in the 1920s was characteristically put only in the hands of women. While European cultural traits were generally seen as positive and only temporarily held in abeyance by poverty and factors related to immigration, the "Mexican problem" was seen as more intractable and hereditary. For example, preceding the aforementioned description of Italian Catholic traits was this analysis of the problem of religious work in the Mexican community:

> We have two kinds of immigrants, European, through New York and Boston, and Mexican across the border. Follow up work on the latter is extremely difficult and often impossible. They are here by the thousands, unheralded and unknown. The women use various names, sometimes their maiden names, sometimes their husband's name. They have compound

names and may use either half. The children have a choice of several names and use them indiscriminately. They often have no address other than an indefinite district, living as they do in huts and shacks, barns, tents, or anything they can find in undeveloped districts. They also roam from place to place with variations of fruitpicking season and industrial occupations. They retain some of the old Indian traits.[44]

Another factor in promoting a distinct vision of the Mexican Catholic tradition as inferior was the confrontation between the church and state in Mexico which erupted again during the Revolution and persisted throughout the 1920s. Catholic priests, particularly the French and Spanish-born, were targets of Mexican revolutionaries, especially Francisco "Pancho" Villa, who operated in Chihuahua and throughout northern Mexico. Many clerics fled north during the Revolution and found themselves in the United States for extended periods of time. By August 1914, for example, the Archbishops of México, Michoacán, Oaxaca, Durango, and Linares and the Bishops of Sinaloa, Saltillo, Aguascalientes, Zacatecas, Guadalupe, Tulacingo, Chiapas, and Campeche were all living in San Antonio.[45] By 1928, the Archdiocese of Los Angeles estimated that 100 Mexican priests continued to reside within its jurisdiction.[46]

Although a few of these priests were assigned to serve Mexican immigrants in Los Angeles, most waited for the earliest opportunity to return safely to Mexico. Even when called upon to participate in clerical duties in the United States, these urbane, typically Spanish-born priests had little connection with the mostly working-class, rural migrants. Their presence, however, did encourage the United States Catholic hierarchy to speak out consistently against religious persecution in Mexico. Part of Cantwell's own appeal to Los Angeles Catholics to alleviate the plight of Mexican newcomers was an admonition that "we, in Los Angeles, so close to the Mexican border . . . cannot be indifferent to the dreadful persecution which is now being waged not only against the Catholic Church but against the most fundamental principles of Christianity."[47] Increasingly, therefore, the Catholic tradition of charity toward those persecuted for their religious beliefs came to dominate the activities of the Church toward the expatriate Mexican community.

The efforts made on behalf of Mexican Catholics, however, were not always welcomed by other ethnic groups represented in the Los Angeles Catholic Church. Indeed, Cantwell himself consciously kept the extent of programs aimed at Mexican immigrants hidden from the majority of Anglo Americans in the diocese. By 1929, the category of "immigrant welfare," which had so totally dominated earlier reports of Catholic welfare work, had disappeared from the annual reports of the Catholic Welfare Bureau.[48] He knew full well that those paying for such activities might react against the larger expenditures on behalf of Mexicans. Indeed, a 1937 editorial in the official Los Angeles diocesan paper, *The Tidings,* rebuffed Catholics who insisted that the Church address issues of social welfare. The editorial stated that the Church was not concerned with "merely human worldly objects . . . . Christ did not found the Church to be a mere humanitarian institution . . . . She has, in fact, plenty to do to attend to her own business."[49] In consistently pressing for progressive Church involvement in working-class lives, individuals like Father Lucey were clearly the exception, not the rule, among the Church hierarchy.

While hiding the costs associated with providing welfare for the impoverished Mexican community in the late 1920s, the Church hierarchy did begin to allow Mexican Catholics to organize for themselves under Church auspices. New Catholic organizations began to form which were ethnic-specific, usually holding their meetings in Spanish and having minimal contact with similar Anglo Catholic organizations. One of the largest of these was the male Holy Name Society, although most of the groups were female service organizations such as the "Mexican Ladies of Charity." As indicated by one report, many pastors "are convinced that these [ethnic organizations] help to hold the Mexican people together and arouse them to more zealous cooperation."[50]

Tolerance toward ethnic organizations within the Catholic Church, coupled with the continued financial support of welfare activities in the Mexican community, meant that most immigrants remained at least nominally Catholic in the United States despite the aggressive efforts of various Protestant denominations in the Mexican community. The Catholic Church was largely successful in stemming the tide toward Protestant conversion among Mexican newcomers to Los Angeles. Although an absolute correlation cannot be proven, most scholars assume that no more than 10 percent of the Chicano population of the 1920s and 1930s became Protestant, and the actual proportion was probably closer to 5 percent, the percentage of practicing Protestants within the Chicano community today.[51]

These figures do not tell the full story, however. Protestants produced a substantial group of community leaders which the Catholic Mexican constituency could not rival. Although often restricted to segregated churches, homegrown Chicano Protestant ministers emerged as important figures during the 1920s, in contrast to the almost nonexistence of Chicano priests and nuns (except for the few that remained after fleeing Mexico during the Revolution). In addition, Protestant ministers were particularly adept at promoting lay leadership among their congregants that translated itself into non-religious, community-wide activity. One highly respected Baptist lay preacher in Belvedere, for example, financed a school to promote Mexican culture and the Spanish language among American-born Chicanos.[52]

Another example of the promotion of Chicano leadership was the work of Dr. Vernon McCombs, Superintendent of the Latin Mission of the Methodist Episcopal Church. McCombs acknowledged that his involvement with this community came from "a feeling of responsibility . . . to develop the Mexican people morally and spirtually to become assimilated into American culture."[53] Because of the nature of this missionary work, however, McCombs's assimilationist philosophy led him to identify potential Mexican leaders, train them under Methodist auspices, and encourage them to develop followings among other Mexicans. Francisco Quintanilla, a former soldier with Pancho Villa's army in Mexico, was thus chosen by McCombs for training at the Boy's Institute in Gardena while still in his early twenties. After twelve years of work in the Mexican community of Watts, Quintanilla had 250 members in his congregation and 400 pupils enrolled in his Sunday school.[54] Catholic lay leadership, by contrast, was always limited by lack of resources and autonomy, and the unwillingness of the Catholic Church to make lay participation more central to the religious mission of the Church in southern California.

Irrespective of all of their efforts, neither Protestant nor Catholic churches adequately met the needs of Mexican immigrants entering the United States in the early twentieth century. Both groups began their efforts by casting their programs under the rubric of "Americanization"—an enigmatic, yet often nativist, ideology which failed to appeal to the majority of Mexican newcomers. Most immigrants believed that they would eventually return to Mexico, and even those who had decided to settle permanently in the United States rarely wanted to give up their ethnic heritage and cultural values. Mexicans were wise enough to accept medical services and employment opportunities without abandoning their cultural values.

This reality pushed both campaigns toward a realization that through a promotion of ethnicity their efforts to create religious allegiance of the newcomers would be more successful. While this newfound tolerance for diversity was premised on the separation of ethnic organizations from those involving Anglo religious practitioners, it did allow for the growth of Mexican religious leadership, particularly among Protestant denominations. More important, however, this shift provided the organizational apparatus by which Mexican immigrants could adapt to a new religious sensibility in the United States.

Through the prism of religion, it is possible to generalize about the adaptation of Mexican immigrants in Los Angeles and to speculate regarding the creation of a distinctive Mexican American culture. For most, life was dominated by secular activities, even among those who might consider themselves religious in outlook. The sheer number of leisure activities offered to immigrants made church-sponsored activities seem lackluster by comparison. Moreover, village structure in Mexico gave the Catholic Church a hold over communal life that it could hardly sustain in the urban centers of this country. Already having moved away from the Church as an institution in Mexico, immigrants brought with them a decidedly personal and familial view of religion, one often tinged with anti-clericalism and anti-authoritarianism.

Even though Catholic officials felt that they had an advantage over

Protestant Americanization efforts because they did not require Mexicans to give up their faith, in fact the North American Church exhibited a form of Catholicism which appealed little to the Mexican populace. Consequently, Mexican immigrants displayed an increasing alienation from the Church as an institution. Anti-clericalism remained high and may have escalated among the Mexican American population because Irish American priests were rarely found to be sensitive enough to the needs of a working class people from a different culture. As one immigrant woman put it: "I don't believe in the sanctity or in the purity of the priests, or that they are invested with superhuman powers. To me they are men like all the rest. That is why I don't pay any attention to their preachings."[55] In the United States, these sentiments could be expressed with greater impunity, especially if directed at clerics in Mexico.

With so many other things to do in a metropolis such as Los Angeles, more and more Chicanos drifted from formal religious practice, even though they continued to consider themselves Catholic. Typical was the response of Julio Cortina, when asked by a WPA interviewer in 1937 what religion he was: "My religion? Well, you know that most of us Mexicans believe we are Catholics, but this is because our parents were Catholics; as far as I am concerned, I haven't visited a church more than four or five times in the same year, and this since I left home."[56] One study showed that while close to 80 percent of the Mexican population remained at least nominally Catholic, only half of that group—or about 40 percent of the total population—actually participated in any Catholic rituals, such as attending mass.[57] A census of California priests in the 1920s confirmed these findings, establishing that fewer than 40 percent of Mexican women and children in their respective parishes fulilled their "Sunday duty," while the percentage for Mexican men was a dismal 27 percent.[58]

As these last numbers indicate, taken out of its community-based context in Mexico, religion appealed less to men. The Mexican Catholic Church had often helped to centralize village life, playing a crucial role in community communication. As one immigrant reminisced, "sometimes I remember the big church I used to go to with my parents, and after the services early in the morning, every Sunday, the many places we went visiting with relatives and friends. And when I was older the chats with the gang around the plaza in front of the church."[59] In the United States, with its myriad churches and growing secularism, participation in Catholic ritual rapidly became an empty shell. At times, male religious practice became uniquely solitary. "I am Catholic but the truth is that I hardly follow out my beliefs," responded one twenty-eight-year-old man. "I never go to the church nor do I pray. I have an amulet which my mother gave to me before dying. This amulet has the Virgin of Guadalupe on it and it is she who always protects me."[60]

Public religion thus gradually became feminized in the Los Angeles Mexican community. In addition to much of the religious work among

Mexicans in Los Angeles being allocated by the Catholic Church to both religious and lay women, Mexican families themselves began relegating the religious component of a family's activities to their women. The Arce family had been in the United States for six years and, like many others, went separate ways on Sunday. "My mother and I go to Church almost every Sunday," explained Elenita. "My brother Miguel never goes, but he believes in God and in religion even though he doesn't carry it out. My sister Sarita goes to Church very regularly and we pray every night."[61] Another husband explained that he was "very Catholic" in Mexico and "went to church there every Sunday with my old woman but here she goes alone."[62] Parents often made it clear that they wanted their daughters to carry on the Catholic faith.[63]

Occasionally, the struggle to earn a livelihood kept even women away from church on Sundays. "I pray at night but I hardly ever go to Mass for I don't have time. Sundays are the days when there is the most work for many people. I have to be fixing the baths, receiving and delivering laundry, taking care of my child, preparing the meals and the rest."[64] In addition, the presence of so few Catholic churches in East Los Angeles made regular attendance a chore for many working-class people. In Belvedere and Maravilla Park, for example, there were only three Catholic churches to serve the fifth largest Chicano population in the nation as late as 1932, while eight small Protestant churches competed for devotional allegiance.[65] For others, it was simply the attraction of other activities. "We quit going to church," explained sixteen-year old Manuel, "because we used to play baseball on Sunday mornings."[66]

In addition, many Mexican immigrants, weary of the fanaticism they encountered in Mexico and the religious competition evident in the United States, became especially tolerant of a variety of different faiths. Doña Clarita, for example, felt that "Mexico ought to be like the United States where all religions are allowed."[67] One native of Guadalajara, after admitting that he had attended Catholic school in Mexico, continued, "but that doesn't mean that I blind myself. I respect the beliefs of all other people and I believe that what is worth most is work and honesty."[68]

The inability of the Catholic Church to provide enough facilities, along with the absence of Chicano priests and a Spanish liturgy, pushed many Mexican immigrants toward practicing a Catholicism rooted in their own homes. The widespread presence of home altars and personal displays of religious sentiment was the urban equivalent of the "folk Catholicism" so often chided by critics of Mexican rural practice. "You may find a Catholic picture in every room of our house, but you will seldom see us in the religious services at the parish," explained one Mexican immigrant.[69] These expressions of spirituality, usually carefully supervised by the matriarch of a family, were the base of a new religious consciousness shaped in Los Angeles, as it was developing throughout the Southwest.

In addition, religious expression increasingly took on an ethnic dimension, with public displays of the Virgin of Guadalupe and other Mexican religious symbols at festivals and events often replacing more orthodox American observance. Not all ethnic celebrations, however, were easily transferred to the barrios of the United States from the tight village communities of Mexico. One Mexican social worker displayed his disappointment with one particular transfer of ritual:

> On Christmas Eve in old Mexico it is customary among the families to hold a *tamalada* after the Midnight Mass. One family will make great preparation for such an occasion and will invite all its relations and perhaps one other family to be present at the festivities. . . . The *tamaladas* here are not of that type. Instead of a private home, they are held in a rented dance hall. Everyone goes. There is not much order and a great deal of confusion. The girls sometimes smoke and drink, which would never be tolerated in Mexico. The older people consider this a desecration of a sacred custom.[70]

For better or worse, Mexican Catholic rituals changed in Los Angeles to reflect the new communal realities of life as ethnic Americans.

By the early 1930s, Catholic officials themselves realized that the Church was more likely to keep Mexicans involved in its workings if it encouraged ethnic organization. The 1934 Annual Catholic Social Service Report indicated that three different organizations had been formed as "Catholic Action" groups. The Mexican Young Women's Association (or Juventud Catolica Feminina Mexicana) assembled 1400 young women into 51 different groups throughout Los Angeles. Young men had a similar organization, involving 30 groups and 1300 men. This focus on adolescents had to be expanded for women interested in remaining with a Catholic organization after marriage, however, as indicated by a new Mexican Women's Federation of four parish-based groups, involving 200 individuals. Religion, one might say, became ethnically stratified in Los Angeles as a result of the unmet spiritual needs of the Mexican immigrant population.

These developments point to the fact that a new religious sensibility had set in among the Catholic Mexican and Mexican American population in Los Angeles which relied less on the Church hierarchy for direction and more on their own ethno-religious organizations. While many immigrants stayed away from the Catholic Church altogether, another group attempted to mix their own version of Catholicism with a new-found ethnic identity as Mexican Americans. They took their cues from their own understanding of what it meant to be Mexican in the United States, no longer from an expatriate elite tied to the Mexican consulate. In this fashion, they were willing to participate with American Catholic officials under the proper conditions, and therefore pushed Catholicism in Los Angeles toward a recognition of the particular needs and concerns of Chicano Catholics. Evidence of this new religious foundation is best found in the developments surrounding the 1934 procession in commemoration of the Virgin de Guadalupe on December 8, which turned out, with 40,000 participants, to be the largest demonstration of any kind in the history of Los Angeles.[71]

Beginning in 1928, local Mexican Catholic organizations had organized an annual procession to honor the Virgin of Guadalupe, Mexico's most prominent Catholic symbol. Since 1930, that procession had been under the auspices of the Mexican branches of the Holy Name Society of Los Angeles. José David Orozco, a local radio announcer and travel agency owner, had organized forty chapters of the Santo Nombre, as it was known by Mexicans, to promote a "Mexican consciousness" and instill ethnic pride among Mexican Catholics. The Guadalupe procession was a crucial part of the blending of ethnic and religious consciousness promoted by Orozco and his uncle, exiled Guadalajara Archbishop Francisco Orozco y Jiménez, for the Mexican expatriate community in the United States. By 1933, the procession had grown so popular that it attracted more than 20,000 participants.[72]

Political developments in Mexico in 1934, however, changed the entire context of the event. Lázaro Cárdenas had been elected president of Mexico and vowed to reduce the influence of the Catholic Church further in the homeland. In the United States, daily reports surfaced in both the English- and Spanish-language press indicating widespread attacks on clergy and religious institutions. For many in the United States, including a certain portion of the Mexican Catholic population, these developments were evidence of a "vicious socialism" which sought to make Mexico "atheistic."[73]

By late fall of 1934, Orozco, as leader of the Mexican division of the Holy Name Society, called on Mexican Catholics in Los Angeles to use the upcoming Virgin of Guadalupe procession as a form of protest against the new persecution of the Catholic Church in Mexico. American Catholic officials, long at odds with the Mexican government over these issues, encouraged this development by asking all Catholics to participate in the procession for the first time. Orozco himself appeared on the cover of *The Tidings,* Los Angeles' Catholic newspaper, dressed in a charro outfit, ready to command his Catholic marchers.[74] The Mexican consul in Los Angeles at the time, Alejandro Martínez, denounced these plans, saying that the parade was organized by "a group of persons . . . already well known as the traditional enemies of the economic, social, and cultural progress of Mexico." He called on all "true patriots of Mexico" not to participate in the upcoming march.[75]

Despite this condemnation, over 40,000 individuals showed up in Boyle Heights on the afternoon of December 8 to march from Monte Carmelo Chapel to the grounds of the Los Angeles Orphan Asylum on Boyle Avenue. Included at the beginning of the procession were "Indians in tribal dress," followed by various American boy scout troops, college marching bands, and groups of Italian, Japanese, and Polish Catholics. The heart of the procession, however, remained the Spanish-speaking units under the direction of Orozco, including separate divisions for men and women and another for various floats. A reporter from *The Tidings* made it clear that few Anglo Americans in the city had ever before seen this type of religious expression:

> It was not an ordinary parade. The rosary beads slipping through rugged fingers, the hymn to Our Lady of Guadalupe on a thousand lips, the streamers and banners proclaiming faith in the Most Holy Name of Jesus and asking Our Lady's intercession, deeply impressed those who had come out of curiosity and, perhaps, those, too, who felt deterred from participating by a false sense of expediency.[76]

Having reached the grounds on Boyle Avenue, Bishop Cantwell proceeded with a benediction, followed by a reenactment of the Virgin's appearance to the Indian Juan Diego in 1531. Father Isadore then addressed the crowd in Spanish, exclaiming "Viva Cristo Rey" at the conclusion of his remarks. Throughout the afternoon, a "Spanish dinner" was being served at St. Elizabeth Church, on nearby Opal Street, for those who had grown hungry.

By the end of the day it was clear that Mexicans in Los Angeles had just participated in an event created by a new sense of ethnic and religious identity. Using all the symbols of Mexico, these Chicanos were not afraid to denounce Mexican government action and to go against the wishes of the local Mexican consulate. By 1934, most of those participating knew that they were likely to remain in Los Angeles, though they were certainly not willing to abandon their sense of *Mexicanidad* nor of Catholicism. They had, in fact, pushed the local Catholic Church to acknowledge, even if for just this day, their own particular needs and desires. At the same time, they had expressed their own unique sense of Mexican national identity. In response to Consul Martínez's denunciation of the marchers, Father Vaughn, a popular radio priest in Los Angeles, put the questions on the minds of Mexican Catholics best when he proclaimed on the airwaves:

> It might be well to consider the meaning of that word patriotism. It connotes love of country. But what is meant by country? Is it the ground we walk on, or our fellow citizens, or the government, its institutions and its laws?[77]

Vaughn concluded that patriotic love of country was conditional and that Mexicans, even if they continued to "love the ground of Mexico and love their fellow countrymen below the Rio Grande," had stopped caring for the government and politicians of Mexico because of attacks on its religious institutions.[78]

The Holy Name Society, for its part, took the unusual step of endorsing a letter written to President Franklin D. Roosevelt blasting the Los Angeles Mexican consulate for violating "our constitutional principles of religion and expression."[79] In this action, these Mexican Catholics

were clearly expressing themselves as Mexican Americans, protected under the U.S. Constitution but still interested in affecting policy in Mexico
toward the Catholic Church. Unlike Catholic officials in Mexico, local
Mexican Catholics could afford to see this animosity subside with the
passage of time, the changing of personnel, and the lessening of tensions.
In fact, within five years José David Orozco himself was serving as the
president of consulate-sponsored ethnic festivities in Los Angeles.[80]
While Mexican politics had served as a source of friction for a time in
Catholic Los Angeles, bonds of ethnicity and identity proved to be a
more long-lasting legacy.

Familiar Sounds of Change: Music and the Growth of Mass Culture

Just south of Los Angeles' central Plaza lay the area known throughout the city as the main arena for activities of leisure in the Mexican community of the 1920s. Sundays were not only a big day for religious practice; they also were big business days for the area's movie theatres, gambling dens, and pool halls—all of which dominated the streets to the south. The constant sound of Mexican music—music that ranged from traditional Mexican ballads to newly recorded *corridos* depicting life in Los Angeles—was everywhere. A burgeoning Mexican music industry flourished in the central and eastern sections of the city during the 1920s, largely hidden from the Anglo majority.

The diminished role of organized religion in the day-to-day life of Mexican immigrants was coupled with increased participation in secular activities. In Mexico, most public events in rural villages were organized by the Catholic Church, with few other opportunities outside the family for diversion. Los Angeles, however, offered abundant entertainment of all sorts. These amusements were generally part of a rapidly growing market in leisure which targeted working-class families during the 1920s. Money spent on leisure-time activities easily outstripped donations to the Church, revealing much about the cultural changes occurring in the Mexican immigrant community.[1] Chicano entrepreneurs responded to the emerging ethnic mass market in cultural forms, even though that market was often dominated by outside advertising and controlled primarily by non-Mexicans. Still, the presence of a growing ethnic market in Los Angeles provided room for many traditional practices to continue, some flourishing in the new environment, but most being transformed in the process.

This chapter will explore the intersection between the growing mass market in cultural forms found in Los Angeles and the leisure-time activities of Mexican immigrants. The various actors who helped shape the creation of a market aimed at providing Mexican immigrants with products, services, and activities that somehow connected with the ethnic self-identification and collective culture will be identified. The complicated nature of this exchange can best be described, however, by looking at one particular arena of cultural interaction. Music, specifically the creation of a Spanish-language music industry and market in Los Angeles, provides one of the best windows for viewing this nexus of cultural transformation in detail.

The Plaza itself continued to cater to single males, offering pool halls, dance rooms, bars, and a small red-light district. Protestant reformers, therefore, consistently viewed Plaza residents as prime targets for moral rejuvenation. In addition, many small, immigrant-owned eateries were located in the area which catered to a male clientele often unable or unwilling to cook for themselves.

A description of a dancing club frequented by single males during this period indicates the extent of the intermingling between sexes and nationalities in the Plaza, a situation which concerned reformers. Located on Main Street, the club "Latino" was open every night except Sunday from 7:30 p.m. to 1 a.m., although it did most of its business on Saturday night. Inside and out, the hall was illuminated by red, white, and green lights, the colors of the Mexican flag. Entrance to the club cost 25 cents, and tickets were 10 cents apiece to dance with women. The female employees were mostly immigrant Mexicans or Mexican Americans, although Anglo American, Italian, Filipino, Chinese, and Japanese women also were available. The band, however, was made up of black musicians and played only American pieces. Mexican immigrant men, dressed in working-class garb, danced "Mexican style" to the American songs; a ticket was required for every dance; and the women partners earned 5 cents per dance. In one corner of the dance floor a Mexican woman sold sandwiches, tacos, pastries, and coffee.[2]

As Los Angeles Mexicans moved away from the Plaza and the community became more familial in structure, different diversions predominated. Some customs were carried over to marriage from single life. For example, a federal survey reported that three-quarters of Mexican families in Los Angeles continued to spend an average of $14 a year for tobacco. Almost two-thirds read the newspaper on a regular basis. Increasingly Mexican families began to purchase other forms of entertainment which could be enjoyed by all ages and in the confines of one's home. Over one-third of the families in the Los Angeles study owned radios, often buying the equipment "on time" for an average of $27 a year. A smaller number (3%) owned phonographs, and only 4 percent owned musical instruments. Expenditures for vacations, social entertainment (other than movies), and hobbies were rare.[3]

During the 1920s, many American manufacturers and retailers discovered a fairly lucrative market in the local Mexican immigrant community. Despite the clamor for Mexican immigration restrictions, these producers understood that Los Angeles contained a large and growing population of Spanish-speaking immigrants. By 1930, some national products were advertised in the Spanish-language press, and increasingly large distributors sponsored programs in Spanish on the radio.[4] Among products heavily advertised in *La Opinión* during this period were cigarettes, medicinal remedies, and recordings to help immigrants learn the English language.

Even more widespread were appeals to Mexican shoppers by certain downtown department stores. In 1929, for example, the Third Street Store advertised in *La Opinión* by asking, "Why are we the store for Mexicans?" The answer stressed the appeal of special merchandise, prices, and service. Located near the Plaza, offering generous credit, the store had apparently already become a favorite in the Mexican community.[5] This kind of ethnic appeal fostered competition among some of downtown Los Angeles' largest retailers. Another department store even offered free "Cinco de Mayo" pennants to any Mexican who purchased its merchandise.[6]

Many of the mass-produced consumer goods in the 1920s were specifically marketed with an appeal to youth. This appeal had profound consequences for Mexican immigrant families. Older children who entered the work force often earned enough to become more autonomous. Adolescents and young adults were often the first to introduce a Mexican family to certain foods, clothing, or activities that were incompatible with traditional Mexican customs. For example, younger Mexican women began to use cosmetics and wear nylon stockings. Young men were more likely to seek out new leisure-time activities, such as American sports or the movie houses. Second-generation youth were often the first in their families to see a motion picture. At times, experimentation led to intergenerational conflict, with much tension revolving around consumer purchases and the control of earned income.

Despite some initial reservations, most Mexican parents joined other Americans in the 1920s in a love affair with motion pictures. Ninety percent of all families in the Los Angeles survey spent money on the movies, averaging $22 a year per family. In San Diego, a government committee investigating local economic conditions observed that "as in American families, movie tickets were an essential feature of these Mexican families' spending ways except under pressure of a special need for economy." In addition, the committee presumed that some working children retained a portion of their wages to spend on movie tickets.[7]

The movie industry in Los Angeles aided Mexicans in retaining old values, but also played a role in cultural change. On the one hand, films produced in Mexico made their way into the many theatres in the downtown area in the late 1920s catering to the Mexican immigrant population. These supplemented American- and European-made silent films which were aimed by their promoters at an often illiterate immigrant population. Sound was not introduced until 1929, so that throughout the decade of the 1920s, movies stressed visual images and presented few language barriers for the non-English speaker.

Since their inception in the nickelodeons of eastern seaboard cities, American films consistently contained storylines intentionally made for the immigrant masses.[8] Messages tended to be largely populist and democratic in tone. Plots stressed the commonality of all Americans. The children of Mexican immigrants were especially intrigued by the open sexuality depicted on the screen. The experience of sitting alone in a darkened theatre and identifying with screen characters, as Lary May has argued, could feel quite liberating.[9]

What made American-made films even more appealing was the appearance of actors and actresses who were Mexican by nationality. Although Ramón Navarro and Lupe Vélez were introduced to audiences in the early twenties, the arrival of Dolores del Río in 1925 brought Mexican immigrants flocking to the box office. The attraction was not simply the desire to support a compatriot; it was also generated by the close proximity of the movie industry. *La Opinión*, for example, the city's leading Spanish-language periodical, regularly followed the Hollywood scene, paying particular attention to the city's rising Latin stars. As citizens of Mexico themselves, the newspaper's editors were quick to condemn stars who distanced themselves from their national origins, while praising others, like del Río, who showed interest in preserving their Mexican identity.[10]

While the motion-picture industry displayed one aspect of the impact of consumerism on immigrant cultural adaptation, opportunities for other entrepreneurs to make an ethnic appeal emerged during this period. Ethnic marketing, usually considered a recent phenomena, in fact has long-standing roots in this era. While huge American corporations consolidated their hold on a national mass market of goods during the 1920s, much room was left for local entrepreneurs to seek sub-markets that catered to the interests and desires of particular groups. In many ways, the standardization of messages brought about by large-scale advertising created new avenues for ethnic entrepreneurs. Since few national advertising agencies were located in Los Angeles or in the American Southwest, little attention was paid by national corporations to distinctly regional appeals. This void was filled by Mexican and non-Mexican entrepreneurs who realized that money could be made by servicing the large and growing Mexican population in the city.

As early as 1916, small Mexican-owned businesses advertised in Spanish-language newspapers.[11] These establishments were generally store-front operations which allegedly provided items that were "typically Mexican." El Progreso Restaurant on North Main Street, for example, claimed that it cooked food in the "truly Mexican style." Similar restaurants were frequented by the large Mexican male population around the Plaza. Other businesses attempted to bring Mexican products into the

Los Angeles market directly. La Tienda Mexicana, on San Fernando Street, carried herbs and cooking supplies which were generally unavailable elsewhere. Down the street, a clothing store, the Sastrería Mexicana, was less successful in its appeal to ethnic taste in dress.[12] It was one thing to continue to put Mexican food in your stomach and quite another to continue to dress in "traditional" Mexican garb on the streets of Los Angeles.

By 1920, large, well-financed operations dominated the Mexican retail business. Their advertisements regularly appeared in the city's Spanish-language periodicals for the next two decades. Farmacia Hidalgo, run by G. Salazar and located at 362 North Main street, declared that it was the only store "positively of the Mexican community." Farmacia Ruiz was founded by an influential Mexican expatriate and quickly gained much status in the immigrant community. Over the next ten years, it was frequented by several candidates for the Mexican presidency, most notably José Vasconcelos.[13] Mauricio Calderón, another emigrant from Mexico, would soon dominate the Spanish-language music industry in Los Angeles. During this decade he established the Repertorio Musical Mexicana, an outlet for phonographs and Spanish-language records, which he claimed was "the only Mexican house of Mexican music for Mexicans." Finally, two theatres, the Teatro Novel and the Teatro Hidalgo, located on Spring and Main streets respectively, were already in operation in 1920, offering both silent films imported from Mexico as well as live entertainment.

A host of rival Mexican-owned firms gave these early businesses much competition. Advertisements usually stressed that their particular establishment was the most "genuinely Mexican" of the group. The Farmacia Hidalgo went so far as to place an Aztec eagle on some of its products to insure "authenticity." A new and important enterprise was the Librería Lozano, providing Spanish-language books to the literate Mexican community and owned by Ignacio Lozano, the editor of *La Opinión*. Not surprisingly, Lozano heavily advertised in his own paper.

In addition, the 1920s witnessed the emergence of Mexican professionals who also targeted their fellow countrymen for patronage. A small, but significant group of doctors, dentists, and lawyers from Mexico set up shop in Los Angeles, and their advertisements stressed that their training had been conducted in the finest Mexican universities.[14]

Mexican entrepreneurs, however, were not the only individuals in Los Angeles who appealed to the Mexican consumer; non-Mexicans also tried to capitalize on the growing ethnic clientele. Leading this effort was the medical profession, particularly women doctors and physicians from other ethnic groups not likely to develop a following within a highly male-dominated, Anglo Protestant profession.[15] Most of these physicians were located near the Plaza area, particularly along Main Street, an area which provided direct access to the immigrant population. Female physicians held special appeal as specialists for women, capitalizing on the sense of propriety among immigrant women. "Doctora" Augusta Stone, for example, advertised as a specialist for "las señoras," and was among the first to use the phrase "Habla Español" in her advertisements. Dr. Luigi Gardini, an Italian American physician, also advertised in Spanish-language newspapers in 1916. Asian American physicians, however, were the largest group of non-Mexican professionals to appeal to Mexican immigrants, largely stressing their training in herbal medicine, an area not unfamiliar to rural Mexicans. Among them was Dr. Chee, who characterized himself as "Doctor Chino" in 1920, and Dr. Y. Kim, who boasted the combination of a Yale degree and a speciality in Oriental herbal treatments.

The growth and increasing economic stability of the Mexican immigrant community in Los Angeles made these appeals profitable. While the Mexican middle class remained small and relatively insignificant, the large working-class community was quickly developing east of the Los Angeles River. Lack of capital and professional training in the Mexican community made it difficult for most Mexicans to take direct economic advantage of this growth. Yet their cumulative purchasing power did allow for the growth of certain enterprises which catered to the unique backgrounds of Mexican immigrants, while creating new modes of ethnic expression.

One of the most important of these enterprises was music. Although the musical legacies of different regions in Mexico were significant, traditions were both reinforced and transformed in the environment of Los Angeles. As a diverse collection of immigrant musicians arrived from central and northern Mexico, often via south Texas, they stimulated the growth of a recording industry and burgeoning radio network that offered fertile ground for musical innovation.

Of 1,746 Mexican immigrants who began the naturalization procedure, 110 were musicians (6.3% of the total), making them the second largest occupational group in the sample, well behind the category of "common laborer."[16] Although 80 percent of the musicians did not complete the process, their ample presence among those who initiated the naturalization process indicates their willingness to remain in the United States. Unlike working-class musicians of Mexican descent in Texas, it appears that many Los Angeles–based musicians were willing to consider changing their citizenship.[17] If, as Manuel Peña has claimed, musicians do function as "organic intellectuals" for the working class, challenging American cultural hegemony while expressing the frustrations and hopes of their social group, then the experiences of Los Angeles musicians indicate a complex, if not contradictory, relationship with American cultural values.[18]

Compared with the larger sample of Mexican immigrants, musicians were more likely to have been born in the larger cities of the central plateau in Mexico, particularly Guadalajara and Mexico City. Over 25 percent of Mexican musicians in Los Angeles came from these two cities alone, compared with 10 percent of the entire sample. Other towns in central Mexico, such as Zacatecas, Guanajuato, Puebla, and San Luis Potosí, were also well represented in the musical community. Unlike the larger sample, northern states were generally underrepresented among musicians, except for the state of Sonora, which accounted for 9 percent of the performers. In central Mexico, the states of Jalisco and Guanajuato and the Federal District alone produced over 41 percent of all Mexican musicians in Los Angeles.[19]

The musical traditions brought to the United States from these locales were varied. The mobility within Mexico caused by economic upheaval and violence related to the revolution had pushed many rural residents, including folk musicians, to seek shelter in towns and cities. There, previously isolated folk music traditions from various locations were brought together, and musicians also encountered the more European musical tastes of the urban upper classes. One study of street musicians in Mexico City during the 1920s, for example, found twelve different regional styles performing simultaneously on the corners and in the marketplaces of the capital. One could hear mariachis from Jalisco, *canciones norteñas* from Chihuahua, troubadors from Yucatán, *bandas jarochas* from Veracruz, and marimba groups from Chiapas and Oaxaca.[20]

If there was one particular musical style which stood out from the rest in popularity during this period, it was certainly the *corrido*. A prominent student of this genre has called the *corrido* "an integral part of Mexican life" and the creative period after 1910 its "most glorious epoch."[21] During the Mexican Revolution, almost every important event, and most political leaders and rebels, became the subjects of one or more *corridos*. Pedro J. González, who later emerged as the most well-known Mexican musician in Los Angeles, remembered composing *corridos* with seven other soldiers fighting with Pancho Villa in secluded mountain hideouts during lulls between battles. None was a trained musician, but each used the opportunity to criticize each other jokingly for past misfortunes or to immortalize some heroic deed through song.[22] As these *corridos* made their way into Mexico's urban centers, they were codified and transformed from folk expression to popular songs.[23]

The *corrido*'s continued popularity during the 1920s in areas far away from its folk origins can be explained by particular characteristics of its style which made it appealing as an urban art form. First, the urban *corrido*, like the *canción ranchera*, embodied what was a traditional music style from the countryside, while adapting it to a more commercially oriented atmosphere. It reminded those who had migrated from rural areas of their provincial roots, and gave urban dwellers a connection to the agrarian ideal which was seen as typically Mexican.[24] Second, most *corridos* appealed to a Mexican's nationalist fervor at a time when the pride of Mexican people, places, and events was flourishing. Several observers have identified the period between 1910 and 1940 as one of "national romanticism" in Mexican cultural affairs, extending beyond music to literature and mural painting. *Corridos* produced in the United States often exalted "Mexicanism" at the expense of American culture, but even those composed within Mexico paid inordinate attention to promoting Mexican cultural identity.[25]

Finally, the *corrido* was an exceptionally flexible musical genre which encouraged adapting composition to new situations and surroundings. Melodies, for the most part, were standardized or based on traditional patterns, while text was expected to be continuously improvised. A vehicle for narration, the *corrido* always intended to tell a story to its listeners, one that would not necessarily be news but rather would "interpret, celebrate, and ultimately dignify events already thoroughly familiar to the *corrido* audience."[26] As such, *corrido* musicians were expected to decipher the new surroundings in which Mexican immigrants found themselves while living in Los Angeles. Its relation to the working-class Mexican immigrant audience in Los Angeles was therefore critical to its continued popularity. As one L.A.-based composer explained, "The *corrido* is a narrative viewed through the eyes of the people—its subject almost always follows the truth."[27] This adaptive style was particularly well suited for the rapidly expanding Los Angeles Mexican community of the 1920s and the ever-complex nature of intercultural exchange in the city.

The first commercial recording of a *corrido* in the United States was "El Lavaplatos." Performed in Los Angeles on May 11, 1926, by Los Hermanos Bañuelos as a duet with guitar accompaniment, the song was apparently originally written by Pedro J. González.[28] The *corrido* describes a Mexican immigrant who dreams of making a fortune in the United States but, instead, is beset with economic misfortune. Finally, after being forced to take a job as a dishwasher, the narrator bemoans: "Goodbye dreams of my life, goodbye movie stars, I am going back to my beloved homeland, much poorer than when I came."[29]

Most Mexican composers and musicians had firsthand knowledge of working-class life in Los Angeles; not only were they products of working-class homes, but most continued in some form of blue-collar occupation while struggling to survive as musicians. Pedro J. González, for example, worked as a longshoreman on the San Pedro docks before being "discovered," and the two musicians who played with him, Victor and Jesus Sánchez, were farmworkers.[30] The vast majority of Mexican musicians never were able to support themselves as full-time artists. One composer of *corridos*, for example, worked in a cement plant, a lumber yard, an oil refinery, the railroad, the telephone company, agricultural fields, and at the Biltmore Hotel while composing songs during the 1920s and 1930s.[31] Several who applied for American citizenship listed additional occupations with authorities.[32] A similar situation existed among Texas conjunto musicians. According to Manuel Peña, they "played and earned just enough to satisfy a few—not all—of their economic needs. There simply were not enough dances during a week for full-time employment: Saturday and Sunday were practically the only days for celebrating."[33]

Los Angeles during the 1920s, however, presented more possibilities for earning a livelihood as a musician than any other location outside of Mexico City, or perhaps San Antonio. To begin with, the Los Angeles metropolitan area contained a huge Spanish-speaking population, second only to Mexico City itself. By 1930 the Chicano population in the city of Los Angeles was larger than any other in the United States. The potential audience for Mexican music was enormous. Since most of these residents were recent migrants from Mexico, they often longed for tunes from their homeland. Others had come from south Texas, where the Spanish-language musical tradition was strong and widespread.[34] In fact, one writer claimed in 1932 that more Mexican music had been composed in the United States than in Mexico.[35]

One stimulus to the Mexican music industry was the explosion of Chicano theatre in Los Angeles during the 1920s. Over thirty Chicano playwrights moved to the city during the decade, producing shows ranging from melodrama to vaudeville. The Spanish-speaking population of the region was able to support five major theatre houses from 1918 until the early 1930s: Teatro Hidalgo, Teatro México, Teatro Capitol, Teatro Zendejas (later Novel), and Teatro Principal. In addition to these five which featured programs that changed daily, at least seventeen other theatres housed Spanish-speaking professional companies on a more irregular basis.[36]

Many of these theatres alternated vaudevillian-style shows with Mexican- or Hollywood-made silent films (three shows a day, four on weekends) during the 1920s. Both live performances and silent movies required musical accompaniment. Theatres, therefore, provided relatively stable employment to a diverse collection of musicians throughout the 1920s. The lack of formal training among many of the musicians did not necessarily hamper them, since playing on the streets often helped them prepare for the spontaneity and improvisation required for this type of performance.

The presence of a large number of middle-class Mexican expatriates also created a market for formally trained musicians who could read music. They performed for a type of theater which featured drama from Spain with orchestral accompaniment, similar to the more refined entertainment among the middle classes in Mexico City. While never enjoying the mass appeal of movies and vaudeville, this European-style performance did provide employment for other musicians from Mexico.[37]

A more disparate, yet still lucrative market for Mexican musicians existed among the streets and informal gatherings of Los Angeles. During Mexican patriotic festivals and the Christmas season, musicians had larger audiences, more exposure, and greater potential for earnings. From these "auditions," Mexican groups were often recruited to play for weddings and other ethnic festivities. Moreover, a market for "traditional" Mexican music also existed among some Anglo residents of Los Angeles, often to provide a nostalgic backdrop to the distinctive "Spanish" past of the city. Pedro J. González, for example, often entertained at parties conducted by city officials and the police department.[38] Another *corrido* composer, Jesus Osorio, was able to make a living as a singer combining work in Olvera Street booths, private gatherings, and in the small theatres and cabarets along Main Street.[39]

The emergence of Hollywood as the leading movie-making capital in the United States during the 1920s stimulated a flourishing recording industry in the city that began to rival New York's. Both these developments boded well for Mexican musicians in Los Angeles, although prejudice, union discrimination, and the lack of formal training kept many out of regular employment in the entertainment industries in the western part of town. Still, by providing the music in English-speaking theatres or working as studio musicians, some were able to break into the larger music business in Los Angeles.[40] Even the possibility of such employment—"the dream of a life in Hollywood"—was enough to attract some performers from south of the border.

Thus musicians from Mexico flocked to Los Angeles during the 1920s, becoming a significant segment of the Mexican cultural renaissance of that decade. Unlike the Harlem Renaissance, where black writers and entertainers were often sponsored by white patrons, this Chicano/Mexicano renaissance was largely supported by Mexican immigrants themselves and existed far out of the sight of the majority of Angelinos. The presence of large numbers of Mexican musicians in the city not only preserved the sights and sounds familiar to Mexican immigrants; it also created an environment of cultural experimentation where traditional music was blended with new methods. In short, musicians often served as social interpreters who translated and reflected the cultural adaptations that were taking place among the Mexican immigrant population as a whole.[41] In fact, one astute observer of *corridos* in Los Angeles recognized that this music often served to "sing what they cannot say":

> Mexicans are so intimidated by the government officials, even by social workers, and so timid on account of the language difficulty that it is almost unheard of for a Mexican to express his opinion to an American. Here, however, he is speaking to his own group and an emotional outlet is offered in the writing of *corridos* on the subject so well known to every Mexican. He is reasonably sure that only Mexicans will ever hear his *corrido*.[42]

Despite their economic and cultural connection to the greater immigrant population, Mexican musicians displayed different patterns of migration and settlement. As a group, they were among the first of the migrants to see the advantages of settling in Los Angeles, with many arriving in the city around World War I or in the early 1920s. Moreover, they usually arrived as adults, crossing the border at an average age of twenty-four years during the 1910s and an average of thirty-two years during the 1920s. The profession was also dominated traditionally by men. Only four of the 110 musicians in my sample were women, and two of these listed their occupation as singers. One study conducted in 1939, for example, found no *corrido* written by a woman in the music shops of Los Angeles.[43] Few avenues of opportunity were available to women in the Spanish-language music industry in Los Angeles, although several women—including the singer Lydia Mendoza—did make names for themselves during the 1930s.

By necessity, most Mexican musicians lived west of the river, even after many of their people began to venture into East Los Angeles. The recording and film industry was located in the western part of the city, and Mexican cultural life continued to be centered around the plaza and downtown areas of the city until the 1940s. Most Spanish-language theatres were located on Main Street, and opportunities for steady income depended on the patronage of the audiences that gathered around the Placita or in downtown restaurants. During the 1930s when so-called "Latin" music clubs were established, they too were located in the downtown area until after World War II. Lalo Guerrero, who arrived in Los Angeles in 1937, remembered that not until economic opportunities around the Plaza declined did many Mexican musicians decide to move to East Los Angeles:

> Since the clubs in the westside started dying off . . . the musicians that had not wanted to come to East L.A., because they thought it was a step down, . . . were practically forced to come back because there was not too much happening on the westside, 'cause the latin scene had passed.[44]

The data confirm this pattern of residence for musicians: 60 percent in the naturalization sample lived around the downtown area, including 20 percent that resided near the Plaza. This compares with less than 39 percent of the overall sample who lived downtown, and 9 percent whose residence was located in the Plaza area.[45] Steady income and the opening up of the electric railway into East Los Angeles gave some musicians the opportunity to move to Belvedere or some of the other communities east of the river. Pedro J. González, for example, moved to Belvedere after residing close to the Plaza.[46]

This residential pattern is an indication that life as a musician did not usually provide the glamour and security which many associated with the entertainment field. Even the advent in the 1920s of an ethnic mass market centered in Los Angeles, which prompted American recording companies and local entrepreneurs to search for Mexican musical talent, meant only short-lived economic returns. Rapid exploitation of the talents of musicians brought quick profits to upstart recording companies but left most Chicano performers—even those who developed a loyal following—with limited resources to show for their newfound fame.

When Los Hermanos Bañuelos first recorded "El Lavaplatos," they ushered in the commercial recording of Mexican music. Already, several large American recording companies such as Vocalion, Okeh (a subsidiary of Columbia), Decca, and Bluebird (RCA) had begun to produce "race" records, featuring black folk music. These companies now realized the potential ethnic market among Mexicans, and sought out Chicano musicians and singers from Texas to California. Many of the early recording sessions took place in temporary studios located in Los Angeles hotels, where a steady stream of performers were expected to produce a finished product in one or two "takes."[47]

To most musicians, the $15 or $20 they earned per record seemed substantial for a few hours' work, especially when compared with the wages they earned as laborers or the limited income from playing on the streets. Yet these tiny sums were a pittance relative to the hundreds or thousands of dollars any single recording could earn, even with records selling for 35 cents apiece. Musicians rarely earned sufficient income to feel secure as recording artists. Offering only "contracts" that were usually verbal agreements consisting of no royalties or other subsidiary rights, the recording companies profited handsomely from this enterprise.[48] Similar contractual agreements were made with Chicano artists as late as the 1940s, even though the pay scale had moved up to $50 a side or $100 per record.[49]

Local ethnic middlemen played an important role in identifying talented musicians and putting them in contact with recording companies. In Los Angeles, one important liason was Mauricio Calderón, owner of the music store Repertorio Musical Mexicana, established on Main Street around 1920 to feature records and phonographs produced by Columbia Records. According to Pedro J. González, Calderón was in charge of everything in Los Angeles that related to Mexican music. He recruited talented musicians by advertising in the Spanish-language press, and kept an ear out for the latest musical trends among the city's performers and audiences. Not only did Calderón make money by serving as a go-between between American companies and the Mexican artists, but he also held a monopoly on the area-wide distribution of these recordings through his store. A standard practice of the time for such businesses was to sell phonographs as well as records, and stores such as Calderón's profited as well from these items. In fact, Calderón's store, located at 418 North Main Street near the Plaza, regularly promoted itself by using a loudspeaker mounted in front on the store playing the latest *corrido*. A small group of men regularly stood in front of the store, listening intently and enjoying the music. Another popular promotion tactic was to give away records with the purchase of a Victrola.[50]

American laws prohibited the importation of records from Mexico, a fact which greatly stimulated the recording industry in Los Angeles. In addition, Mexican companies were not allowed to record in the United States. These restrictions severely crippled the music industry in Mexico, while creating a vast economic opportunity for American companies and ethnic entrepreneurs. When Mexican recordings were finally admitted during the 1950s, interest in immigrant and native-born Spanish-language talent evaporated quickly, and many Chicano musicians were left without an outlet in the recording world. In fact, some labels which had showcased Mexican artists, such as Imperial, began concentrating on black rhythm and blues artists, such as Fats Domino and T-Bone Walker.[51]

During the 1920s and 1930s, however, a vibrant environment for Mexican music existed in Los Angeles. Another factor in creating this cultural explosion was the advent of the radio. During the 1920s, commercial radio was still in an experimental era where corporate sponsors and station managers tried to discover how best to make radio broadcasting profitable and enlightening. For most of the decade, the radio was seen as a way of uplifting the masses, of bringing elite American culture into the homes of common laborers.[52] By the end of the decade, however, advertising and corporate economic interests dominated the airwaves. This transformation created a market for Spanish-language broadcasts. Although many Anglo Americans continued to believe that only English should be heard on the nation's airwaves, the goal of reaching Spanish-speaking consumers silenced their opposition.

American radio programmers scheduled Spanish-language broadcasts during "dead" airtime—early morning, late night, or weekend periods which had proven to be unprofitable for English programs. Pedro J. González remembers first broadcasting from 4 to 6 a.m. on Station KELW out of Burbank. He often scheduled live music, including many amateur musicians and singers from the community.[53] While Anglo Americans were rarely listening at this hour, many Mexican immigrants tuned into González's broadcasts while they prepared for early morning work shifts. González's daily shows provided day laborers important information about jobs as well as cherished enjoyment to workers who toiled all day.[54]

Corporate radio sponsors in the mid-1920s were quick to understand the profitability of ethnic programs. Large advertisers such as Folgers Coffee used airtime to push their product in the Spanish-speaking market. More often, local businesses appealed to Mexican immigrants to frequent their establishments. In Los Angeles, radio broadcasting soon became a highly competitive industry. By selling blocks of airtime to foreign-language brokers, marginally profitable stations could capture a ready-made market. During the late 1920s, the hours dedicated to Spanish-language broadcasts multiplied. González's program was expanded until 7 a.m., and additional hours were added at lunchtime and in the early evening. Chicano brokers such as Mauricio Calderón profited handsomely as they negotiated with stations, paying them a flat rate during cheap broadcasting time, which they then sold to businesses advertisements.[55]

Key to the success of Spanish-language broadcasting was its appeal to the thousands of working-class Mexican immigrants within the reach of a station's radio signal. Radio, unlike *La Opinión* and other periodicals, reached Mexican immigrants whether or not they could read. In addition, the content of radio programming focused less on the tastes of the expatriate middle class and more on those of the masses. A 1941 analysis of Spanish-language programming found that over 88 percent of on-air time (outside of advertisements) was dedicated to music, with only 4 percent used for news.[56] Programming was dominated by "traditional" music from the Mexican countryside, rather than the orchestral, more "refined" sounds of the Mexican capital and other large urban centers. "The corrido, the shouts, and all that stuff was popular" with working people, remembered González. Although some bemoaned the commercialization of the *corrido* tradition and its removal from its "folk tradition," most Mexican immigrants found this transformation to their liking because it fit well with their own adaptations to urban living.[57]

The potential power generated by this mass appeal was so substantial that it not only threatened the cultural hegemony of the Mexican middle class in Los Angeles but also worried local Anglo American officials. González himself was the target of District Attorney Buron Fitts, who in 1934 had the musician arrested on trumped-up charges. Earlier, Fitts had attempted to force González off the air by getting federal authorities to rescind his broadcasting license. Along with other government authorities, Fitts believed that only English should be heard on the radio and that only American citizens should have the right to broadcast. As a result, many radio stations curtailed their Spanish-language programs during the early 1930s, often because of the continued harassment directed at ethnic broadcasters and the imposition of more strigent rules for radio licensing.[58]

These restrictions in the United States encouraged the growth of Spanish-language broadcasting in Mexico. Although many American stations continued to reserve Spanish-language blocks, entrepreneurs based just across the border capitalized on the potential market on both sides by constructing powerful radio towers capable of reaching far-flung audiences. Increasingly, individuals unable to be heard on American-based stations moved their operations to Mexico. It proved much harder for American authorities to control the airwaves than the recording industry. Mexican immigrants could now listen to radio programming from Mexico itself, ironically often featuring music performed by U.S.-based Mexicans.[59]

The economic crisis of the 1930s curtailed much of Mexican cultural activity in Los Angeles. First, deportation and repatriation campaigns pushed almost one-third of the Mexican community back to Mexico, effectively restricting the market for Spanish-language advertising campaigns. Second, the enthusiasm of American companies for investing in "experimental" markets that did not insure a steady flow of income understandably cooled. The Mexican immigrant community itself had fewer resources to support cultural activities, given its precarious economic situation. Since expenditures on leisure-time activities were the first to be reduced during times of need, many families cut back drastically on attendance at musical events or the purchase of radios and phonographs. Many theatres in the community shut down during the Great Depression.[60]

Movies and other forms of cheap, cross-cultural entertainment continued to thrive in Depression-era Los Angeles. Simply because of the economics of scale, Hollywood was able to continue to produce entertainment accessible to families at every economic level. In addition, the introduction of sound to motion pictures made it more difficult to sustain a steady Spanish-language audience with Mexican imports, since the Mexican film industry had difficulty throughout the transition of the 1930s.[61] English talking-pictures, on the other hand, had a wider, and therefore more secure audience. The advent of sound coincided with the rise of the second generation of Mexicans in this country, more likely to be as fluent in English as in Spanish. Increasingly, changing demographics and limited economic resources stunted the growth of the ethnic market. A new era in Mexican/Chicano cultural activity began.

Although commercial activity was slowed during the Depression, Mexican cultural life did not die out in Los Angeles. Indeed, aspects of cultural life were altered dramatically, reflecting the changing composition and nature of the Mexican/Chicano community. Musical activity, for example, became less dependent on *corrido* story-telling (which required the ability to understand Spanish lyrics) and more concentrated in dance clubs. La Bamba night club, at Macy and Spring streets, and La Casa Olvera, adjacent to Olvera Street, were only two of many small clubs which opened during the decade. Dancing, of course, did not require a working knowledge of Spanish, and had appeal well beyond the Mexican immigrant population.[62]

Second-generation youth, in particular, flooded the dance clubs during the 1930s. Social commentators of the period commented on the "dance craze" that had seemingly overtaken adolescents and young adults in Mexican American families. One such nineteen-year-old, known only as Alfredo to his interviewer, boastfully explained this "craze":

> I love to dance better than anything else in the world. It is something that gets in your blood. Lots of boys are that way. I go to five dances a week. I can't wait for Saturday night because all the time I am thinking of the dance. It is in my system. I could get a job playing my trumpet in an orchestra but then I couldn't dance. I quit school because I got plenty of everything they teach, but dancing.[63]

This new "dance craze" did not often sit well with Mexican immigrant parents. Even when participation was closely chaperoned in school clubs and community centers, public dancing seemed to offend the sensibilities of decency among older Mexicans. Increasingly, however, it became difficult for parents to withstand the effect of peer pressure on their children, as evidenced by the words of one mother in the early 1930s:

> Juanita has joined a club and now she wants to learn to dance. That is what comes of these clubs. It is wrong to dance and my Juanita wants to do it because the others do. Because everybody does it does not make it right. I know the things I was taught as a girl and right and wrong cannot change.[64]

Although the vast majority of musicians and clientele in each of these establishments were Mexican, the music demonstrated a wide variety of American and Latin American styles. Cuban music was especially popular in the latter half of the decade, with many orchestras specializing in the mambo. The Cuban style was popular throughout Latin America, and this trend filtered into Los Angeles through traveling bands and musicians. Regular groups that played in these clubs all included Mexican songs in their repertoire.[65] In addition, English-language music increasingly became popular among American-born youth. Many Mexican immigrants bemoaned this turn of events, as evidenced by the comments of one unnamed señora:

> The old Spanish songs are sung only be the old people. The young ones can sing the "Boop-da-oop" like you hear on the radio but they can't sing more than one verse of *La Cruz*. Do you know *La Cruz*? It is very beautiful. It is about our Lord carrying the cross. It is sad. In Mexico we would all sing for hours while someone played a guitar. But here, there are the drums and the saxophones.[66]

Undoubtedly, a more eclectic and diverse musical life than in former decades emerged among the Mexican/Chicano community in Los Angeles. In fact, Los Angeles probably offered a richer environment for such leisure-time activity than any other city in the American Southwest.

This diversity of choice in musical styles and taste not only created a more experimental environment for musicians themselves but also reflected developments in Chicano culture as a whole. Clearly, the control of the individual over his or her own cultural choices paralleled the growth of an ethnic consumer market. In a consumer society, each Mexican immigrant alone, or in conjunction with family, embraced cultural change—consciously or unconsciously—through the purchase of material goods or by participation in certain functions. Neither the Mexican elite nor the Anglo American reformers intent on Americanization could completely determine the character of these private decisions. Instead, an unsteady relationship between American corporations, local businesses, Mexican entrepreneurs, and the largely working-class community itself influenced the range of cultural practices and consumer items available in the Spanish-language market. If appeals to Mexican nationalism could be used to sell a product, then so be it. Although barriers to the ethnic market were constructed by local officials, particularly during the Great

Depression, change in economic circumstances and in cultural tastes of the population had the most important impact.

Appeal to the tastes of youth also created subtle power shifts within the Chicano community. In Mexico, few outlets were available to young people for influencing cultural practices in an individual village or even one's own family. The American metropolis, on the other hand, gave Mexican youth an opportunity to exercise more cultural prerogatives merely by purchasing certain products or going to the movies. Rebellion against family often went hand in hand with a shift toward more American habits. This pattern was stimulated by the extent to which adolescents and unmarried sons and daughters worked and retained some of their own income. As the second generation came to dominate the Chicano population by the late 1930s, their tastes redefined the community's cultural practices and future directions of cultural adaptation.

Behind the vast American commercial network lay an enterprising group of ethnic entrepreneurs who served as conduits between the Mexican immigrant population and the corporate world. These individuals were often the first to recognize cultural changes and spending patterns among the immigrant population. Individuals such as Mauricio Calderón and Pedro J. González were able to promote Mexican music in entirely new forms in Los Angeles because they had daily contact with ordinary members of the Los Angeles Mexican community. Although they found tangible financial rewards in their efforts, they also served an important role in redefining Mexican culture in an American urban environment.

CHAPTER 9

Workers and Consumers: A Community Emerges

I dreamed in my youth of being a movie star
And one of those days I come to visit Hollywood.
One day very desperate because of so much revolution
I came over to this side without paying the immigration.
What a fast one, what a fast one, I crossed without paying anything.

On arriving at the station, I ran into a friend,
Who gave me an invitation to work on the track.
I supposed the track would be some kind of a store.
And it was to repair the road where the train ran.
Oh my friend, oh my friend, how he took me to the track.

When I became angry with the track, he invited me again,
To the picking of tomatoes and the gathering of beets.
And there I earned indulgence walking on my knees,
About four or five miles they gave me as penance.
Oh what work, and so poorly paid, for going on one's knees.

My friend who was no fool, continued giving them a bad time.
And on completing (enough) for his fare, he returned to his land.
And I earned but a trifle, and I left for Sacramento,
When I had nothing I had to work with cement.
Oh what torment, oh what torment, is that famous cement.

Toss some gravel and sand in the cement mixer,
Fifty cents an hour until the whistle blows.
Four or more of us strained at that famous pulley.
And I, how could I stand it, I was better off washing dishes.
How repentant, how repentant, I am for having come.

—"El Lavaplatos" (The Dishwasher), ca. 1930

The song "El Lavaplatos" depicts many of the obstacles workers from Mexico encountered in the United States. As a *corrido,* or folk ballad, the story is constructed around humorous lyrics which portray the failure of the immigrant dream. Instead of living as a Los Angeles movie star would, the young man finds life in America made up of temporary, low-paid, and difficult jobs in which geographic mobility is key to both survival and the retention of dignity. Pedro J. González claimed to have composed the tune after meeting several celebrated generals of the Mexican Revolution who were working as dishwashers in Los Angeles during the 1920s.[1]

Most studies of the work experience of immigrants in this period have, like González's *corrido,* noted that low wage labor with little possibility of economic advancement characterized the Mexican experience.[2] Indeed, the economic opportunity for Mexican immigrants was bleak, especially during the Great Depression which profoundly affected the economic outlook of Mexicans who had arrived during the 1920s. This chapter will explore the significance of these limited opportunities in framing the cultural adaptation of Mexican immigrant in Los Angeles. Although work was difficult and social mobility circumscribed, Mexican workers did construct communities which indicated their commitment to staying in the United States. Home ownership and buying patterns of these settlers, despite limited funds, indicate that a new Mexican American sensibility emerged when workers refused simply to move on to find employment.

From the moment they stepped on American soil, most Mexican workers were treated as little more than mobile, low-cost employees. Herded onto railroad cars, male Mexican immigrants saw their migration within the United States dominated by the large corporate employers of the American Southwest who funneled Mexicans toward jobs for which—at least in this period—only Mexicans were deemed suited. Jumping off the rolling train of migratory labor and seeking better work in the city was clearly an act of defiance for the worker. For the employer, however, it represented only another telling example of how Mexicans were "unreliable" and had to be controlled as laborers. Restricted to low-paying jobs, often subjected to oppressive working conditions, most Mexican immigrants were rarely in a position to improve their economic position in this country.

Contact with middle-class American values, therefore, was minimal. The vast majority of Mexican immigrants did not leave the working class. Even those that were able to obtain white-collar employment often continued to live in communities made up largely of working-class ethnics. How can we best describe their adjustment to American culture in the face of little economic advancement?

My research indicates that Mexican immigrants displayed very distinct patterns of cultural adaptation. Acculturation of Mexican immigrants in Los Angeles occurred primarily within the confines of the working class, often in the barrios in which they lived. Yet cultural transformation did occur. Even if we acknowledge that the United States in the twentieth century has not been as conducive to immigrant social mobility as was previously thought, the adaptation that took place among immigrants and their children who remained in the working class must be understood better.

A good place to begin is by examining the nature of the work experience itself. The historian Herbert Gutman, in his pathbreaking essay "Work, Culture, and Society in Industrializing America, 1815–1919," used a Mexican *corrido* to characterize one type of confrontation between an immigrant with preindustrial work habits and American industrialized labor. Although Gutman argued that the tune "El Enganchado" ("The Hooked One") "celebrated the disappointments of immigrant factory workers," in fact it lamented the inability of a skilled Mexican laborer based in Chicago to get factory work. "I am a shoemaker by trade," says the Mexicano, "but here they say I'm a camel and good only for pick and shovel. What good is it to know my trade if there are manufacturers by the score, and while I make two little shoes they turn out more than a million?"[3] Like the song's composer, most Mexican immigrant laborers were concentrated in jobs outside the factory during the 1920s.[4]

The kind of work obtained by Mexican immigrants in the agricultural, railroad, and mining industries did not resemble modern factory labor in several important ways. The "rule of the clock" which, according to Gutman, characterized factory labor discipline often did not apply in these industries. Outside of a controlled environment, if not in the "open air," Chicano workers could be driven to longer hours based on piece-rates and completion of tasks, rather than on hours in the day. At times, the tasks were limited only by daylight. Since workers lived on or near work sites, their day did not end with a clear separation from the work environment. Instead of requiring the development of internal discipline within each worker, Mexican laborers were often controlled through gangs driven by a foreman. Intimidation, pressure, and even violence were more characteristic than time discipline and economic incentive. This feudal system of labor dominated the industries in which Mexicans were concentrated throughout the first three decades of the twentieth century.

The same employers who maintained harsh conditions and low wages for Mexican workers often complained that Mexicans showed no loyalty to the arduous work they were assigned or to the companies for which they worked. "The average Mexican likes the easy way to payday," said one Los Angeles railroad official. "He will work hard but does not like to be held responsible or likes to worry after the day is done. . . . As soon as he has a little money he lays off and goes to spend it. He does not come back till he has [to]."[5] Another frequent complaint was that the Mexican worker was generally slower than other laborers.

Mexicans laborers had their own explanation for employer attitudes. One particularly astute observer complained about working next to "bol-illos"—a derogatory term for Anglos—because they had learned to play the game:

When the foremen is there they work perhaps harder than we do. We go our pace and that makes the foreman think we are slow. But the minute the foreman goes away the bolillos all stop work and they do nothing till he comes back. The minute they see him they start to work like mad. . . . Then if we rest on the shovel for a minute they tell the foreman we are loafing and why don't we get our time for being loafers and foreman fires us.[6]

Mexican workers were quick to develop their own critique of a work ethic that seemed contrived only to serve the interests of employers. These laborers realized that the jobs they were asked to perform in the United States were often the most arduous and hazardous. "I know Mexicans who do the hardest work in the cement mills that would kill elephants and that work under unsanitary, unhealthy conditions that no man—no white man—would allow any other white man to tolerate," said one man with a mixture of disgust and pride. "But it is all right! We are Mexicans. We can stand it. If one of us becomes sick, dies or gets tuberculosis there will be more Mexicans to take our place."[7] It was not difficult for workers to suspect that American employers cared more about profit than the livelihood of any individual Mexican.

Ironically, Mexicans were often lauded by their employers for their supposed submissiveness in the face of utter degradation and horrible working conditions.[8] According to many observers, these "values" were brought by the Mexicans from their homeland. Dr. George Clements, head of the Los Angeles Chamber of Commerce during the 1920s and 1930s, gave his own version of *indigenismo* in detailing the relationship between race, culture, and industrial wages:

> The Mexican is an Indian and must be considered so. He is undergoing active evolution and we must always take this thought into consideration in dealing with him. His wants are few and his habits, while docile, are not in harmony with western civilization, and he so recognized it and was willing to abide by it. To pay him an exorbitant salary only meant to cater to his extravagance; to pay him a living wage and add to his future comforts seemed to be the only way in which to handle him.[9]

This interpretation clearly served the economic interests of the U.S. employers. Pre-industrial values, as interpreted by employers, functioned rather efficiently in the context of migratory, low-wage labor. Maintaining the flow of workers from Mexico, therefore, became a prime political consideration of pre-Depression employers in the Southwest.[10]

Los Angeles, for Mexicans as for others, promised at least the possibility of a greater range of choice, possibly higher pay, and certainly a more exciting environment than many rural locales. Yet a number of studies indicate that employment in Los Angeles usually turned out to be a disappointment for many Mexican workers. Ricardo Romo found that over 91 percent of Mexican men in the city in 1917–18 had blue-collar occupations, with close to 70 percent involved in unskilled labor.[11] Pedro Castillo estimates that approximately 90 percent of Mexican male workers were involved in blue-collar labor in both 1910 and 1920, while

Table 12. Occupational Structure for Mexican Male
Naturalizers

Occupational rank	Number	Percent
High white-collar	66	4.4
Low white-collar	302	20.2
Total white-collar	368	24.6
Skilled blue-collar	406	27.1
Semiskilled blue-collar	206	13.8
Unskilled blue-collar	441	29.4
Total blue-collar	1,053	70.3
Not employed/unknown	77	5.1
Total	1,498	100.0

Source: Declarations of Intention, Naturalization Documents, National Archives, Laguna Niguel, California.

Albert Camarillo has discovered similar figures for other southern California cities in the early twentieth century.[12] Using marriage certificates, Douglas Monroy charted similar employment figures for the Depression decade.[13]

My analysis of almost 1500 Mexican males who filed their first naturalization papers reveals a few important differences. Only 70 percent of the men in this sample held blue-collar jobs (see Table 12). An unusually high proportion, 20 percent, had low white-collar occupations, which includes clerical or sales positions, along with a substantial group of musicians whom I classified as white-collar.[14] Moreover, less than 30 percent were unskilled workers, while over 27 percent were skilled blue-collar laborers. Though still heavily concentrated in blue-collar occupations, the distribution of jobs in this sample as a whole is higher on the occupational ladder than previous historians' figures.

The most obvious reason for this discrepancy is that Mexicans who were skilled or white-collar laborers disproportionately chose to naturalize themselves, while unskilled and semi-skilled laborers were more likely to return to Mexico or to defer naturalization. Even though the most cited occupation among the naturalizers was "common laborer," and blue-collar workers made up a substantial majority of those who took out first papers, those in white-collar jobs in the city were much more likely to want to settle in Los Angeles permanently. Since other studies derive from census materials, city directories, and marriage certificates, they better characterize the entire Mexican immigrant population in Los Angeles at any given time. My sample, on the other hand, is a more accurate representation of those who decided to make their permanent home in the United States, particularly in the city of Los Angeles.

One might assume that the men who sought naturalization were exactly those who had achieved higher economic status during their sojourn in Los Angeles. Although this group of men as a whole held higher status jobs, they did not necessarily achieve them through the process of economic mobility while they resided in Los Angeles. Combining economic data with other information available in the naturalization documents, it is clear that the men who listed professional and skilled blue-collar occupations had arrived in the city with either substantial capital or the skills needed to maintain their economic position. In other words, few had moved up the ranks from unskilled or semi-skilled positions. Of the 66 individuals who held professional occupations, for example, the majority arrived in their thirties or forties—a strong indication that they had achieved professional standing elsewhere before arriving in the city. Many of these doctors, dentists, and businessmen had earned their degrees or made their fortunes in Mexico, and had fled their native country during the revolution to settle permanently in the United States. These individuals often made up the elite class that associated itself with the Mexican consulate office and joined in the growing merchant and professional organizations of Mexican Americans in the 1920s and 1930s.

Skilled blue-collar workers were heavily represented by immigrants from the northern states, where skilled work was often available in the railroad shops and mining towns of Mexico. Increasingly during the 1920s, and despite pressure by racist union officials and unsympathetic employers to keep them out of skilled positions, these workers found employment in small-scale shops that dotted the Los Angeles industrial landscape.[15] A small but not insignificant group had moved up to skilled work from lower-status blue-collar occupations, but in general this group consisted of individuals who obtained their training elsewhere. In the 1930s, this group of skilled blue-collar workers would serve as the base of a new Mexican American working-class movement in East Los Angeles.

The low-status white-collar workers are important on a variety of levels. First, musicians, who made up the second largest occupational group, skew the figures for this category. Musicians were placed in the white-collar category because they worked as independent entrepreneurs in the streets and restaurants of Los Angeles.[16] Second, a clear pattern emerges if one focuses on the second major component of this category: salespeople and clerks. These workers tended to arrive in the United States as children, and therefore their experiences more clearly reflect those of American-born Chicanos. They undoubtedly spoke English and were able to obtain employment on the basis of their familiarity with American life.

Unskilled laborers, who made up one-third of the sample, tended to arrive in Los Angeles as young adults, just as had the majority of the Mexican immigrant population. They came from all parts of Mexico (with the exception of under-representation from Mexico City), but, unlike many of their fellow blue-collar workers, they decided to remain permanently in this country. These 441 individuals were the most repre-

Table 13. Occupational Mobility Among Male Naturalizers

Rank on initial declarations	Rank on Final Petitions						
	NE	LBC	MBC	HBC	LWC	HWC	Total
High white-collar (HWC)	2 (5.9)	2 (5.9)	0 (0.0)	2 (5.9)	6 (17.6)	22* (64.7)	34 (100%)
Low white-collar (LWC)	3 (2.5)	7 (5.9)	6 (5.1)	9 (7.6)	87 (73.7)	6 (5.1)	118 (100%)
Skilled blue-collar (HBC)	5 (2.6)	26 (13.5)	7 (3.6)	143 (74.5)	10 (5.2)	1 (0.5)	192 (100%)
Semiskilled blue-collar (MBC)	3 (2.8)	18 (16.5)	72 (66.1)	12 (11.0)	3 (2.8)	1 (0.9)	109 (100%)
Unskilled blue-collar (LBC)	9 (5.1)	127 (72.2)	12 (6.8)	24 (13.6)	3 (1.7)	1 (0.6)	176 (100%)
Not employed (NE)	15 (40.5)	7 (18.9)	3 (8.1)	3 (8.1)	4 (10.8)	5 (13.5)	37 (100%)
Total	37 (5.6)	187 (28.1)	100 (15.0)	193 (29.0)	113 (17.0)	36 (5.4)	666 (100%)

Source: Declarations of Intention and Petitions for Naturalization, National Archives, Laguna Niguel, California.

*Italics denotes those that remain at the same level.

sentative of Mexican immigrants in the urban United States—poorly paid workers employed at construction sites and railroad yards, or in homes as domestics and gardeners. Since most of their compatriots in similar economic circumstances refused to adopt American citizenship, their decision to apply for naturalization was careful and conscious.

Economic mobility, however, did not seem to explain their decision. A comparison of the occupations listed by immigrants on their initial naturalization papers with those listed on the final papers reveals how little economic mobility these applicants experienced (see Table 13). Those applying for naturalization were required to wait at least two years before filing the second set of papers, and many waited far beyond that period. (The median number of years between filing dates was about four.) Not surprisingly, over 70 percent of the 666 Mexican male workers showed no change in occupational level in this period. For the rest, downward mobility was more likely than upward movement. Of the 192 skilled workers, for example, eleven had moved up to white-collar occupations by the time they took out their final papers. Some 33 workers, however, had moved down to unskilled or semi-skilled labor, and five had no job at all.

These figures indicate that only a few individuals were able to move from blue- to white-collar occupations. Besides the eleven skilled workers, four semi-skilled and four unskilled workers also moved up. Nine others who had been out of the labor market, mostly as students, also took white-collar employment by the time of their final naturalization papers. Security for Mexican white-collar workers was scarce, however. Twenty-six white-collar workers "fell back" to blue-collar categories, including four professionals also unable to remain out of blue-collar employment. Even if one looks only at workers in pre-Depression Los Angeles, the figures remain fairly consistent. This bleak pattern corroborates Ricardo Romo's analysis of occupational mobility among Mexicano workers from 1917 to 1928.[17]

Even if coming to Los Angeles did not translate into greater occupational mobility, Mexican workers did find more stable employment possibilities in the city. Moving out of highly seasonal and migratory jobs in agriculture and railroad work was the goal of many immigrants, particularly as they married and began to have their own families. Like city resident Mr. Martínez, who had regularly left the city each year to pick sugar beets in Idaho, many laborers refrained from migratory work once their families were reconstituted.[18] Though Los Angeles did not provide stability through a single employer for Mexicans, it was a place where one could find relatively stable work. Though most Mexican workers remained in the secondary labor market, it was easier to find niches in the economy where work was readily available for family members in Los Angeles than in other, more rural locales.

Patterns of residential settlement increasingly reflected differentiation within the Chicano community itself. The relative stability of work opportunities allowed individual barrios in Los Angeles to take on particular characteristics which reflected the economic makeup of their residents (see Table 14). Though laborer was the most common occupation in each barrio (with the exception of the downtown community), different areas of Mexican/Chicano settlement were more likely to attract individuals engaged in skilled or white-collar jobs. Other barrios remained dominated by unskilled workers.

The Plaza area continued to attract recent migrants, most of whom were involved in unskilled labor. These single, unattached males created a climate of occupational and geographic instability in the area as they filled the city's inexpensive motels, boarding houses, and house courts. In the Plaza they competed for space with other recent migrants, most notably single Italians. Musicians represented the only significant white-collar group in this location, living near and playing in the downtown streets and restaurants. The accompanying North Main district shared most of the Plaza's composition, although it contained fewer *solos* and a greater percentage of families. This entire area was dominated by Mexican migrants who periodically, and sometimes permanently, returned to Mexico to visit family and friends.

Immediately to the east of the Plaza lay the area in which railroads had come to dominate the industrial scene (see Map 9). Like the Plaza, this barrio, stretching along both sides of the Los Angeles River, also

Table 14. Occupational Rank by Neighborhood

Neighborhood	NE	LBC	MBC	HBC	LWC	HWC	Total
Plaza Area	7	59	15	31	38	7	157
	(4.5)	(37.6)	(9.6)	(19.7)	(24.2)	(4.5)	(100%)
North Main district	7	32	5	11	7	2	64
	(10.9)	(50.0)	(7.8)	(17.2)	(10.9)	(3.1)	(100%)
Downtown area	9	26	17	30	29	2	113
	(8.0)	(23.0)	(15.0)	(26.5)	(25.7)	(1.8)	(100%)
Rest of Central Los Angeles	42	85	53	99	100	25	404
	(10.4)	(21.0)	(13.1)	(24.5)	(24.8)	(6.2)	(100%)
Railroad area	4	40	14	30	13	10	111
	(3.6)	(36.0)	(12.6)	(27.0)	(11.7)	(9.0)	(100%)
Lincoln Heights	5	26	16	27	14	4	92
	(5.4)	(28.3)	(17.4)	(29.3)	(15.2)	(4.3)	(100%)
Brooklyn Heights	8	20	12	24	16	0	80
	(10.0)	(25.0)	(15.0)	(30.0)	(20.0)	(0.0)	(100%)
Boyle Heights	13	29	20	41	23	5	131
	(9.9)	(22.1)	(15.3)	(31.3)	(17.6)	(3.8)	(100%)
Belvedere	23	51	36	52	30	9	201
	(11.4)	(25.4)	(17.9)	(25.9)	(14.9)	(4.5)	(100%)
Outside East/ Central L.A.	37	75	32	65	72	15	296
	(12.5)	(25.3)	(10.8)	(22.0)	(24.3)	(5.1)	(100%)
Total	155	443	220	410	342	79	1,649
	(9.4)	(26.9)	(13.3)	(24.9)	(20.7)	(4.8)	(100%)

Source: Declarations of Intention and Petitions for Naturalization, National Archives, Laguna Niguel, California.

sustained a heavy concentration of unskilled laborers, along with many single males who eventually returned to Mexico. The industrial shops which dotted this region, however, increasingly brought skilled laborers to the area, including carpenters, bricklayers, boilermakers, mechanics, and machinists. Many of the shops and restaurants that catered to the Mexican community were located here, since rental space was far less expensive than in the Plaza. Consequently many of the proprietors of these establishments lived in this district, thereby producing the largest proportion of high white-collar workers (9%) of any barrio.

To the south and west lay smaller, more integrated districts. I have noted elsewhere that these neighborhoods were most likely to contain intermarried couples and single women living alone. Additionally, up to one-quarter of the Mexicans living here had white-collar jobs, the highest percentage in any region of the city. The only barrio not dominated by common laborers was downtown, for example, around the All Nations Church. There, musicians were the most numerous occupational group; laborers came next, with clerks following closely behind.

Map 9 Mexican Residences in Railroad Area
Source: Naturalization Documents, National Archives, Laguna Niguel, California.

Native-born Chicanos and Mexicans who migrated to Los Angeles as children comprised most of the sales and office clerks who lived in this region. Despite widespread prejudice against hiring Mexicans in downtown offices, many young Mexican women learned English and basic skills in night school in order to obtain positions as stenographers and clerks. These women often accomplished this feat while working full-time in blue-collar jobs in clothing or canning factories.[19] More than any other area, this was a zone of social mobility.

In time, however, large sections of residential neighborhoods south of downtown disappeared. Rezoned by the city for industrial purposes during the early 1920s, factories and warehouses increasingly replaced many of the modest houses and apartments of the district. Early in the decade, reformers affiliated with the All Nations Church worried that this industrial plan would keep city officials from improving the area's schools and playgrounds, even though hundreds of ethnic children desperately needed educational and recreational facilities. Although the Great Depression slowed the movement of industry into this area, socially mobile immigrants during the late 1920s could no longer find the same opportunities for integrated living as they had earlier in the decade.[20]

To the east of the river lay barrios that eventually emerged as the cornerstones of Chicano East Los Angeles. At this point in time, however, Boyle Heights, Brooklyn Heights, and Lincoln Heights still housed a variety of ethnic groups. Mexican settlement here was slow, due mostly to the fact that home ownership was a characteristic of these communities. Close to half of all the Mexican laborers who lived here in the 1920s and 1930s were skilled or semi-skilled workers, having achieved the economic stability which allowed them to purchase homes. High-density Mexican residence did not occur until the late 1930s and 1940s when Jews, Italians, Japanese Americans, African Americans, and other ethnics moved out of East Los Angeles in substantial numbers.

Further east, however, the Belvedere community rapidly became Chicano (see Map 10). With the completion of a Pacific Electric streetcar line into Belvedere in the mid-1920s and the division of land in the area into tiny, often unimproved, lots, Chicanos of all economic levels were able to move into Belvedere by the second half of the 1920s. Belvedere— officially outside the Los Angeles City limits—quickly became the most economically representative barrio in the region. Containing one-quarter unskilled workers and one-quarter skilled laborers, the blue-collar component of its population dominated community life. By 1930, it had already emerged as the fifth largest area of Mexican settlement in the United States, preceded only by the cities of Los Angeles, San Antonio, El Paso, and Laredo.[21] Family-oriented and working class, Belvedere in many ways symbolized the future of East Los Angeles.

Despite being concentrated in low-paying jobs, Mexican immigrants established themselves during the 1920s as home owners, particularly east of the Los Angeles River. The 1930 census revealed that 18.6 percent of Mexican-origin residents of Los Angeles owned their own homes, while Belvedere's rate was even higher at 44.8 percent. Ownership was particularly characteristic of native-born Chicanos, although the immigrant generation in Los Angeles also increasingly owned their own homes. Compared with other urban areas, Los Angeles' rate of home ownership among Mexicans was high.[22]

Home ownership for Mexican immigrants, as for other ethnic newcomers, allowed communities to develop a quality of permanence. For example, in the 1920s when the traditional center of Mexican life in Los Angeles was being displaced by the growth of downtown businesses

Map 10 Mexican Residences in Belvedere
Source: Naturalization Documents, National Archives, Laguna Niguel, California.

around the Plaza, many Mexican immigrants chose to move elsewhere. Increasing numbers sought a less vulnerable location, and found this stability in East Los Angeles.

Home ownership did not necessarily imply moving up the occupational ladder. A comparative study of Mexican homeowners in San Diego, for example, revealed that they did not differ appreciably from renters in terms of occupation, income, or family size. They did, however, tend to spend more of their income on housing than those who rented. The average Mexican family in this study spent $225 per year, or

16 percent of their income, on housing, while those still paying off mortgages averaged over $400 a year. Once loans were paid off, however, homeowners averaged under $100 a year on housing costs.[23] Buying a home, therefore, was a considerable investment for most blue-collar laborers, both in financial terms as well as in commitment to a particular barrio.

Several historical studies, however, question the wisdom of this investment. In his pathbreaking study of social mobility, Stephan Thernstrom discovered what he termed widespread "property mobility" among immigrants in Newburyport, Massachusetts, with upward of 60 to 80 percent of workingmen who stayed in town for more than a decade able to own their own homes.[24] According to Thernstrom, this largely Irish blue-collar population usually achieved their goal by sacrificing mobility out of the working class. In both this study of Newburyport and a subsequent one on Boston, Thernstrom argued that home ownership often precluded investment in education and training which would have allowed Irish immigrants or their children to move out of blue-collar jobs.[25]

Daniel Luria has argued similarly from a Marxist perspective. His 1976 study found that home ownership, although appearing to lead to greater economic equality from 1880 to 1910, in fact masked increasing social inequality. Homes, unlike other forms of property wealth, did not translate into what he termed "the social form of capital." They did not, he maintains, contribute to an economic leveling of society. Indeed, according to Luria, owning a home could be a distinct liability which limited further investments and inhibited geographic mobility during periods of economic downturn.[26]

Consideration of the emotional, psychological, and cultural rewards of home ownership, however, is often lost in these historical discussions. Olivier Zunz argues in his work on Detroit that "owning one's home was more an ethnocultural phenomenon than one of class . . . ."[27] This was certainly true for Mexican immigrants. Home ownership in a barrio symbolized an adaptation: permanent settlement in the city. Permanent settlement must be viewed in the context of patterns of geographic mobility that had characterized the Mexican condition even within Mexico during the Porfiriato. Furthermore, employers in the United States spoke with disdain of Mexicans as "homing pigeons," willing to forgo stability in order to follow the seasonal crops across the agricultural landscape or repair railroad track wherever it was needed. They believed only Mexicans were willing to tolerate the migratory nature of such labor and were willing to return dutifully to Mexico when the season was over. Within the context of the social and economic roles delineated for Mexican immigrants, buying one's own home was an act of defiance and a form of self-assertion.

Some Mexicans in southern California were even able to combine home ownership and migratory labor in order to make extra income.

"During the summer we rent our houses to other Mexicans who come there [to Whittier] to work," reported one fruitpicker traveling through the San Joaquin Valley. Owning their own automobiles, these working-class entrepreneurs selected higher-paying jobs in agriculture, and rented their houses during their absence. When they came back to the Los Angeles area to pick oranges at the peak of the season, they brought back tidy earnings to combine with the rent they collected from their tenants.[28]

Home furnishings provided further indication that Mexican immigrants were in the United States to stay. The rapid expansion of consumer credit in the 1920s made it easier for Mexican immigrants to come into contact with American consumer items, allowing cash-poor individuals to purchase "on time." Almost half of the families in the San Diego sample, for example, were paying for furniture on installment.[29] The most common labor-saving devices purchased by Mexican families were irons and sewing machines, items which had already been widely used in Mexico. Rarely did Mexican families buy washing machines or telephones, although these became more common among the American working class during the 1920s.[30]

Despite these assertions of permanency, home ownership did not always afford the same security to Mexicans as it did to European immigrants. In his work on Detroit, Olivier Zunz found that Polish and German immigrants largely controlled their own housing market. Many had built modest single-family dwellings themselves and owned their homes outright. This immigrant-controlled market existed independent of the formal housing market in other sections of Detroit.[31] Although in Los Angeles as in Detroit single-family dwellings dominated the urban landscape, virtually all housing in the City of the Angels was tightly controlled by a small group of realtors. Mexican Americans could not influence the circumstances of the market, and found themselves increasingly segregated or consciously manipulated toward residence in less desirable areas. Belvedere, for example, located in unincorporated Los Angeles County, was developed in the 1920s as an exclusively Mexican community largely because developers could ignore city statutes concerning lot size, sewage, and other services.

Working-class families who wished to move to East Los Angeles invariably needed more than one income to afford housing and the other amenities which characterized east-side living. These supplemental wages often provided the income necessary to purchase a home and to make payments on the credit used to buy furnishings.

The desire to settle in East Los Angeles, therefore, often forced families to alter "traditional" notions of work, gender, and childrearing in the American context. Among the 187 married couples in my sample for which information on the spouse's employment status is available, almost 40 percent of wives worked for wages outside the home. In two of the barrios on the east side, Brooklyn Heights and Lincoln Heights, the percentage of married women who worked increased to 73 and 57 percent, respectively.

There are indications, however, that Chicano families viewed married women working as a temporary expedient and often reversed the pattern once families settled securely in East Los Angeles. The portion of married women working in Belvedere and Boyle Heights was much nearer the city-wide proportion of 40 percent. In addition, unlike nineteenth-century Chicana work patterns described by Griswold del Castillo, American-born Chicanas were less likely than Mexican immigrant women to work once married. In the naturalization sample, 80 percent of Chicanas married to Mexican immigrant men did not work, while almost half of the Mexican immigrant women did work outside the home. Since Belvedere and Boyle Heights contained a substantial population of native-born Chicanos—estimated to be 23.7 percent in 1930—this evidence points to a reassertion of traditional attitudes toward sex roles once families were formed among the second generation.[32]

Occupational patterns for married women also differed in some important respects from those of men. Although 62 percent labored in blue-collar work, this figure was lower than the 70 percent of men. Moreover, fewer women were employed in skilled positions, while greater percentages were involved in semi-skilled and low-status white-collar positions. In fact, almost one-third of married women in the labor force found themselves in low white-collar jobs, often working as sales or office clerks in downtown businesses.

Temporary or seasonal ventures into the wage-labor market characterized Mexican female employment. This highly elastic work pattern allowed Mexican families to supplement the male wage earner's salary with extra income during times of need or in anticipation of future purchases. One study conducted in nearby San Diego revealed that almost half of the 100 couples surveyed in 1930 included working wives, although married women rarely engaged in full-time, year-round employment. Instead, women worked in seasonal industries such as fruit packing or part-time jobs such as laundering or sewing, where work that could be brought into the home. Paul Taylor discovered that 48 percent of Mexican women who worked in Los Angeles industries did not want to work, saw employment as temporary, and, in fact, had gone to work only because of financial need.[33]

Despite the variable nature of women's participation in the work force, their work experience provided a significant cultural learning experience. Unlike many men, women were often less racially segregated on the job. Mexican immigrant women were therefore exposed to a variety of ethnic and religious traditions at the workplace. In addition, the extra income that their work generated often went toward the purchase of commodities—automobiles and radios, for example—that themselves opened the home to greater contact with American life.[34]

Once families relocated and matured, older children often replaced mothers as the supplemental wage earners in the family. A 1927–28 California survey reported that 35 percent of families had working children, most part-time. Another study of 100 Mexican families in San Diego in 1930 revealed that when children worked, their wages often doubled that provided by their mothers. Almost all the children sixteen and older worked, although fewer than half were actually self-supporting. Still, the earning power of older children seemed to outpace that of working mothers substantially. Often, these adolescents and young adults secured jobs in the same places of employment as their parents, and employers viewed them as steadier, more productive employees than their elders. As families in East Los Angeles grew larger and children grew older, many could be expected to take the place of their mothers in the labor force.[35]

The tenuous economic position of Mexican immigrant workers meant that the entire family income had to be carefully monitored and distributed. The average Mexican family brought in between $80 and $100 a month in 1930, although this amount might have varied tremendously month to month given seasonal employment opportunities. Some 80 percent of family income came from the main male wage earner, except, of course, in instances where the husband was incapacitated or permanently unemployed. Besides women and children, other sources of income included the contributions of relatives living in the household. Unlike other immigrant groups, Mexican families typically took in few boarders who were not relatives.[36]

How Mexican families spent the money they earned is an important indicator of the changing cultural values of immigrants. Food, clothing, and housing accounted for roughly 70 percent of the budget of a typical Mexican immigrant family in 1930. The rest of a family's income went toward expenses considered secondary, including leisure-time activities, transportation, insurance, and savings. This distribution, of course, varied tremendously, and proved much more elastic than the money spent on basic necessities.[37]

Among working-class Chicanos, the chief item in the family budget was food. Studies indicate that food alone accounted for over one-third of a Mexican family's total expenditure, about $500 a year. Most reports indicated that the Mexican American diet resembled that of old Mexico. Most food money was used to purchase dry goods, particularly corn meal for tortillas and beans. Some 95 percent of the families surveyed by one researcher served tortillas regularly, with beans being a daily staple in 72 percent of the families. Women (who prepared virtually all of the meals) increasingly bought commercial meal for tortillas rather than grind their own. This certainly was a change from Mexico, where corn mills often stood idle while Mexican women labored daily with their own *metates*.[38]

Meals were largely consumed at home, except for the occasional lunches of day laborers or school children. Single male migrants, however, ate out much more often. The appeal made by food establishments to garner business was usually couched in ethnic terms. Restaurants and grocery stores often depicted their fare as "típica Mexicana"—typically Mexican. Descriptions of chilis and prepared meals often included elaborate commentary on how similar this food was to meals one could remember from home.[39] Thus, even food invoked traditional values and longings.

Larger and poorer Mexican families often suffered from inadequate nutrition. During difficult times, most families tried to insure that children remained well fed while parents ate last and least. Coffee continued as the chief drink among adults, with milk being strictly designated for children. Despite their abundance in California, fresh fruits and vegetables were expensive and eaten only occasionally as families struggled creatively to add nutritional items to their diets. Although rare in Mexico, meat was added to the diet in the United States when family income expanded. Dietary budgets were occasionally augmented by purchases of slightly stale breads or wilted produce. Many families who owned property kept a garden or raised chickens and goats to provide additional supplements.[40]

Though resistant to changes in diet, the Mexican immigrant adapted dramatically to new clothing styles. Clothes, which had to be worn publicly and more easily reflected economic circumstance, were carefully selected to mirror the new environment. Comprising about 13 percent of a typical family's budget, or $175 a year in 1930, dress allowed much room for decision-making and cost-consciousness among all family members. Mexicans in the United States rapidly discarded traditional styles, abandoning muslin trousers, open sandals, straw hats, calico dresses, and even *rebosos*. El Paso, Texas, the main entry point for immigrants to Los Angeles, reported a brisk business in selling "American" clothes to recently arrived Mexican immigrants. Advertisements for clothing in Spanish-language periodicals rarely appealed to Mexican tradition and custom. Like the ads in the English-language newspapers, clothing retailers invariably depicted their offerings as the latest and trendiest in fashion. A few reported their clothing as durable and long-lasting, a wise investment for the cost-conscious.[41]

Most Mexican men in Los Angeles regularly wore a cheap cotton shirt with overalls or pants, cap, and a sweater. Women of all ages typically wore cotton house dresses. Each family member also had at least one "good" outfit, although families tended to skimp on clothes in this category. Men waited four years before replacing a suit, while women often owned a few "better" dresses, hats, and stockings for street wear. Children's clothes were a problem since rapid growth often negated parental attempts at saving money. The amount of money spent on clothing obviously differed according to circumstances of employment and residential location, yet most Mexican families dressed simply and conservatively.

Although considered a luxury by many contemporary observers, one item that many Mexican families purchased during the 1920s was an automobile. Car ownership was not confined to the economically better-off. One-quarter of the families surveyed in the San Diego study owned automobiles. Most had secured secondhand cars for a relatively small sum. Upkeep on these vehicles, however, could be substantial. In fact, the Heller Committee conducting the San Diego survey criticized poor Mexican families, arguing that "automobile owners seemingly were economizing on food in order to buy gasoline." It did not appear to the committee that those who owned cars spent any less on transportation than those who used public transit.[42]

Although today Los Angeles residents view car ownership as a necessity, during the 1920s the city possessed a relatively extensive public transportation network which included trolley car stops in most Mexican immigrant neighborhoods. Service, however, was usually shoddy, and rides could prove to be rather expensive, particularly if more than one wage earner in a family traveled by trolley to work.[43] Depending on family circumstances, automobile ownership in Los Angeles could be more economical than public transporation. A car also gave families greater flexibility to find employment, especially to venture out to nearby agricultural fields to pick fruits or vegetables. This type of temporary employment often proved to be an important supplement to income generated by urban jobs, particularly because entire families could engage in this seasonal activity. Automobiles thus afforded a high degree of geographic mobility in the city and environs.

Second, auto ownership made it easier to visit relatives along the border or in Mexico without having to depend on the railroad. Visits on the train could be prohibitively expensive, especially if one was forced to stay for any length of time in a border town waiting for a connection. Automobiles also gave families the opportunity to get out of the city and enjoy the still unblemished beauty of the surrounding area. The automobile thus quickly became a vitally important commodity in Los Angeles, for immigrant families as well as native-born citizens.

The growth of a stable community of Mexican immigrants east of the Los Angeles River symbolized the important transformations that were occurring within the Chicano culture of Los Angeles during the 1920s. More integrated into the American consumer economy, Mexican immigrants nonetheless were adamant about controlling the direction of life in the barrios themselves. Increasingly they bought property and set down roots in Los Angeles. These communities, however, were not immune to the larger economic forces at work in the city. The barrio, despite its appearance as an isolated Mexican enclave, was still largely politically and economically controlled from the outside.[44] Certainly living within the barrio allowed Mexican immigrants a measure of control over their future. Yet economic and political developments outside the barrio continued to impinge on these decisions.

Few during the expansive 1920s could have predicted the enormous impact that the economic crisis of the next decade would have on Mexican immigrant families in Los Angeles. A decade of rapid demographic expansion, cultural growth, and adaptation was followed by a time of crisis. The Mexican/Chicano community in the city was put on the defensive as the events of the Great Depression rapidly altered the normal patterns of immigrant adaptation which had unfolded in previous decades. In their responses to this crisis, a new ethnic identity among Mexican Americans in Los Angeles took shape.

During the 1920s, the Mexican government futilely attempted to curtail emigration to the United States by issuing official emigration documents and restricting their use. This is the legal emigration paper of Artemio Cerda Duarte, issued on August 11, 1926. *(Courtesy of the Security Pacific National Bank Photograph Collection, Los Angeles Public Library)*

Various organizations encouraged Mexican immigrants to participate in American patriotic festivities in order to promote Americanization. This 4th of July celebration was sponsored by the International Institute of Los Angeles in 1923. *(Courtesy of the Security Pacific National Bank Photograph Collection, Los Angeles Public Library)*

The Mexican government, along with the local Mexican immigrant community, regularly sponsored festivities to commemorate Mexican national holidays. This 16th of September celebration of the 1940s featured a queen and her court on stage *(background)*, both Mexican colors and an American flag, a Spanish dancer, and a Persian rug. *(Courtesy of the Security Pacific National Bank Photograph Collection, Los Angeles Public Library)*

Elaborate weddings, such as this one from the 1920s, often signified a decided commitment to remain in the United States. *(Courtesy of the Security Pacific National Bank Photograph Collection, Los Angeles Public Library)*

Marriages between Mexicans and non-Mexicans increased with the greater interethnic contact possible in an urban area like Los Angeles. Leo Montenegro poses here with his wife Pauline and their son Leo, Jr., in front of their home on E. 65th Street in 1920. *(Courtesy of the Security Pacific National Bank Photograph Collection, Los Angeles Public Library)*

In Los Angeles, religious ritual increasingly became the responsibility of female members of the family. Here Aurora Moreno celebrates her first Holy Communion in the early 1920s. *(Courtesy of the Security Pacific National Bank Photograph Collection, Los Angeles Public Library)*

During the 1920s, Pedro J. González formed "Los Madrugadores" with other recent Mexican immigrants to southern California. In this publicity photo for KMPC radio, González displays the flair that would make him one of the favorite Spanish-language radio personalities in Los Angeles in the late 1920s and early 1930s. *(Courtesy of the UCLA Chicano Studies Research Library)*

¡FIGURANDO AHORA!
Durante el Programa Regular de Variedad
"FLOOR SHOW"
TOMANDO LA PARTE PRINCIPAL
"El Tenor de Alma Gitana"
ISIDRO LOPEZ
en
El Templo Maya de la Costa
"MOCTEZUMA INN"
403½ N. FIGUEROA ST.   L. A.

A significant number of Los Angeles residents took advantage of the opportunities available for providing Spanish-language entertainment to the immigrant community. This is Isidro Lopez's business card from 1934–35, featuring his regular stint as a singer at the "Moctezuma Inn" in the central Plaza area. *(Courtesy of the Security Pacific National Bank Photograph Collection, Los Angeles Public Library)*

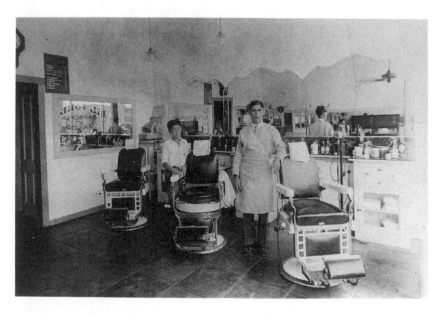

Businesses catering to an immigrant clientele solidified the cultural presence of the Mexican American community in East Los Angeles during the 1920s. This barber shop, located on Ford Boulevard in Maravilla Park, was ready for business in 1929. *(Courtesy of the Security Pacific National Bank Photograph Collection, Los Angeles Public Library)*

Employment opportunities in white-collar office work slowly increased for American-born Chicanas, along with more readily available blue-collar jobs in cannery and garment work. This woman was the office manager of the Beauchamp Penmanship Company on Broadway Street in downtown Los Angeles in the early 1920s. *(Courtesy of the Security Pacific National Bank Photograph Collection, Los Angeles Public Library)*

Ownership of an automobile provided increased mobility and opportunities for leisure away from the urban center. Here Julia and Eloise Arciniega enjoy the snow in Wrightwood in 1930 with a few of their friends. *(Courtesy of the Security Pacific National Bank Photograph Collection, Los Angeles Public Library)*

Rose Pesotta (on the right) returned to Los Angeles in 1941 to aid garment workers on strike. Here Pesotta comforts Jennie Lechtman, an arrested garment striker. *(Courtesy of the Los Angeles Times Collection, UCLA Department of Special Collections)*

One of the leaders of *El Congreso,* Josefina Fierro de Bright, confers with her fellow labor organizers Francisco Macin and Salvador Carrillo on May 25, 1942, to discuss Mexican participation in the wartime industries of World War II. *(Courtesy of the Hearst Newspaper Collection, Special Collections, USC Library)*

Mexican Americans increasingly demanded their rights as citizens to protest working conditions and low wages. In this 1941 picket, 300 C.I.O. demonstrators marched in support of street railway employees. *(Courtesy of the Hearst Newspaper Collection, Special Collections, USC Library)*

In the World War II era in Los Angeles, Manuel Ybarra (1942) and Mary Martel (1944) attired in their finest clothes. *(Courtesy of the Security Pacific National Bank Photograph Collection, Los Angeles Public Library)*

THE MEXICAN

VOICE

AUGUST 1939 VOL. 2 No. 3

The cover of the August 1939 edition of *The Mexican Voice,* published by the Mexican American Movement (MAM) to promote Mexican American progress through education. *(Courtesy of the YMCA Room, Whittier College)*

The arraignment of the young men accused in the 1942 Sleepy Lagoon murder case. The defendants are *(left to right)* Ruben Robert Peña, Daniel Verdugo, Edward Cranpre, Manuel Delgado, Jack Melendez, Joseph W. Valenzuela, Andrew Acosta, Joe Herrera, Richard Gastelum, Lupe Orusso, and Benny Alvarez. *(Courtesy of the Hearst Newspaper Collection, Special Collections, USC Library)*

Marcos and Zeferino Ramírez, pioneer residents of Belvedere, celebrating their sixtieth wedding anniversary in 1958. *(Courtesy of the Hearst Newspaper Collection, Special Collections, USC Library)*

PART FOUR

AMBIVALENT AMERICANISM

I worked in the U.S. of A. since 1904 with different companies. I registered in the world war in Johnson, Arizona, Cochise Co. I have never given my services to the Mexican government nor to Mexican capital. I have worked all of my life, since I was 19 years of age in the U.S. of A., and that is why I wish to return to the country where I am entitled to live with my children so that they be educated in the schools of your country and not in Mexico.

—Pablo Guerrero, repatriate from Los Angeles, 1934 [1]

We're tired of being pushed around. We're tired of being told we can't go to this show or that dance hall because we're Mexican or that we better not be seen on the beach front, or that we can't wear draped pants or have our hair cut the way we want to. . . . I don't want any more trouble and I don't want anyone saying my people are in disgrace. My people work hard, fight hard in the army and navy of the United States. They're good Americans and they should have justice.

—Alfred Barela, Mexican American teenager, 1943 [2]

We're not awakening ourselves for Mexico nor the United States, but for ourselves.

—Paul Coronel, President of the Mexican American
Movement (MAM), 1938–39 [3]

CHAPTER 10

Where Is Home?:
The Dilemma of Repatriation

Every morning at the crack of dawn in 1931, María Olazábal watched at her kitchen window as hundreds of Mexican men and women ventured out in search of work. As entire industries collapsed during the Great Depression, Mexican workers were among the first to feel the effects of unemployment. Desperation rapidly overcame the Belvedere barrio. Night after night, Mrs. Olazábal saw workers return to their homes exhausted and empty-handed with little strength or money to feed their families. "The power and gas companies are shutting down the services to these people because they cannot pay, and it is frightening to see the misery endured by people ready and willing to work," she complained in a letter to *La Opinión*. As a recent resident of Belvedere, she was grateful that her husband was still making ends meet as a grocer.[1]

Concerned about the poverty around her, Mrs. Olazábal decided to help the less fortunate in her community. With the assistance of a few other Mexican women along North Rowan Avenue, Olazábal organized a group that made and sold tamales at cost to the barrio's unemployed. Though the women preferred to give the food away, they could not afford to do so; besides, they felt direct charity might hurt the pride of their compatriots. Olazábal boasted of the quality of the service, assuring her customers that the tamales were "made with total cleanliness." To link themselves with their friends and neighbors, the group called itself the Cooperative Society of Unemployed Mexican Ladies.[2]

María Olazábal was one of hundreds of Mexican immigrants in Los Angeles who aided those left stranded without jobs during the 1930s. During the Depression, many were forced to make difficult adjustments regarding their future in the city, caught in economic circumstances beyond their control. While Mexicans attempted to cushion the deprivation in the barrios, county officials and local businessmen, largely focusing on those without jobs and on welfare, developed campaigns to rid southern

California of Mexicans. They believed the hard times made it imperative that the scarce jobs and resources be reserved for American citizens.

This chapter addresses the responses of Mexicans to economic crisis and the mass repatriation campaigns of the 1930s. Their responses reflected fundamental changes in Chicano cultural development in the city. Faced with limited choices, Mexicans made crucial family and individual decisions shaped in part by governmental policy in the United States and Mexico. Though the impact of outside forces was clearly significant, their personal choices eventually had a powerful effect on group ethnic identity. Approximately one-third of the Mexican community in Los Angeles returned to Mexico during the decade.[3] The majority who stayed in Los Angeles became ambivalent Americans, full of contradictory feelings about their place in American society.

After a period of phenomenal industrial growth since the turn of the century, Los Angeles, with its rapidly increasing population, found itself ill-prepared for the economic bust. A week before the stock market collapse in October 1929, city officials, brokers, and bankers had attended ground-breaking ceremonies for a new stock exchange building on Spring Street. Despite an emergency conference on unemployment in mid-December 1929, most officials felt that the next decade would bring continued prosperity to the area.[4]

But by 1930, most city residents realized the widespread economic depression had deep roots and the nation would not easily recover. Many lost their jobs; thousands of families found themselves without income. Nationally, unemployment rose to four million by January 1930, to five million by September, and to eight million by the spring of 1931. By 1933, one-third of the work force in the United States—15 million people—were out of work, a fivefold increase from 1929.[5]

In Los Angeles, the census of 1930, conducted in April, reported that 50,918 city residents were unemployed, or just under 10 percent of all gainful workers. Especially hard-hit were the skilled building trades—electricians, stone masons, and tile layers. Though, proportionally, Los Angeles had fewer unemployed than industrial cities such as Cleveland, Buffalo, and Detroit, by the end of 1930 one out of every five Angelenos could not find work. One study showed unemployment in Los Angeles peaking at 41.6 percent in 1933.[6] Even those with jobs severely felt the impact of the Great Depression. Wages fell by one-third in the United States, as both wages and hours were cut. In Los Angeles County, average wages declined 38 percent between 1926 and 1932.[7]

Because of their seasonal employment in agricultural work, Mexicans were among the first in Los Angeles to experience the consequences of the Depression. Agriculture was one of the earliest casualities of the economic crisis. Although agricultural employment in the city proper mirrored the overall employment picture, farm work in the San Joaquin Valley and Imperial Valley declined even before the stock market crash. The value of California farm products dropped precipitously from $750 million in 1929 to $372 million in 1932, with wages falling from 35 cents an hour in 1928 to 14 cents in 1933. Agriculture could no longer absorb urban workers who increasingly needed to supplement their low wages from unskilled industrial jobs. This development caused Mexican workers to depend even more heavily on wages earned in urban Los Angeles.[8]

Finding work in the city, however, became more and more difficult. By April 1930, one of every seven Mexican laborers was unemployed, a figure almost twice as high as that of any other ethnic group in Los Angeles. Moreover, as Anglo Americans found themselves without work—particularly after several months of unemployment—they began to exert pressure on city employers to hire only "citizens" for work that had normally or occasionally been limited to Mexicans.[9] By February 1931, *La Opinión* reported that laundries, factories, stores, and construction companies regularly replaced Mexican workers with Americans. Ethnic background, rather than strict definitions of citizenship, seemed to prevail in determining this form of discrimination. An ex-sergeant in the United States Army, born in the U.S., described how he was denied jobs because he was Chicano:

> During the last three months, I have been getting up very early; I dress up and go downtown or uptown to the construction sites where the supervisors know me and always have given me a job. Soon the supervisors come out and tell the people that are waiting to get a job to line up on one side, all the white people, and on the other side the ones that are not. Because I am of dark complexion I stay with the people of my race and of course, do not get hired because the supervisor has the order to hire only the "white people" and that is what he does.[10]

In August 1931, the California state legislature enacted a law making it illegal for any company doing business with the government to employ "aliens" on public jobs. The Alien Labor Act was a form of legislative discrimination that displaced many Mexican workers from construction sites, highways, schools, government office buildings, and other public works projects. The Mexican consul in Los Angeles estimated that this law immediately excluded more than 900 Mexicans from work in the city alone.[11]

Many Mexican families were forced to turn to public and private charities for help in surviving unemployment and economic deprivation. Yet here, too, discrimination became the norm. During the 1920s, Mexicans constituted about one-fourth of all city residents who received some form of public assistance. Erratic employment opportunities for men, coupled with dangerous work conditions that often left unemployed women in charge of large families, placed Mexicans among the groups in Los Angeles most likely to need periodic economic assistance. But during the Great Depression more Anglo American residents also found themselves in difficult circumstances. The result was increasing pressure on public officials to give preference in welfare allocations to American-born heads of households. By 1931 public works projects financed by local monies likewise barred employment of aliens.[12]

As unemployment climbed, almost all new relief was allocated for Anglo laborers who had lost their jobs. In Los Angeles County, the number of welfare cases jumped from 18,650 in the fiscal year 1928–29 to 25,913 in 1929–30, and to 42,124 in 1930–31. Expenditures sky-rocketed from $1,690,450 in 1928–29 to $2,469,520 in 1929–30, and to $4,209,729 in 1930–31. Yet the percentage of Mexicans on relief steadily decreased from 21.5 percent in 1928–29 to 15.8 percent in 1929–30, and to 12.5 percent in 1930–31, despite widespread impover-ishment in the Mexican community. The role of the federal government in public assistance was minimal during this period, and, thus, local offi-cials determined this inequitable allocation of resources.[13]

Of the private groups dispensing assistance, the Catholic Welfare Bu-reau was the largest in the city, and handled most of the needy Mexican families unable to receive aid elsewhere. The bureau's unemployment re-lief steadily grew from $63,719 distributed among 3,211 families in 1928–29 to $112,883 among 9,172 families in 1930–31. The bureau's funds, however, came largely from Los Angeles' Community Chest, which had been founded in 1924 to coordinate relief efforts among reli-gious groups in the city. In better times, the Chest raised substantial funds from private sources. During the Depression, however, these re-sources evaporated, and the Chest grew utterly dependent on public funds. Between 1929 and 1932 roughly 94 percent of its monies came from either the city or county. As a consequence the Catholic Welfare Department was subject to the same financial pressures regarding the distribution of its funds; thus it too began to discriminate among ethnic groups. In 1931, for example, food allowances for "American" families were reduced by 10 percent, while allocations for Mexican families were cut by 25 percent.[14]

Many Mexican residents of Los Angeles responded to the worsening economic conditions and growing pattern of discrimination by returning to Mexico. Although most of the early repatriates came from Texas bor-der towns, during the winter of 1929–30 a sizable group of Los Angeles residents departed for their homeland. This first group was usually not destitute; many returned with automobiles and furniture accumulated by hard years of work and saving in the United States. A number of early returnees were single male white-collar workers who had never intended to stay permanently in Los Angeles. As office and sales clerks they had enjoyed modest economic success in the 1920s, and left the United States when the possibilities for making extra income vanished. As one resident of Los Angeles in this period remembered, the average Mexican always had it in his heart to return to Mexico. With economic opportuni-ties dwindling in the U.S. and Mexico calling, many decided simply to head back.[15]

This group of early repatriates resembled those identified by Manuel

Gamio in his classic study of Mexican immigrants in 1927. He discovered a great deal of movement back to Mexico with returnees bringing with them a host of consumer items, from agricultural tools to phonographs.[16] In fact, Gamio's investigators interviewed many southern California residents who had been enticed back to Mexico by their government's promise of agricultural land. These individuals left the United States with hopes of turning their small savings into larger enterprises in Mexico. Though ready to acknowledge the economic benefits of their stay, they also often deeply resented their treatment by American employers and local officials. José Castillo, for example, planning to return to Acámbaro, Guanajuato, never liked the customs in the United States, and expressed his desire to see his grandchildren raised as "good Mexicans."[17]

As the winter of 1930 approached, an increasing number of Mexican residents of Los Angeles decided to head south. All along the border consular officials reported large caravans of returnees from throughout the Southwest and Midwest. Some 2700 repatriates crossed through Nuevo Laredo during the first half of December, while 800 more were counted leaving through Laredo and Nogales on a single day in early January, 1931. By the end of 1930, nearly 10,000 Mexicans had crossed the border. But while hundreds had left Los Angeles, they were quickly replaced by new arrivals who had fled California's small rural towns expecting to find better public relief in a larger city.[18]

The growing financial burden of local relief led many officials in Los Angeles to look for scapegoats. Resentment and suspicion of "alien" Mexicans on relief increased as limited resources for the unemployed grew more scarce. President Hoover's attitude encouraged such feelings. While Food Administrator during World War I he had enthusiastically recruited Mexican farm workers to maintain wartime production, but in 1930 Hoover denounced Mexicans as one of the causes of the economic depression—"they took jobs away from American citizens"—and he initiated plans to deport them. Thus, a host of factors coalesced during this crisis which culminated in the depopulation of the Mexican community in Los Angeles by as much as 30 percent between 1930 and 1935.

Several historians have described the Mexican repatriation/deportation programs of the 1930s. In the first book-length study on this subject, Abraham Hoffman focused on Los Angeles County officials' role in sponsoring trains that returned Mexicans to the border from 1931 to 1934. Another perspective on repatriation is provided by Francisco Balderrama. Utilizing Mexican consulate records, he chronicled the role of the consulate office in Los Angeles, demonstrating how Mexican officials in the United States encouraged repatriation efforts while consistently fighting against abuses in the processes of deportation and repatriation of Mexicans. Finally, in another study, Mercedes Carreras de Velasco described the role of various branches of the Mexican federal government in repatriation efforts, particularly concentrating on the attempt to promote agricultural cooperative communities among the returnees. These studies and other evidence from the period provide insight into how the actual planning, coordination, and implementation of repatriation campaigns were conducted in Los Angeles and throughout the nation.[19]

What historians of repatriation have not yet examined is how the loss of thousands of family members, friends, and neighbors affected the social identity of those individuals who stayed in the United States. Moreover, we know little about repatriation's impact on the cultural development and adaptation of Mexican American communities. The departure of nearly a third of all Mexicans from Los Angeles had profound consequences among Chicanos.

To understand the emotional and psychological impact of the repatriation period, one must first examine which groups encouraged repatriation in Los Angeles. Communications between local officials, the business community, and federal authorities in the Labor Department who were responsible for immigration control set activities in motion. In late 1930 newly appointed Secretary of Labor William Doak began his work with a promise to rid the country of the "four hundred thousand illegal aliens" he believed were taking jobs away from American citizens. In a desperate attempt to rescue the Republican party from responsibility for the economic crisis, Doak charged the Immigration Bureau (which at the time was under the Department of Labor) to ferret out these "thousands." However, the task could really not be accomplished solely by federal authorities because the tiny Border Patrol, founded in 1924, was largely responsible for administering border checkpoints, while few other federal officials were assigned immigration responsibilities. Los Angeles, for example, had only thirty-five immigration agents assigned to it in 1931.[20]

Los Angeles officials were more than willing to aid in the effort. Charles P. Visel, newly appointed head of the Los Angeles Citizens Committee on Coordination of Unemployment Relief, devised a scheme in January 1931 to publicize a visit to the city by the regional immigration director in order to frighten local "aliens" into returning to their native country. In coordination with Secretary Doak and Colonel Arthur Woods, national coordinator of the President's Emergency Committee for Employment, Visel sent out a publicity release on January 26 to all newspapers in the city, especially the foreign-language press, which emphasized the upcoming campaign to rid the city of all deportable aliens. This plan eventually included a high-profile sweep of the Plaza district on February 26, 1931, in which four hundred individuals were detained, yet only seventeen people—eleven Mexicans, five Chinese, and one Japanese—were taken into custody.[21]

Fewer than 300 Mexican aliens were actually deported by federal authorities during this entire campaign.[22] The scare was successful, however, in encouraging Mexicans of varying legal status—including American-born citizens of Mexican descent—to contemplate leaving. As far as Mexicans' legal status was concerned, the laws were relatively new, contradictory, and largely intended for European immigrants expected to enter the country at designated seaports. In addition, various exemptions had been granted Mexican laborers which further complicated the issue. The net effect of the confusion was the rampant abuse of authority by those charged with implementing the law and the widespread distrust of American officials by Mexicans in Los Angeles.

The scare tactics were so effective in alarming the Mexican population that local businessmen began to worry about losing an abundant, reliable supply of cheap labor beyond the immediate crisis. The Los Angeles Chamber of Commerce, chaired by George Clements, pulled back from their initial support of repatriation and called on political officials to restore calm in the Mexican community. The ties between local urban industrialists and rural landowners in the San Joaquin Valley and Imperial Valley were strong, so despite widespread unemployment in the city, the business community was adamant about protecting the large pool of Mexican workers. What business leaders wanted was an orderly program that would lessen the burden on local welfare agencies without disturbing the availability of Mexican workers needed to complete the harvest at minimal wages.[23]

Meanwhile, welfare officials had begun discussing strategies for reducing the relief rolls as early as the fall of 1930. Much of this dialogue occurred among Los Angeles County officials, since the county administered a large percentage of welfare relief in the region, particularly among Mexicans. By January 1931, welfare directors had approached the County Board of Supervisors with a plan to pay the train passage to the border of those Mexican residents on relief. Officials estimated that one-way train fares were far less expensive than maintaining Mexicans on relief rolls.[24] The plan was adopted by the end of the month, and the first trainload of county-sponsored repatriates left Los Angeles on March 23, 1931.

Paradoxically, the Mexican government's representative in Los Angeles was crucial to the success of repatriation efforts. Along with local Mexican businessmen, Consul Rafael de la Colina had protested against the Immigration Bureau's efforts to scare Mexicans out of the city with threats of deportation. In his view, however, voluntary repatriation was another matter entirely. By late 1930, the consular staff had received frequent inquiries from disenchanted, unemployed barrio residents wanting assistance to return to Mexico. On January 29, when local officials briefed Colina about their evolving plans to launch a repatriation campaign, the consul argued that the proposed program should be extended to Mexicans outside the county as well as to those who did not receive welfare benefits. On March 8, 1931, when plans for county sponsorship of repatriation trains was announced, *La Opinión* called the deal a "great victory" for Colina. A month after the program began, he appeared before the Los Angeles Chamber of Commerce to allay fears that continued movement back to Mexico would hurt local industry. He argued that repatriation was the "only alternative for many unemployed Mexicans."[25]

The consulate's strong advocacy of repatriation was motivated by the growing economic misery of Mexican immigrants in Los Angeles. Also underlying his support was the Mexican government's long-standing desire to see immigrants return to Mexico, particularly after they had acquired experience in the American labor force. The presence of so large an emigrant community across the border had rankled Mexican officials since the revolution. Mexican officials viewed the mass immigration as international mutiny amidst growing nationalistic fervor. They also sincerely believed the migration had been a labor drain that had depopulated the northern states. "México llama a sus hijos"—or "Mexico calls out to her children"—was the headline one southern California resident remembered from a flyer encouraging repatriation put out by the Mexican consulate office.[26]

In 1931, many destitute Mexicans in Los Angeles also began to believe that a return to their homeland would be prudent. But long spells of unemployment and debt left many Mexican residents unable even to pay the costs of transportation. During the winter of 1930–31, the Mexican consulate responded by arranging a reduced rate for Mexican repatriates to El Paso via the Southern Pacific railroad. Requests for additional help, however, continued to pour into the consulate offices. County plans to use relief funds to pay for passage to the border were welcomed by those who had organized more informal programs. In addition, welfare committees set up by the consulate office began to divert a significant percentage of their funds toward repatriation. One historian estimates that as many as 1500 individuals who left Los Angeles in April and May of 1931 had their passage paid for by the Comité de Beneficencia Mexicana in Los Angeles.[27]

Compared with those who left the city before formal deportation and repatriation campaigns began, Mexicans who departed after 1931 were more likely to be low-paid blue-collar workers.[28] They were destitute, unemployed for many weeks or months, and usually on relief. In fact, what often distinguished those who decided to go to Mexico from other working-class Mexican immigrants who did not, even within their extended families, was whether or not members of the family were employed and had been forced to go on relief. The Bureau of County Welfare had little trouble finding willing participants during its first year of organized repatriation; as many as three to five thousand residents on relief had requested county aid for transportation back to Mexico. Heads of families who had planned on waiting out the economic downturn now realized that their decision had simply allowed precious resources to dwindle. This was their chance to start anew in their native land, even if their return was prompted by desperate circumstances.[29]

Unlike the earlier group of repatriates who had exhibited some measure of optimism, these returnees were deeply troubled. They had brought back little to show for the years of hard work in the north. One observer noted that "the majority of the men were very quiet and pensive" on the trains, while "most of the women and children were crying."[30] Moreover, they resented their treatment in the United States, feeling that they deserved more than an unfriendly send-off after many years of toil. Some had lived in the city for over a decade and had developed close attachments to neighbors and neighborhoods in East Los Angeles. Moreover, they were anxious about the uncertainty of their future. To ease their anxiety, the only Mexican American student at Occidental College in 1931 remembered dressing in Mexican costume, singing songs in Spanish (particularly the farewell song, "Las Golondrinas"), and expressing gratitude to the departees for their role in building this country from the train platform. A few Anglo American church officials and parishioners also organized collections of food and clothing for the repatriates.[31]

Many, if not most, experienced little improvement in their status once back in Mexico. While early repatriates had brought resources with them that could be put to good use in starting over, later returnees often found themselves unable to translate their American experience into tangible economic results in Mexico. As one woman who repatriated to Mexico City said:

> How can we do anything? We are so poor. Surely many have learned useful skills there, but what good does that do here when they come back without anything, no tools, no work, nothing at all, not even to eat. What help can *repatriados* like that be?[32]

In fact, people returned to places where they had familial and other ties rather than to areas with greater economic possibilities. Despite government attempts to encourage repatriates to settle in agricultural colonies, Mexican officials estimated that fewer than 5 percent did so. Approximately 15 percent settled in large urban centers such as Mexico City and Guadalajara, but the economic opportunities there were also limited. The majority, probably close to 80 percent, went home to familiar villages, often returning to the place of their birth.[33]

The response of repatriates to their native villages varied tremendously. Paul Taylor discovered that many returnees to Arandas, Jalisco, easily reintegrated themselves into the community's life, quickly abandoning customs and dress that they had acquired in the United States. Many felt quite positive about their experience in the north and wanted to return when economic conditions improved, though Taylor surmised that this attitude may have resulted from the fact that Arandas produced very light-skinned individuals with few Indian features—many of whom might have remained immune from the worst of American prejudice.

James Gilbert, a sociology student at the University of Southern California, traveled throughout central Mexico in 1934 interviewing over

100 repatriates. He took a much broader sample than Taylor, and uncovered greater problems of adjustment, particularly among those who had returned with few resources. Most had difficulties obtaining employment or land to resume agricultural endeavors, and the single colony he investigated was fraught with environmental and economic difficulties.[34]

Within a repatriated family, it became clear that different members had very distinct adjustment problems. Women often had to readjust to more austere housekeeping conditions. "Here it is harder," said one woman who was living in a small village with her husband. "Cooking is more difficult. There we had gas ranges, but not here, and we used flour while here it is *maíz*." One young girl had to accustom herself to waking up at 4 a.m. in order to prepare corn tortillas for the family meals for the day, even though she herself refused to eat any not made of flour nor to use the tortilla as a replacement for a spoon. Another complained that the lack of running water forced her to bathe "in a little tub of water in the middle of the patio exposed to the four winds." Restrictive mores were particularly oppressive to young women, one of whom complained that she never went to dances or went out with boys because of local customs. "If you do, everyone starts talking, and you are regarded as a lost person. They won't have anything to do with you."[35]

Older children also had difficulty adapting to their new surroundings, having grown up accustomed to amenities in the United States. One woman who repatriated back to a family ranch from southern California at age seven remembered laughing at the first adults she encountered wearing huarache sandals, large sombreros, and white cotton pants she called "pajamas," only to find herself crying months later when her own clothes and that of her siblings ran thin and they had to begin dressing like other Mexicans. This same girl, however, taught other Mexican youngsters at the ranch how to dance the tango and play American baseball and basketball, earning the nickname "La Norteña" in the process. Many others did not know how to read or write Spanish, so they were held back in school. Often children who had been born and raised in the United States dreamed of returning long after their parents had already decided to remain in Mexico.[36]

The most successfully adjusted repatriates were usually those who had spent the least amount of time in the United States and had been most isolated within barrios. For them, life in Mexico consisted of familiar surroundings. Not surprisingly, the most skilled and the most Americanized repatriates—the very people the Mexican government hoped would bring progress to the villages—became the most discontented. They looked for the first opportunity to return to the United States, and often felt like social outcasts in their native land. Those that gave up "American ways" had an easier time of adjustment, yet failed to distinguish themselves in a way that would bring progress to the entire community.[37]

One significant exception to this pattern—but one totally unintended

by the Mexican government—was the emergence of agrarian radicalism
in certain areas of the central plateau due, in part, to the influence of
repatriates. A study of *agraristas* in Lagos de Moreno, Jalisco, revealed
that many of the early leaders of that region's agrarian movement were
repatriates who returned from the United States with growing families;
they demanded communal farm lands promised by the Mexican government as part of its post-revolution land redistribution policy. They had
enjoyed higher wages while in the north and some had been involved in
unions; in Mexico, they refused to return to sharecropping or debt peonage. In addition, some individuals, because of their exposure to other
religious beliefs in the United States, tended to oppose priests who hin-
dered agrarian reform. Some had gained basic literacy skills in the United
States, a talent which catapulted them into leadership positions among
the villagers. As one of these *agraristas* explained: "You see, there is pre-
cisely something about the United States which awakens me. . . . We
saw in the United States that progress comes from work . . . and we
remembered that here, the rich men don't work, they just exploit the
poor."[38]

In Mexico City, a different form of political protest emerged among
the repatriate community. In late 1932, the newly formed La Unión de
Repatriados Mexicanos (the Union of Mexican Repatriates) pressured
the Mexican government to halt repatriation efforts until the promises
they made to the returnees were fulfilled. They asked Mexican officials to
recognize the "painful reality" of the repatriates' economic condition.
They demanded, moreover, a stop to the "deception" by the National
Committee on Repatriation which seduced potential repatriates with "a
thousand promises of improvement and aid to all Mexicans who returned
to their native country." On April 19, 1933, La Unión sent a letter to
La Opinión in Los Angeles asking the newspaper to print the enclosed
photographs of their condition and publish the news that "they had returned only to die of hunger and to inspire pity at the doors of charitable
organizations, where they receive only one meal a day."[39]

This was not the first news to reach Los Angeles that repatriates were
suffering in Mexico. As early as November 1931, *La Opinión* published
a report from Aguascalientes that repatriates—some originating in Los
Angeles—had been abandoned without money or help, often unable to
find family or friends after years of separation. Other reports confirmed
the problems experienced by the repatriates. A group considering repatriation from Anaheim, California, for example, was warned by a former
resident who had returned to Abasolo, Guanajuato, that "the situation
in our Mexico was distressing because there was no job openings." In
the spring of 1932, Mexican consulates throughout the United States
warned those considering repatriation to forgo expectations of locating a
job in Mexico and to come only if they could not find work in the
United States. In early 1933, a tragic story of a repatriate from California
who resettled in Morelia, Michoacán, circulated throughout the Los

Angeles Mexican community: despondent over his family's economic misery, he killed his wife and his four children with a hammer and then committed suicide.[40]

Increasingly, a large number of repatriates began to consider returning to the United States.[41] Los Angeles County officials, for example, received a letter in May 1934 from a former resident whose family had been repatriated in 1932 through the county-sponsored program. Pablo Guerrero wrote from the border city of Mexicali, Baja California, to request legal U.S. immigrant status for he and his family; ironically, all five children were born in the United States and, therefore, were legal citizens of the nation.

> I worked in the U.S. of A. since 1904 with different companies. I registered in the world war in Johnson, Arizona, Cochise Co. I have never given my services to the Mexican government nor to Mexican capital. I have worked all of my life, since I was 19 years of age in the U.S. of A., and that is why I wish to return to the country where I am entitled to live with my children so that they be educated in the schools of your country and not in Mexico.[42]

Most repatriates, including Pablo Guerrero, found county officials unsympathetic. In fact, United States federal authorities at the border were instructed to turn back any Mexican who had been repatriated under the sponsorship of a charitable organization or government agency and who requested reentry. The stamp of "L.A. County Charities" on the back of a voluntary departure card precluded immigrants' return to the United States. Many were forced to reenter this country illegally, risking capture even if their families included American-born children.[43]

still the case, despite the laws on the books

Instead of reconsidering their plans, county officials continued to promote repatriation long after it was clear that the situation in Mexico was, in many ways, worse than in the United States. The county program became more difficult to implement by mid-1933 as fewer Mexicans were willing to consider repatriation as an alternative to their economic woes. Evidence of this slowdown was reflected in fewer and fewer county-sponsored repatriation trains. Until April 1933, Los Angeles County had organized shipments of repatriates at approximately two month intervals. After this time, there was an abrupt decline in the number of departures. For example, the shipment in early August, 1933, contained only 453 people as compared with an average of 908 people in the thirteen previous shipments. The subsequent train in December 1933 contained even fewer repatriates. A six-month hiatus followed before the next, and final, county-sponsored train left Los Angeles.[44]

There were a number of reasons which account for the decline of interest in repatriation among Los Angeles Mexicans. First, and probably most important, many of those inclined to return to Mexico had already done so by mid-1933. After four years of economic depression and more than two years of continued encouragement by local American and Mexican officials to repatriate, the Mexican community in Los Angeles had lost thousands of individuals whose commitment to stay in the United States, both in economic and psychological terms, was relatively weak. The single male migrants to the city were among the first to leave, since they had fewer familial obligations and generally had not invested in real estate. Recent migrants to the city, including newcomers from both Mexico and rural California, were less well established in Los Angeles and likely had greater difficulty securing jobs in the midst of the Great Depression. Those that remained in the city in 1933 tended to be members of a family unit, to be property owners, and to be residents in the city for at least a decade.[45]

Second, the dismal reports about conditions of repatriates in Mexico that circulated by word of mouth and in the Spanish-language press undoubtedly discouraged further repatriation. Moreover, by 1932 the Mexican consulate in Los Angeles no longer encouraged repatriation except under the most exceptional circumstances. The departure of Consul Rafael de la Colina in March 1932, a staunch proponent of repatriation, no doubt also diminished the call for return to Mexico. Joaquín Terrazas, Colina's immediate successor, reported to *La Opinión* in May that Mexican nationals should stay in the United States even if they had poorly paid jobs. The economic situation in Mexico was so difficult, he reported, that repatriates should have little expectation of finding work in their native land. Consul Alejandro Martínez, who replaced Terrazas the following January, went so far as to tell a federal official that "there is no agreement between the United States and the Mexican Government" concerning repatriation, and that Mexican nationals who desire to return may do so "but the Mexican Government feels as long as they voluntarily left Mexico the government is under no obligation to help them return." According to a Los Angeles official, the Mexican government was increasingly reluctant to support repatriation because local communities in Mexico complained about the burdens which destitute repatriates created for their towns and cities.[46]

The third reason interest in repatriation had waned was the inauguration of President Franklin D. Roosevelt in March 1933. This change in administration brought about a transformation in the Mexican community's outlook toward their future in this country. Immediately after Roosevelt assumed the presidency, *La Opinión*, which generally focused on Mexican politics, told its readers not to be alarmed by the immediate fiscal crisis: "The confidence and solidarity of the North American people in this emergency situation are admirable." Much of the praise centered on the federal controls that Roosevelt promised for revitalization and the presidential theme of cooperation among the population during times of crisis. When compared with the disastrous policies of Herbert Hoover, Roosevelt's early presidency appeared to Mexicans to foreshadow not only economic recovery but a reduction in racial discrimination as well.[47]

In very practical terms, the policies of the Roosevelt administration rearranged the distribution of relief for Mexicans in Los Angeles. Federal

Table 15. Applications for Naturalization
by Three-Year Intervals

Years	Number	Percent
1904–6	1	0.06
1907–9	4	0.23
1910–12	3	0.17
1913–15	25	1.44
1916–18	66	3.79
1919–21	206	11.84
1922–24	152	8.74
1925–27	279	16.04
1928–30	249	14.31
1931–33	192	11.03
1934–36	374	21.49
1937–39	189	10.86
Total	1,740	100.00

Source: Declarations of Intention and Petitions for
Naturalization, National Archives, Laguna Niguel, California.

assistance to the state and county contained provisions which prohibited
discrimination in the allocation of direct relief on the basis of legal status,
while raising the level of public assistance for all. Moreover, federal relief
funds could not be used to transport aliens out of the county. On the
other hand, the Federal Civil Works Program welcomed only American
citizens or aliens who had taken out their first papers for naturalization.
Nevertheless, not only did Mexican residents of Los Angeles now have
greater possibilities for direct relief, they were motivated to file for naturalization because of the new provisions of work relief (see Table 15).
The possibility of surviving the economic crisis in Los Angeles increased
substantially with the Democratic administration in Washington.[48]

Los Angeles County officials who had committed themselves to a
policy of repatriating or deporting Mexican aliens looked askance at these
developments. Largely in the hands of the Republican party, government
agencies in southern California resisted the federal intervention of the
New Deal program on many different levels. Since most relief in California before 1933 had been distributed by private agencies, new federal
rules demanding that government funds be handled by public agencies
created great turmoil. At both city and county levels, the establishment
of public control ushered in mismanagement, unnecessary bureaucracy,
and blatant political patronage. Lorena Hickok, an unofficial observer of
relief efforts for Harry Hopkins, Roosevelt's administrator of relief,
wrote in the fall of 1934 that relief in California was "the damnedest
mess . . . a wretchedly inefficient business."[49]

Local officials, bent on solving Los Angeles' economic woes in their

own way, doubled efforts to entice Mexicans to leave. Earl Jensen, Superintendent of Charities, admitted to an "intensive recruiting drive on the
part of the Welfare and Unemployment Relief Districts" to get Mexicans
to sign up for the December 1933 trip which netted only 120 relief
cases, or 412 individuals. The failure of that effort hastened officials to
offer other incentives to repatriate, such as a modest cash allowance for
families once they disembarked from the train in Mexico. Although this
incentive produced a slight increase in the number of Mexicans departing, it largely failed.[50]

This third phase of repatriation produced the most overt examples
of abuse and manipulation, and certainly increased the level of racial discrimination by local officials against Mexicans. Growing resentment
against repatriation by Mexican Americans created conflict between them
and local officials determined to continue the movement of Mexicans
from southern California. This conflict occurred on a day-to-day basis,
especially between county social workers and Mexican families seeking
relief dispensation. John Anson Ford, who was elected onto the County
Board of Supervisors in December 1934, remembered that during his
tenure welfare officials, though lacking the legal authority to do so, pressured Mexican nationals to return to Mexico, trying to convince them
that they had to go back. A longtime Mexican resident recollected that
as a little girl she picked up margarine, peanut butter, other groceries,
and clothes from a county warehouse. Her father, after being pressured
to give up his Mexican citizenship, decided to return to Mexico rather
than continue to suffer the uncertainties of poverty and county welfare.[51]

Señor Natividad Castañeda, like most others who left during this
third phase of repatriation, entered the depression decade as a skilled
worker, proud of his craft and able to earn good wages during more
prosperous economic times. He, and others like him, had arrived in the
city before 1923, and had been in the United States for much longer. By
1933, tens of thousands of Mexicans had lived in the United States for
over two decades, many having arrived as adolescents and young adults.
They had often bought property east of the river in Brooklyn Heights or
Belvedere, married, and decided to raise their families in Los Angeles. In
many ways, repatriates of this third phase closely resembled those who
refused to return to Mexico during the Great Depression. What often set
them apart, however, was a particular misfortune that made family survival extremely precarious and forced them to reevaluate their previous
decisions. Castañeda's family, for example, finally agreed to repatriation
only after the mother fell ill and died of tuberculosis, and the family
home was foreclosed by the state.[52]

Pressure to repatriate and harassment in allocating relief was combined with the manipulation of the welfare system to serve the interests
of local employers. For example, after the El Monte Berry strike of 1933
was settled, the County Charities Department investigated the predominantly Mexican work force to determine who would soon be out of

work. They then placed an undercover agent among the workers and his task was to persuade people to return to Mexico, thus ridding the county of potential Mexican troublemakers and welfare relief recipients.[53] Local efforts could be hampered by federal guidelines, however, because accepting federal funds bound officials to distributing aid to all the unemployed, including strikers. The federal government's increased relief to Mexicans actually strengthened their bargaining power as field laborers, because the amount received in aid in 1933 was approximately the same amount per month as wages earned by agricultural workers. In fact, income from federal assistance programs created more security than seasonal farm labor.[54] This situation angered local businessmen, as the comments of Los Angeles Chamber of Commerce President Clements suggest:

> The Mexican on relief is being unionized and is being used to foment strikes among the few still loyal Mexican workers. The Mexican casual labor is lost to the California farmer unless immediate action is taken to get him off relief.[55]

By 1935, farm owners, in fact, working in concert with the Los Angeles Chamber of Commerce, were instrumental in changing state and federal allocation of resources. In May, the California Relief Administration and the WPA agreed to drop workers from their relief rolls if they refused agricultural work. From August to October, 75,000 workers were denied relief in order to force them to work in California fields and processing plants. In addition, relief recipients in the Belvedere barrio, along with Mexicans from throughout Los Angeles, were sent out to work in agriculture. Between 1935 and the beginning of World War II, relief rolls in southern California were expanded or contracted depending on the seasonal labor needs of agricultural interests in the state.[56]

The maltreatment of Chicanos on relief, particularly the pressure put on residents by county officials to repatriate, deeply affected those Mexicans who stayed in the city. Many Chicanos recall vivid memories of the Great Depression. Antonio Soto, for example, interviewed during the 1970s, remembered that Mexicans in the 1930s were indiscriminately picked up and sent back to Mexico. "Even if they were citizens they had no rights and were treated like animals and put in cattle cars."[57] Mexicans who stayed behind also retain memories of relatives, neighbors, and fellow workers who departed under questionable circumstances. Those who remained in the United States realized that they, too, might have easily been deported or repatriated had it not been for the benevolence of a trusting neighbor, a child's extra income, or the family vegetable garden in the back yard.

By 1935, the Mexican community of Los Angeles had been substantially transformed by the effects of depression, deportation, and repatriation. First, a profound demographic shift resulted from the loss of one-third of the Mexican residents of the city, and, consequently, the internal composition of the Mexican population was altered. The departure of young, single, Mexican-born men who had greatly influenced the environment in the central Plaza area made more prominent the role of family units in the evolving culture of Chicanos in Los Angeles. Since housing around the Plaza area was geared primarily toward single men, during the Depression movement into East Los Angeles by the Mexican community accelerated. In addition, the return to Mexico by clerks and other low-level white-collar workers contributed even more to the homogenous profile of Mexicans as a low blue-collar work force.

Perhaps most important for the future of the Chicano community, the net effect of the repatriation of single men and young Mexican families was to quicken the demographic shift toward second-generation dominance. For the Mexican origin population in California as a whole in 1930, the ratio of native-born to foreign-born was 91 percent. By 1940, however, the native-born now dominated by a ratio of 164 percent. The midwestern states, also sites of widespread repatriation, witnessed a similar shift.[58] Within the span of five years, what had been largely an immigrant community before the Depression became one dominated by the children of immigrants. This generational shift had profound implications on Mexican American ethnic leadership and cultural identity in Los Angeles.

The major outcome of repatriation was to silence the Mexican immigrant generation in Los Angeles and make them less visible. As construction on the new Union Train Terminal alongside the Placita in 1934 began, the presence of the Mexican immigrant community diminished further in the downtown area. Reminders of a vibrant Mexican immigrant life disappeared for the larger Anglo American population. The ethnic diversity which had in the past so profoundly marked the city was now becoming more segmented as movement of Mexicans into East Los Angeles gained momentum. Increased residential segregation, decreasing inter-ethnic contact, and concerted efforts on the part of local officials to rid Los Angeles of its Mexican population resulted in Chicanos becoming an "invisible minority."

Nothing epitomized the redefined status of Mexicans in Los Angeles better than the movement to restore the Los Angeles Plaza area, including Olvera Street and the city's oldest standing structure, Avila House. In 1926, Christine Sterling, a San Francisco–born woman of English descent, approached Harry Chandler, publisher of the Los Angeles *Times,* with the idea of rejuvenating the site of the city's founding. Two years later when the Avila House was scheduled for demolition, Sterling had gathered enough support to bring her "Plaza Beautiful" campaign to fruition. In addition to raising $30,000 for the restoration, a much larger program for the incorporation of the Plaza—involving some of the leading citizens of the city—was set in motion over the next few years. These citizens hoped to make the Plaza a major tourist attraction, featuring Olvera Street as a "picturesque Mexican market place."[59]

Ironically, restoration was completed at the very moment when thousands of Mexicans were being prodded to repatriate. The lesson was clear: Mexicans were to be assigned a place in the mythic past of Los Angeles—one that could be relegated to a quaint section of a city destined to delight tourists and antiquarians. Real Mexicans were out of sight and increasingly out of mind. Physically farther away from the center of power, Mexican immigrants remained close enough to provide the cheap labor essential to industry and agriculture. Repatriation removed many, but others continued their struggle for survival east of the river. Their children, however, made it much harder for the Anglo American community to designate Mexicans as relics of the past. These young people, born and educated in the United States, demanded to be included in the city's future as Mexican Americans.

CHAPTER 11

Forging a New
Politics of Opposition

On October 13, 1933, workers in the garment trade filled the Embassy Auditorium in downtown Los Angeles. They gathered to discuss the future of an International Ladies' Garment Workers' Union (ILGWU) strike that had been launched by Local 96 the day before. Local 96 itself was less than one month old. Spirits ran high and strike leaders felt optimistic. But what began as a mass meeting demonstrating unity quickly degenerated into a disappointing exposé of the racial and gender tensions which permeated the labor movement in Los Angeles. The 1600 male, largely Anglo and Jewish cloakmakers, represented by Local 65, announced that they had reached an agreement with sweatshop owners earlier in the fall which included union recognition and a closed shop. Consequently, they would not strike. The 1500 female dressmakers of Local 96, consisting mostly of Mexican and a smattering of Anglo, Italian, and Jewish workers, took this position to be a repudiation of their massive organizing efforts of the last three weeks—efforts that had resulted in the registration of over one thousand new members.[1]

This campaign had been led by Rose Pesotta, a recently arrived ILGWU organizer sent from New York. Pesotta was a Russian Jewish immigrant dressmaker herself. Local ILGWU leadership, exclusively male and white, had considered the Mexican and Mexican American women workers as largely unorganizable. But the women proved them wrong. Using bilingual appeals in handbills and over the radio for the first time, they had constructed a local which directly challenged not only the dressmaking industry but also the ILGWU local leadership. With union members already being fired from their jobs and aggressive pickets surrounding sweatshops throughout downtown Los Angeles, Local 96 decided to continue their strike—the 1933 Dressmakers' Strike.[2]

This strike not only set the stage for union organizing in Los Angeles during the 1930s but put in motion two decades of organizing Mexican workers in the city under American unions. The 1933 Dressmakers' Strike and the agitations which followed also expose the complex intersection of racial and gender hierarchies in industrial unionism of the period, and highlight the role of exemplary women organizers who, while often fighting sexism within the unions themselves, managed to lead successful efforts to organize Mexican women to improve their lives and the lives of their community. The coming of age of American-born Chicanas and Chicanos who filled fledgling industrial unions that clamored for their rights as workers and as citizens of the United States made possible this new organizational development and reinforced the changing cultural orientations of the Mexican-origin community of Los Angeles. In the aftermath of the repatriation exodus, the immigrant generation that remained saw their political orientation increasingly turn toward issues of economic security in the United States, with their advocates primarily coming from the ranks of American labor unions and the liberal and left political establishment, including American-born citizens of Mexican descent.

During the 1920s, the constant influx of Mexican newcomers and the attraction Los Angeles held for elite refugees had sustained a tightly organized, Mexico-focused community leadership centered around the consulate office. But the Depression and the demographic effects of repatriation greatly diminished the elite's influence over community life. After aiding in the repatriation efforts, the consulate office's importance in grass roots politics greatly decreased after 1935. Historian Francisco Balderrama attributes this decline to the transfer of activist consul Ricardo Hill and a directive from the Mexican government in 1936 warning consular agents to cease their involvement in sociopolitical activities.[3]

Yet even more fundamental to the change in political direction was the rapid transformation of the population from a community dominated by the Mexican-born to one which centered around the American-born. The maturation of Mexican immigrant families in the United States, the cutoff of new Mexican immigration with the onset of the Depression, and the repatriation of thousands of Mexican nationals all contributed to the tilt toward second-generation dominance in the 1930s and 1940s. In the city of Los Angeles, the actual number of Mexican-born residents fell from 56,304 in 1930 to 38,040 in 1940, while the percentage of ethnically Mexican, but American-born members of the overall Chicano community skyrocketed from 45 percent to 65 percent over the same period.[4]

Although most Mexican Americans remained connected to the immigrant experience through family ties and through life in the barrios of East Los Angeles, they now increasingly assumed the mantle of leadership in the community. Even the immigrants from Mexico who got involved in the politics of the late 1930s and early 1940s increasingly came from the ranks of those who had migrated as children with their parents, and who connected intellectually and experientially with the perspectives of the American-born. Coupled with the reorientation brought about by repatriation and the lack of continued migration from Mexico in the decade, the 1930s witnessed the emergence of a leadership decidedly focused on events north of the border.

In addition, the steady movement of Chicanos into East Los Angeles reduced the influence of Plaza-area merchants and professionals who had grouped around the consul to define the local agenda for Mexicans. Unlike San Antonio, where a middle-class Mexican American leadership had begun to develop even before the Depression, Los Angeles did not contain a large or potent enough Mexican American middle class to replace the Mexico-focused leadership.[5] That this American-born generation in Los Angeles came to maturity in the midst of the Great Depression also curtailed possibilities of economic advancement. Thus, the Mexican American community in the southland developed its ideology of dual identity primarily among young people with working-class roots, if not working-class economic positions. It was this generation, one that came of age during the 1930s, that was destined to redefine being Mexican in Los Angeles.

These young people demonstrated a greater willingness to participate in American political institutions. The old orientation toward Mexico no longer engaged their passions. And, increasingly, the immigrant community joined with the American-born generation in economic and political activity that committed them to life in the United States. Moreover, the political activism of this "new" Mexican American community was decidedly identified with the aspirations of skilled and semi-skilled blue-collar laborers. As the most rooted members in East Los Angeles—men and women who had invested in small homes and sought education for their children—they led the way in shaping a Mexican American activism within their neighborhoods and larger community.

Manifestations of this new orientation surfaced most forcefully in the upsurge in labor union activity among Chicanos during the 1930s. Though agricultural unions organized by and for Mexican immigrants had been aided by the Mexican consulate and remained active throughout the decade, what is most striking about the growth of unionization among Chicanos during the 1930s is their participation in *American* labor organizations. Luis Arroyo, for example, traced the development of Mexican activism within the American Federation of Labor (AFL), the Committee for Industrial Organization (CIO), and independent locals in the Los Angeles furniture industry after 1933. Douglas Monroy and Vicki Ruiz have shown that CIO unions, in particular, promoted ethnic leadership by bringing large numbers of Mexican industrial workers into contact with the affiliated locals. Participation in CIO locals often encouraged Mexicans to naturalize for job protection and increased political strength.[6] Rather than detail the particularities of each of the many strikes and union activities of the 1930s, this chapter will examine the role of this new level of activity among Mexican workers in American unions in reshaping ethnic identity and political orientation toward participation in the American political and social arena.

The organization of Mexicans under American unions in southern California, of course, was not new. Since at least the turn of the century, mostly male Mexican workers had organized strikes and pickets in coordinated campaigns with the AFL to improve wage scales and working conditions.[7] Previous to World War I, the AFL had displayed little hostility toward the notion of organizing Mexican workers, particularly when compared with their antagonism toward Asian workers in the same period. In two southern California strikes in 1903, the Oxnard sugar beet strike and the Los Angeles Pacific Electric Railway strike, AFL organizers appeared ready to give assistance to primarily Mexican work forces engaged in labor agitation with their employers. Historian Luis Arroyo credits the small numbers of Mexican industrial workers, the positive impact of Mexican radicals on labor's perception of their fellow countrymen, and the eagerness of Mexicans to join organized labor as contributing factors to this AFL readiness.[8]

Despite the consistent, albeit limited, AFL activity among Mexican workers, radical activists were a more visible presence among Mexican workers before the 1920s. The Partido Liberal Mexicano (PLM) had been organized under anarcho-syndicalist principles in Mexico, but its leaders fled to the United States in 1904. From various locales north of the border, the Flores Magón brothers—Ricardo, Enrique, and Jesús—led efforts to overturn the dictatorship of Porfirio Díaz and later Mexican officials. The PLM worked closely with the Industrial Workers of the World (IWW) throughout the American Southwest. In Los Angeles, as many as 400 Mexicans were members of the IWW, providing the backbone of that organization in the city.[9] This PLM-IWW connection would prove to be the most important link between Mexican and American labor radicals in the first two decades of the twentieth century.

The strength of the IWW among Mexican workers in Los Angeles and throughout the Southwest can be credited to a variety of factors. First, the IWW itself was opposed to any sort of racist policies, and thereby was eager to compare itself with the exclusionary AFL. This rivalry with the AFL also involved a philosophical difference over the nature of labor organizing. While the AFL stressed craft unionization, thereby making it less likely to undertake widespread organizing among Mexican laborers, the IWW focused on industrial unions which would stretch among all workers, no matter the skill level, at a particular work site. Also, given its roots among itinerant westering miners, IWW leaders often romanticized migratory labor, which Mexicans had come to typify by World War I. Finally, various Wobblies played highly visible roles in the Mexican Revolution, enlisting with combatants in northern Mexico in order to participate in what was considered at the time to be a crucial event in the promotion of social revolution around the world.[10]

However, rather than promote prolonged political activity in the United States among Mexicans, the IWW-PLM connection kept labor activists focused on events in Mexico. The PLM clearly used the United

States as a launching pad for activities designed to overturn regimes in Mexico, and although their activity brought about arrests and incarcerations in the U.S., their primary focus was insurrection south of the border. The IWW, for its part, saw revolution in Mexico as a positive sign of the possibility for political transformation and evidence of Mexican laborers' radical potential, but organizers rarely took Mexican workers in the United States as their prime focus. Since both radical organizations usually targeted the current Mexican government for overthrow, the Mexican consulate kept a close watch on their activities and often aggressively attempted to intervene and aid local law enforcement agencies in the apprehension of radical leaders.[11]

With increased migration of Mexican workers to the Southwest after World War I and the stabilization of the Mexican government through an "institutionalized revolution," the cross-national dynamic of labor politics changed considerably in the 1920s. On the international level, AFL leader Samuel Gompers established close connections with Mexican labor leaders in La Confederación Regional Obrera Mexicana (CROM), especially leader Luis Morones, who increasingly played a central role in the direction of the Mexican government during the 1920s. In the United States, Gompers advocated immigration restriction from Mexico before his death in 1924, while local AFL officials in the mid-1920s half-heartedly attempted to organize Mexican workers in segregated locals. These efforts usually took place in industries employing almost solely Mexican laborers, or when the inclusion of Mexicans aided whites in organizing unions in the open-shop city. Attempts were made to establish segregated locals among the painters, the hod carriers, and in the building trades during the mid-1920s. Most organizing campaigns were conducted in accordance with the Mexican consulate, who sought to protect the rights of Mexican nationals working temporarily in the United States.[12]

The Mexican government, for its part, stepped up efforts to encourage unionism among Mexican workers in Los Angeles during the late 1920s, but cast that effort securely under its own control. Partly to counteract the continued presence of radical organizers in the city, Los Angeles' Mexican consulate Alfonso Pesqueria played a central role in forming La Confederación de Uniónes Obreras Mexicanas (CUOM), the Federation of Mexican Workers Unions, in 1928, patterned after CROM in Mexico. Pesqueria encouraged the Federation of Mexican societies, consisting mostly of *mutualistas* and social organizations, to support trade unionism and place disparate groups under a larger umbrella controlled from the consulate offices. CUOM particularly targeted agricultural workers, who were largely first-generation immigrants and seen as more susceptible to radical influences, as in Mexico proper. The larger agenda of the Confederación de Sociedades Mexicanas, like that of the consulate, rested on support for immigration restriction and repatriation, and the development of Mexican schools for American-born Chicanos.

Trade unionism in this context simply became another tool used by the Mexican government for the protection of Mexican nationals living outside of their country and the encouragement of their eventual return to the homeland.[13]

Increasingly, however, segments of the Chicano population outside of the limited framework of CUOM entered industrial labor in Los Angeles and became active in labor union activity. Mexican American women often took the lead in this new union activism of the 1930s. In both garment factories and food processing plants, Chicanas often constituted the majority within this labor force. During the Depression, when men in skilled trades lost their jobs, many older women worked side by side with their daughters. As the Depression wore on, more Mexican women of all ages were drawn into the labor market. Given the severity of male unemployment, the welfare of entire Mexican families was often dependent on female wage labor. For example, in the ladies' apparel industry, 212 establishments employed 6,024 women and minors in 1930. By 1931, this figure had grown to 221 plants employing 6,302 women and children, despite the deepening of the Depression.[14]

According to research by economist Paul Taylor and historian Vicki Ruiz, often the first women to enter the industrial labor force were teenage daughters of Mexican immigrants. Ruiz estimates that from 1930 to 1950, for example, approximately 70 percent of Mexican canning and packing workers were single, largely daughters living at home or striking out on their own. Many started working when their fathers were unemployed or ill, and continued on to improve the family's living conditions or purchase a radio or new furniture. After they had situated themselves in their workplaces and additional income was needed as the men in the family found it more difficult to find employment during the Depression, these daughters would encourage their Mexican-born mothers to join them in the factory.[15]

Though the signs of a new industrial work force were evident, it was only the most resourceful of labor organizers who responded sensitively to this newfound constituency. Rose Pesotta of the ILGWU was one of the first to see potential unionists in this young group. Sent by ILGWU national president David Dubinsky from New York in September 1933 on her first assignment, Pesotta arrived in Los Angeles seeking to take advantage of the recent elevation of Franklin Roosevelt to the presidency and the passage of the National Industrial Recovery Act. The conditions Pesotta found in Los Angeles were among the worst she had encountered in the country. In an industry where 75 percent of the dressmakers were Mexicanas, 40 percent made less than $5 per week, despite the fact that California's minimum wage was $16. The hours were long, and home work was widespread.[16]

The 1933 strike was the culmination of an intensive organizing campaign launched by Pesotta despite the fact that local union leaders such as Paul Berg, secretary of the dressmakers' local, felt Latina workers could not be organized.[17] Pesotta was able to sign up hundreds of Chicana workers by using appeals and announcing meetings in *The Organizer,* the local's bulletin, printed in Spanish and English. The official strike declaration itself would be printed bilingually. A Spanish radio hour, broadcasting twice a week on Wednesday and Saturday, was enlisted to disseminate information on the union and the garment industry through a five-minute talk in Spanish, then translated into English. This recording occurred live in a downtown theatre with an audience of hundreds. When the local theatre owner who had been sending the union's message over the radio waves was threatened, Mexican workers suggested buying time on a station in Tijuana, which would be beyond the reach of American officials.[18] Taking advantage of Mexican workers' bilingual abilities through the radio was an innovation in labor organization of Pesotta's that would become commonplace by the end of the decade.

By all accounts, Mexican women were among the most active participants on the picket lines. They reportedly battled with workers brought in to replace them, and several ended up spending nights in jail for "disturbing the peace." According to Pesotta, singing on the picket lines was a regular occurrence. The favorite strike song was apparently "Solidarity Forever," sung to the tune of the "Battle Hymn of the Republic." Ironically, this "all-American" labor tune was written by a Wobbly activist, Ralph H. Chaplin, in 1915, after experiencing revolutionary Mexico first hand.[19] The spectacle of Mexican immigrant women and Mexican Americans singing these fighting words in unison on the streets of Los Angeles must have been a sight; Pesotta claimed it fascinated crowds of shoppers, who stopped to gawk:

When the Union's inspiration through the workers' blood shall run,
There can be no power greater anywhere beneath the sun,
Yet what force on earth is weaker than the feeble strength of one,
But the union makes us strong.
　　Solidarity Forever!
　　Solidarity Forever!
　　Solidarity Forever!
For the Union makes us strong.[20]

Pesotta was able to organize Mexican women because she appealed to them directly as a fellow dressmaker, an immigrant American, and, probably most important, as a woman. Because the strike lasted past Halloween, Pesotta went as far as arranging a children's party at the union hall with youngsters in costume and performances of native Mexican dances—a strategy until then inconceivable to any of the ILGWU's male organizers. The children were even marched through the streets to where their mothers were picketing, garnering additional favorable publicity for the union.[21] Pesotta wrote back to national leader Dubinsky to explain her success: "WE got them because we are the only AMERICANOS who take them into our organization as equals. They might become the backbone of the union on the west coast."[22]

The Dressmakers' Strike that resulted from the October 1933 meeting at the Embassy lasted four weeks and affected 2000 female workers in 80 factories. The arbitrated settlement that ended the conflict left much to be desired. Although employers theoretically agreed to a thirty-five-hour work week and wages which conformed to the National Recovery Administration's Dress Codes, the inability to gain union recognition and enforcement provisions for the contract left most strikers out of work because of the coming of the slack season.[23] More important than the settlement itself, the strike proved to more skeptical union officials that Latina workers could be stalwart union activists. Moreover, the dressmakers provided the nascent Mexican American community with their first significant experience under an American union. The point is not that the community or the women themselves were "taught" political activism, or even labor activism, by the ILGWU; rather, the experience of these garment workers was the first for Chicanos under the labor politics of the American New Deal and set the stage for the growth of a Mexican American ethnic identity forged in this struggle for workers' rights.

Clearly, Pesotta was unique among American labor leaders in being able to reach Mexican women workers because of her ability to treat them as equals and her willingness to learn quickly about culture and traditions which had been foreign to her before arriving in Los Angeles. But also part of Pesotta's appeal was the fact that the Mexican women themselves seemed ready to engage more directly with an American culture reflected in the American labor movement:

> Some of the women quietly admitted to me that they, too, would like to be Americans. In Mexico, they said, women still had no freedom; a married woman could not vote nor hold a job without her husband's consent, and the father was still the supreme ruler over unmarried daughters until they reached the age of 30. The poor were always overburdened with work, entire families toiling on the plantations owned by the rich.[24]

Union participation, therefore, became recognized by women as a legitimate form of incorporation into the American political scene, because its ultimate goal was the social betterment of the family's economic condition. It also provided an outlet for Mexican women to learn English, regularly interact with non-Mexicans, and voice political protest. As long as such activity remained tied to the context of the family economy, it remained culturally acceptable.

Moreover, these new Chicana workers were among the first members of their community to express a civil rights agenda as American citizens, largely through their participation in the labor movement of the 1930s and 1940s. Along with the male Mexican American labor leaders emerging in Los Angeles in this same period, Chicana laborers combined a tradition of resistance emanating from the Mexican Revolution with a newfound belief in the rights of citizenship bestowed on them by virtue of their birth in the United States. This kind of 1930s "Americanism" was present among other ethnic labor activists of the period and took political form in its strong support for Franklin Delano Roosevelt and the Democratic party's labor agenda. Unlike the middle-class Mexican expatriate vision of "Mexicanos de afuera," which looked to a return to Mexico as an ultimate goal, their mind-set was rooted in a belief in the socio-economic advancement of Chicano families in the United States through labor and political organization.[25] Clearly a new generation of Mexican American workers was coming of age that was willing to press for social improvements, workers willing to respond to organizers emanating from American labor organizations, whether radical or moderate.

American union leadership typically lacked continuity, however, and this led to tensions between Mexican workers and the local ILGWU leadership. When Rose Pesotta left Los Angeles for other situations demanding her organizational abilities after the 1933 strike, she emphasized to national leader David Dubinsky the importance of establishing a Spanish-speaking local there. Although Dubinsky had promised a separate charter for a Mexican local when 600 had enrolled, he did not make good on it when membership passed this figure in 1936. In addition, the male leadership left in charge of Local 96 failed to follow up on Pesotta's earlier efforts to nurture Mexican workers to assume leadership positions.[26] Despite this turn of events, the 1933 Dressmakers' Strike left a lasting legacy of Mexican American participation in American unions.

By contrast, labor activity by Mexican agricultural workers in southern California earlier in 1933 demonstrated the conflict between an organization spawned by American radicals and that under the rubric of the Mexican consulate and a Mexico oriented perspective. Largely ignored by the AFL, the 200,000 agricultural workers in the state, three-fourths of them Mexican, provided fertile organizing territory for more radical labor activists. Already organized around mutual aid organizations or *mutualistas,* Mexican workers bearing the brunt of the depression in California agriculture welcomed both Anglo and Mexican Communist organizers from the Cannery and Agricultural Workers' Industrial Union (C&AWIU). Since the late 1920s, a small cadre of dedicated leftists had been busy establishing contacts with agricultural and cannery workers throughout the state, developing relationships with Mexican, Spanish, Filipino, Japanese, and Anglo workers by offering their expertise, experience, and limited resources. Even with passage of the National Industry Recovery Act, which acknowledged workers' right to bargain collectively, the Communist party continued to provide virtually the only American labor organizers throughout the decade, since agricultural workers were excluded from the protection of Section 7a.[27]

The 1933 El Monte Berry Strike was only the biggest of many agricultural strikes which hit California that year; at least 37 labor conflicts arose between April and December in the fields. El Monte itself was a small agricultural community twenty miles east of downtown Los Angeles with a mixed population of Anglos, Japanese, and Mexicans. Like many of the smaller communities within Los Angeles and Orange counties, El Monte relied on its proximity to urban Los Angeles both for selling its agricultural produce and for producing a ready supply of farm laborers. A mixed work force dominated by Mexican laborers picked approximately 700 acres of berries, which were in turn managed largely by Japanese growers on land almost exclusively owned by Anglos. After being rebuffed in their request for higher wages, workers voted to strike on June 1, 1933, and the action quickly spread throughout the agricultural communities dotting the San Gabriel Valley. Because the berries had to be picked within days of ripening or be lost, the growers quickly attempted to settle. When the strike committee rebuffed them, they immediately brought in scab workers.[28]

By the end of this first week of the strike, the chairman of the strike committee, Armando Flores, requested that the Mexican consul of Los Angeles, Alejandro Martínez, be consulted. Upon arriving in El Monte, Martínez denounced the C&AWIU organizers as "reds," had the leaders arrested, and was able to wrest leadership of the strike away from the organization. In its place, the consul formed a new union along the lines outlined by the earlier CUOM confederation. Vice consul Ricardo Hill and the Anglo attorney for the consulate, David C. Marcus, joined with strike leader Flores to appeal to the general public for support. Both President Roosevelt and ex-president of Mexico Plutarco Calles were solicited. By the third week of the strike, financial contributions arrived from Mexican labor unions, as the strike spread to Santa Monica, Culver City, and Orange County. What had begun as a limited labor walkout quickly became an international incident, as labor unions throughout Mexico began to consider a widespread boycott of products from Japan to protest the intransigence of the Japanese growers. By the second week of July, however, an agreement had been reached between the strike committee and the growers, with the Japanese and Mexican consulates acting as the main intermediaries.[29]

The active role of the Mexican consulate in the 1933 El Monte Berry Strike indicates the continued sway of the office in influencing Mexican nationals to respond to purely nationalist messages. Although CUOM's membership and activities had dwindled with the onset of the depression, many of its former leaders joined in the aftermath of the El Monte Strike with the new Los Angeles consulate Alejandro Martínez and vice consul Ricardo Hill to form a new organization of Mexican farmworkers, La Confederación de Uniones de Campesinos y Obreros Mexicanos (CUCOM). This confederation was clearly organized as a response to the successful organizing campaign of C&AWIU in the fields. CUCOM gave Mexican agricultural laborers an alternative organization which was less radical in its demands and more strictly nationalistic in its goals.[30]

The conflict over union leadership in the 1933 El Monte Berry Strike clearly put Communist organizers at odds with the Mexican consulate. The Communist party did not hesitate to publish headlines denouncing the Mexican consul as a strikebreaker and claiming that Martínez and vice consul Hill had actually sold out the workers they claimed to represent. What particularly outraged Communist organizers were the efforts of the Mexican and Japanese consulates to break the racial solidarity of the union by reorganizing the workers into nationalist camps and negotiating separately.[31] Three years later, during another agricultural strike, this time in Santa Ana, the new consul Ricardo Hill played a central role in pushing an agreement with Orange County citrus growers. The CUCOM, along with several Communist organizers in its ranks, denounced the agreement and made it clear that "the CUCOM must further state for the information of its loyal members, that we will never sign any agreement that ignores the other nationalities and the success of the working class depends on all workers working together toward the common aim."[32]

This transnational position did not prevent the Communist party from utilizing facets of ethnicity when needed to organize workers. On December 15, 1933, for example, the West Coast contingent of the Party launched the first issue of the *Lucha Obrera,* a Spanish-language Communist newspaper intended "as a great step forward towards organizing them [Spanish-speaking workers] together with the other workers for the fight to establish human working conditions, against imperialism, discrimination, and for the overthrow of the capitalist system."[33] What distinguished Communist organizers, however, was their determination to transform the radical tradition Mexican laborers derived from the Mexican Revolution into a new "American" form of radicalism. As a budding leftist organizer in the 1930s, Dorothy Healey remembered speaking about Communism to a group of twenty-four Mexican agricultural workers packed into a small home in the Imperial Valley. Their response was indicative of a longstanding anarcho-syndicalist tradition emanating from the Mexican Revolution: "Of course we're for the revolution. When the barricades are ready, we'll be there with you, but don't bother us with meetings all the time. We know what to do, we know who the enemy is!"[34]

The strike actions also exposed the ideological and cultural divisions evident among the leadership of the Mexican workers themselves. Armando Flores, the chairman of the strike committee, opposed the influence of the C&AWIU organizers and called in the Mexican consulate to shift the focus along more nationalist lines. In contrast, William Velarde, the vice president of the union, supported the alliance with C&AWIU and consequently was accused of being a communist by vice consul Hill. Velarde countered by claiming Hill and Flores had sold out to the growers, but continued to work with the more radical members of CUCOM. Later in the decade, during another strike in Riverside, Velarde observed that "a large proportion of the strikers are American citizens and [vice consul] Hill and [honorary commission leader Lucas] Lucio have no authority whatsover to claims to negotiate for them." [35] Increasingly it became more difficult for the Mexican consulate to take such an active role in a labor dispute uncontested by other actors from the Mexican American community.

In the 1933 Dressmakers' Strike, for example, a small group of female strikers did write to consul Martínez seeking support. Marcela Puente, Dolores Nuño, Dolores de Veytia, and Victoria Martínez took it upon themselves to write directly for assistance:

> Señor Consul, we believe that we have as much or more right to ask for your help as the [El Monte berry] pickers, that the consulate aided in immeasurable ways when they went on strike to demand better wages from their Japanese overseers, and we ask you to give us the same assistance. We have waited for some representative of your office to come and ask us about our work conditions and the victimization we have faced from the authorities, but disgracefully our struggle has been ignored by the consulate office. [36]

Since the consulate had provided food and funds for the striking berry pickers earlier in the year, these women believed that help in "their struggle for the bread to feed their families" should also be forthcoming. They asked consul Martínez not only for moral support and material aid but also for a public denunciation of the Mexican strikebreakers who were crossing the ILGWU picket lines. As loyal unionists, "they were ashamed that the majority of strikebreakers were Mexicans and that these women were therefore their worst enemies." Especially when *La Opinión* published an editorial in full support of the garment strikers, these women clearly came to expect that the Mexican consulate would lend every resource available to it to help in their time of need. [37]

Quickly, however, the complexity of the situation became clear. Ramona González, head of the Latino department of the ILGWU in Los Angeles, wrote the consul hurriedly after reading letters by individual strikers in the newspaper. González wrote guardedly that the garment workers' struggle "found elements from all nationalities, like Mexicans, Russians, Americans, Jews, Italians, Poles, etc., etc., being under the direction of foreign workers groups affiliated with the AMERICAN FEDERATION OF LABOR [emphasis in original] . . . ." She went on to emphasize that even though 50 percent of the striking work force was Mexican, the ILGWU itself had *not* yet asked for intervention on the part of any consulate "because the union is a conglomeration of various nationalities." Although González made it clear that any Mexican who found it necessary to call on the consul for aid and protection was certainly welcome to do so, a formal request from the ILGWU at this point "would only interrupt the negotiations occurring between the appropriate [ILGWU] directors and the manufacturers." [38]

For his part, consul Martínez probably breathed a sigh of relief, having no desire to respond directly to a request for involvement. In addition to meeting the needs of Mexican workers organized under an American union, Martínez would have been faced with a decision to support women engaged in an often violent struggle against fellow Mexicans acting as strikebreakers, with the local police defending the interests of the manufacturers. He wrote to *La Opinión* that "although always attentive to the problems of his countrymen, he was confident that the conditions affecting the Mexican garment workers would improve with the involvement of the National Recovery Act and the government of the United States." Now that the ILGWU had taken an interest in these workers, he felt satisfied playing the role of "sounding board."[39]

Involvement by Mexican workers under the rubric of the AFL (and later the Congress of Industral Organizations (CIO)) altered the dynamics in the delicate relationship between labor organization and national identity. During the 1930s, American labor demands were often centered around fufilling the promises of FDR's New Deal, especially the National Industrial Recovery Act policies on wages and hours and the acknowledgment of workers' right to bargain collectively. This position allowed labor leaders of all ideological persuasions to organize with national support for unionization. But the newfound patriotic rhetoric of labor organization also redefined the meaning of "Americanism" for ethnic workers, including Mexicans. Their struggle was increasingly seen as not only one on behalf of a particular ethnic nationality but also the fufillment of the American promise of equity and cooperation. Racial and ethnic justice and the American dream were thereby intricately interwoven in the rhetoric of labor unions of the 1930s.[40]

American labor organizers such as Rose Pesotta, although ranging in ideological positions from communist to reformist, were crucial in redefining the oppositional culture of Mexican organized workers during the 1930s. Building on histories of Mexican cooperative organizations and memories of radicalism on each side of the border, these organizers placed this history within the context of current labor struggles and encouraged workers to see themselves living out an important American tradition of radicalism. In this manner, the same memories of the Mexican Revolution that the Mexican consulate might try to appropriate for purposes of nationalist activity could be commandeered to represent the highest form of "Americanism." Increasingly, as workers participated in activism that placed them side-by-side with fellow workers of other nationalities, a reformulated notion of radicalism bonded individuals of different ethnicities together, creating new definitions of American political activity.[41]

What Pesotta had started early in the decade was carried on by others later. Throughout the 1930s and 1940s, different women and men would rise up to champion the needs of workers in the Mexican American community, dedicating their lives to helping laborers to speak in a collective voice. Increasingly these leaders would be Latino. As the Depression wore on, other industries besides garment work and agriculture would become battlegrounds over the right of Mexican workers to organize and bargain collectively. New unions would challenge the right of the conservative AFL to speak for all American workers. In Los Angeles, Mexican female and male workers were at the forefront of establishing locals in the newly formed CIO.

In 1935 in the aftermath of increased labor militancy throughout the nation, AFL leaders were confronted at their annual convention with an angry challenge by unionists who felt the AFL had consistently followed, and at times forestalled, direct action by the rank and file. John L. Lewis of the United Mine Workers led the call to "organize the unorganized" with a strategy of industrial unionism which had characterized the growing independent local actions. Rather than rely on associations based on craft skills, Lewis called for activities which would put all workers at a labor site, no matter their particular function, within the same labor organization. Delegates, fearing pressure from "the rabble," voted down this crucial shift in approach. Following the convention, Lewis, along with Sidney Hillman of the Almagamated Clothing Workers and David Dubinsky of the ILGWU, formed the CIO as an opposition movement inside the AFL. Within a year, this labor affiliate would be thrown out of the AFL and the CIO would become an independent labor organization which directly challenged the hegemony of the AFL for the next two decades.[42]

One of the major Los Angeles industries to experience this tumultuous transition to industrial unionism was furniture manufacturing. As the city's population exploded after World War I, local businesses cornered the market for furniture for the multitude of residences built to meet the housing needs of the expanding population in the western states. Furniture manufacturers took advantage of the largely non-unionized, low-paid work force that increasingly was stratified along racial lines. Mexican male workers, and to a lesser extent African Americans, were restricted to the most dangerous, low-paying jobs in the industry, even when they acquired experience that would have allowed them to move up to more skilled positions. Future organizer Frank López, for example, recalled being placed at the end of a conveyor belt full of coiled springs in his first job in 1933 at the Nachman Spring Filled Corporation. Since this was the worst position in the piece-work occupation of sorting springs, López earned only $4.17 for 67 hours of work in his first week on the job. Even with more experience, he was able to increase his wages only from $6 to $8 a week.[43]

The onset of the Great Depression exacerbated racial tensions among the stratified work force, but also created the conditions necessary for movement away from the limited craft unionism that had marked the industry. Communist organizers took advantage of the more positive climate created by the NIRA to challenge the AFL's hold on labor organization in the industry, which had previously been limited to skilled white workers. The creation of the Communist-led Furniture Workers Industrial Union Local 10 in Los Angeles in September 1933 pushed the AFL to engage in industrial unionism, if only halfheartedly. In May 1934, when the Sterling Furniture Company announced a 15 percent pay cut for the mostly Mexican workers who put in the springs for the upholsterers, the entire work force walked off the job. Shocked at this newfound solidarity across race and skill, the employer asked the largely white upholsterers: "What are you fellows fighting for those Mexicans and unskilled workers for? We're not bothering your wages." This first show of solidarity helped rescind the pay cuts and served to crystallize sentiment among furniture workers in favor of industrial unionism and racial cooperation.[44]

It was the development of the Independent Furniture Workers Local 1 in 1933 and 1934, however, which cleared the path for the decisive turn to industrial unionism in the furniture business. Positioning itself between the conservative AFL and radical Communist locals, Local 1, organized by what historian Doug Monroy calls "a motley but seasoned core of unionists," courted both Mexican and white workers in order to strengthen the bargaining position of both. After being incorporated as an AFL affiliate (Local 1561 of the Carpenters Union) in March 1935, it led the first industry-wide strike to hit Los Angeles during the Depression decade. Among the many strikers who grew in experience and assumed central leadership positions during the 1935–36 Furniture Strike were several Mexican Americans. Vice president of Local 1561 Frank López, for example, led a highly successful boycott of non-union furniture sold by retailers such as the May Company and Bullock's.[45]

In December 1937, the now independent CIO chartered the United Furniture Workers of America (UFWA), an act which decisively put workers involved in furniture manufacturing into the ranks of industrial unionists. In Los Angeles, workers affiliated with AFL locals overwhelmingly went over to the CIO in January 1938. Among the leaders advocating this switch was Manuel García, president of Upholsters Union Local 15 of the AFL, yet sympathetic to the strategy of the CIO. Mexican American labor activists also led the movement of Carpenters Local 1561 to the CIO. Frank López, Julius Davila, and Jack Estrada had been elected to important leadership positions in the local after their rise to prominence in the 1935 strike. The transformation of the local's leadership to a younger, more ethnically representative group of individuals in the 1937 elections made possible a wholesale exodus from the AFL. In February 1938, despite a tense confrontation with AFL officials attempting to hold on to the local's office, many members of Local 1561 also joined the CIO-affiliated UFWA Local 576. After a year of intense red-baiting by the AFL, a unity meeting between all locals in April 1939 led to a period of cautious coexistence in the furniture industry. Tellingly at that meeting, it was Manuel García Jiménez, the first president of

UFWA Local 576, who spoke eloquently about the importance of the CIO for furniture workers and advocated unity between all factions.[46]

The appearance of active, vocal Mexican American leaders with CIO-affiliated union locals in Los Angeles was not limited to the furniture industry. Tony Rios, who would later lead the Community Service Organization (CSO) of East Los Angeles, began his activist career as president of the Utility Steel Lodge of the Steel Workers Organizing Committee (SWOC) before World War II. With Mexicans constituting the majority of workers in several foundries, at least two other SWOC lodges, Continental Can and Bethlehem Steel, elected Mexican presidents. Rios immediately launched eighteen grievances on behalf of Mexican workers unfairly held back in lowly positions, some for up to seventeen years.[47]

Similarly, the International Longshoremen's and Warehousemen's Union (ILWU) launched the career of Bert Corona, who would become a founder of several Mexican American political groups from the 1950s to the 1970s. Corona was one of a collection of volunteer Chicano organizers who gathered together employees at drug warehouses under Local 1-26 of the ILWU in 1937, nearly doubling its membership. This achievement among a diversified work force, half of which was Chicano, a quarter Russian Molokan, and the rest African American, Anglo, and Jewish, eventually led to the election of Corona as president of the local in 1941. Before he was inducted into the army in 1943, Corona aggressively signed twenty-six contracts as local president.[48]

Vicki Ruiz's chronicle of the formation of the CIO's United Cannery, Agricultural, Packing, and Allied Workers of America (UCAPAWA), Local 75 in Los Angeles, describes the sense of empowerment Mexican women operatives gained from their participation in shaping cannery life. Working side by side with relatives as well as women from other cultures, these Mexican female employees of the California Sanitary Canning Company (or Cal San) constructed a work culture based on kinship networks, interethnic cooperation, and an "us against them" mentality toward the management. Cal San workers experienced poor working conditions and the constant threat of sexual harassment, and their cooperation in a predominantly female work environment made them ripe for labor organizing.[49]

Enter Dorothy Ray Healey, a twenty-four-year-old Communist working in the Popular Front period as national vice president of UCAPAWA, one of the CIO's most aggressive, decentralized unions, with an expressed interest in recruiting women and minority members. Healey's open, spirited style, along with her experience working with Mexican laborers, encouraged some of the Cal San workers to assist in her organizing efforts. Employee Julia Luna Mount remembered: "Enthusiastic people like myself would take the literature and bring it into the plant. We would hand it to everybody to pay attention."[50] Within three weeks, 400 of 430 employees had joined UCAPAWA, whose organizing strategy exploited the established kin network and focused on meeting with entire families in workers' homes.[51]

Healey's success in initially organizing UCAPAWA was greatly due to the fact that no one pushed the theme of racial equality in this era more than American Communist party organizers. Beginning in 1935, the Party entered what is known as their "Popular Front" period, abandoning their own separate Communist unions to work with fledgling CIO unions and ethnic organizations. When Earl Browder, the 1936 presidential candidate of the American Communist party, rhetorically linked the organization to a radical American tradition of 1776 and 1861 and adopted the campaign motto "Communism is Twentieth Century Americanism," he launched a softening of the Party's hardline stance against other progressives while recasting its message into decidedly flexible terms. Healey represented a tradition of local, more independent, Communist organizers who had long been representing themselves as bridges to ethnic communities by using notions of "Americanism" emanating from an ideology promoting a transnational working-class struggle. The Popular Front period gave official sanction to this approach and made it easier for organizers such as Healey to flourish in multicultural settings.[52]

Like anti-Communist Rose Pesotta, what distinguished the best Communist organizers from the others was their ability to listen to workers' concerns and exhibit a tolerance for different opinions. In describing an attempt to attract Mexican walnut packing workers in Los Angeles, Healey remembered:

> I had been trying to organize a strike around the issue of wages and had not met with much success. What finally brought about a spontaneous walkout on the part of the workers was the fact that as they stood at the tables sorting the nuts, splinters from the legs of the tables would tear their stockings. They were just infuriated by this. It was not the kind of issue that I would have thought up to organize around. In order to be successful as an organizer you first had to acquire the ability to listen to what the workers had on their minds, and then you had to learn to articulate coherently back to them what they already felt in a disconnected or fragmented way.[53]

Although this form of organizing often strayed considerably from Communist principles, it also was a most effective response to a community still steeped in its own radical tradition. As David Roediger, in his review of Robin Kelley's book on Communists in Alabama's African American community, observes, "measuring radicalism not by its ideological purity but by its ability to interact with a received culture to generate bold class organization" is the most productive way of assessing the impact of Communist organizers.[54]

On August 31, 1939, Cal San employees launched a massive strike which crippled the plant and, after three months, was settled in the union's favor. In the course of the strike, UCAPAWA garnered food and support from East L.A. grocers and encouraged big food chains to boycott Cal San products. After this successful action, Luisa Moreno was put in charge of consolidating Local 75 of UCAPAWA in late 1940. She, like Healey, proved instrumental in continuing an UCAPAWA tradition of Mexican women taking leadership posts. During the 1940s, for example, Mexican and Mexican American women held more than 40 percent of all shop steward positions in the union. In 1941, Moreno herself was elected vice president of the national UCAPAWA, an event which solidified her position as the leading Latina labor organizer in Los Angeles and in the United States.[55]

Born in Guatemala to well-to-do parents and sent to boarding school in the United States, Moreno first emigrated as a young woman to Mexico City to work as a journalist for a Guatemalan newspaper. She moved to the United States in 1928 after marrying a Mexican artist. Unemployed and pregnant at the beginning of the Great Depression, Moreno was forced to work as a seamstress in a New York City garment factory near Spanish Harlem. Here she had contact with socialist Puerto Rican workers, an experience that radicalized her and pushed her toward professional labor activism. Eventually abandoned by her husband and left to raise an infant daughter alone, Moreno threw herself into a lifelong career as a labor organizer.[56]

Moreno's first job was with the Needle Trades Workers Industrial Union, where she stayed until the AFL called upon her to organize Italian and Cuban cigar makers in Florida. Like other Latino labor leaders, Moreno became disenchanted with the conservatism of the AFL, and quickly joined the CIO when it was founded in 1936. When she learned about Mexican workers in the Southwest, Moreno asked to be sent to work with striking San Antonio pecan shellers for the militant UCAPAWA of the CIO. Over the next decade, she proved to be a strong leftist and a skilled organizer at a time when bilingual leaders with her experience and abilities were sorely needed.[57]

As the chief organizer of UCAPAWA in Los Angeles after 1940, Moreno encouraged employees to band together to break the discriminatory hiring practices of the Cal San owners, George and Joseph Shapiro. For example, union pressure forced the Shapiros to hire blacks in early 1942. Local 75 also aided other organizing efforts at the California Walnut Growers Association and the Royal Packing plant that processed Ortega chiles. Through these efforts, Moreno made the motto of UCAPAWA—"An Injury to One Is An Injury to All"—ring true. As Dorothy Healey would recall, "a strong sense of national identity held these workers together, but did not prevent them from making common cause with others, like their Jewish and Russian fellow-workers."[58]

Moreno also became the main force behind the first national civil rights conference for Spanish-speaking peoples. Earlier in 1938, she had taken a leave of absence from UCAPAWA, using $500 of her own money to travel throughout the Southwest and organize local committees of the National Congress of Spanish Speaking Peoples—or El Congreso de Pueblos que Hablan Español. Moreno recognized very early on that workers' rights for Mexican laborers could be gained only by also working for the civil rights of Mexican women and men. As she remarked: "You could not organize workers in the face of violence and terror."[59] She envisioned a coalition of CIO unions, various Mexican American and Mexicano organizations, and liberal and left political groups working in unison through El Congreso to protect the rights of Mexican laborers.[60]

Delegates to the first congress held in Los Angeles in April 1939 represented 136 union locals and Latino organizations throughout the United States. Although a few representatives came from eastern Puerto Rican and Cuban organizations, the vast majority were centered among Mexican American associations in the Southwest. A delegate from Mexico proper represented organized labor south of the border, but most representatives were from Los Angeles. In other regions of the Southwest, Chicano union activity had been more sporadic and uneven, so the developed Latino leadership in American unions in Los Angeles provided the core of El Congreso's active membership. In fact, the first meeting of the Congress had to be moved to Los Angeles after red-baiting led to the revoking of the group's permission to meet at the University of New Mexico in Albuquerque.[61]

There is little doubt that El Congreso was part of a Communist "Popular Front" strategy to encourage ethnic minorities in the United States to join them in a fight against racial and class oppression. Although it is virtually impossible to ferret out exactly who in the organization was a Party member, Party publications such as *The People's World* took credit for the achievements of El Congreso. Moreover, it is hard to believe that El Congreso was not patterned after the National Negro Congress, organized by the Party in 1935 so that it "could take the lead in launching a nationwide coalition of black organizations concerned with eliminating racial discrimination, fighting lynching and disfranchisement, and encouraging black participation in unions."[62] Originally conceived on the campus of Howard University (reminiscent of El Congreso's attempts to meet at the University of New Mexico), the National Negro Congress drew from labor and church leaders, middle-class organizations, and independent black intellectuals and professionals, much like the wide spectrum of Latinos who attended El Congreso's first conferences.[63]

More telling evidence of this relationship is the shift in the tone of resolutions concerning fascism and the war in Europe between El Congreso's first conference in late April 1939 and its second conference in December, *after* the signing of the Nazi-Soviet pact. While the first congress passed resolutions which described racial discrimination as "one of the touchstones of fascism" and called for "efforts to block Nazi and

Italian fascist domination of the economies of Central and South America," the second congress pulled back from these condemnations and called for pacifism. Instead, it described the war in Europe as "essentially imperialistic in nature," and warned that "the interests which profit through war in the United States are already encouraging a false patriotism. . . . " The second meeting of El Congreso went so far as to call on those present to "oppose every proposal that may be made to carry the United States towards war," being careful to connect the rise of sentiments toward U.S. involvement with the Allied side in Europe with talk of an invasion of Latin America.[64]

Yet saying El Congreso was a part of the Communist party's "Popular Front" strategy does not really address what went on at these meetings nor in this organization. It is important to note that these particular resolutions which dealt explicitly with fascism came out of the "English-speaking panels" at both the early conferences. Though passed by the body as a whole, each was crafted almost exclusively by Anglo American left activists invited to participate.[65] Not only did the rest of the resolutions deal with more immediate concerns for the Latino delegates, but they point to the fact that Latino organizers of the conference had their own "Popular Front" strategy in the late 1930s. While they welcomed aid and support from all fronts, including white liberals and leftists, they defined their own direction for fighting the oppression they believed was integral to American society. By not excluding Communists from their ranks—both Anglos and Latinos—El Congreso proved to be an inclusive organization, but not one "captured" by any outside group. In fact, the leadership of El Congreso, though clearly a product of labor and left organizations, prided itself on being able to appeal to *all* Latinos, regardless of political affiliation.

For example, at only eighteen, Josefina Fierro de Bright emerged as one of El Congreso's main organizers and its executive secretary throughout most of its history. She was born in 1920 in Mexicali, Mexico, while her parents fled political persecution from the revolution of 1910. Her mother, Josefa, had supported Ricardo Flores Magón, the most radical leader of the Revolution, while her father fought for Pancho Villa. Their political disagreements led to a separation, and Josefina was brought to Los Angeles by her mother in the 1920s. The elder Josefa opened a restaurant catering to Mexican immigrant workers in downtown Los Angeles, while the family itself resided everywhere from Santa Monica to East Los Angeles. Josefina attended eight different schools. Eventually the family joined the agricultural migrant stream into the San Joaquin Valley and settled in Madera, a small town in central California.[66]

Josefina returned to Los Angeles in 1938 to attend UCLA, where she hoped to study medicine and become a doctor. She lived with an aunt who sang at a Latin nightclub, and Josefina found herself balancing studying with watching the floor show. At the club she met a young

Hollywood screenwriter, John Bright, with whom she fell in love and eventually married. The marriage ended her college career but jump-started her work as a political activist. Bright, an active member of the Screen Actors Guild, would eventually be blacklisted in the 1950s. Josefina was encouraged to begin organizing in the Mexican community, and she persuaded Hollywood friends such as Orson Welles, Anthony Quinn, and Dolores del Rio to donate time and money to her causes. Luisa Moreno recognized her work and saw in this teenager a Mexican American woman who could help establish the Spanish Speaking Congress in Los Angeles.[67]

Though not directly a labor organization, El Congreso placed union activity at the center of an organizing strategy for the Mexican-origin community. The platform of El Congreso boldly asserted that "the Trade Union Movement provides the most basic agency through which the Mexican and Spanish-speaking people become organized" and called on all Spanish-speaking people to affiliate with the local union in their industry. Fierro de Bright and other leaders of the congress consistently sponsored and spoke at meetings designed to influence Mexican American workers to unionize. Unity among workers was a central goal, as expressed in the platform: "The main problem of labor is creating greater unity and understanding between the Mexican laborer and the American laborer and having Mexican people see the necessity for fighting for the right to collective bargaining, to organize and strike." This first national Latino civil rights organization advocated many policies that Rose Pesotta had introduced into American unions, measures such as bilingual meetings, publications, and announcements.[68]

To combat discrimination against Spanish-speaking people, the congress called for programs designed to educate the general population about the contributions of Latinos in the western hemisphere and to denounce theories of racial supremacy. These included ethnic studies classes at American universities, the censure of textbooks portraying Latinos as inferior, increased teaching of Latin American history, and the creation of schools and programs to render an authentic portrayal of the historical and cultural background of the Latino in the United States. El Congreso also proposed bilingual education for Mexican American children and the training of more Mexican teachers in American public schools. Central to the battle against discrimination was a plea for the protection of the foreign born, including the cessation of deportation and discriminatory legislation aimed at non-citizens.[69]

Unlike other Mexican American civil rights groups of the period, El Congreso also sought to raise consciousness regarding gender discrimination. Some 30 percent of its membership were women, and many, such as Josefina Fierro de Bright, served in leadership positions. Its official position on women was adopted at its second California state convention:

Whereas: The Mexican woman, who for centuries has suffered oppression, has the responsibility of raising her children and of caring for the home, and even that of earning a livelihood for herself and her family, and since in this country she suffers double discrimination, as a woman and as a Mexican.

Be it resolved: That the Congress carry out a program of organization and education of the Mexican women, concerning home problems; that every Pro-Congress Club establish a Women's Committee as soon as possible; that it support and work for women's equality, so that she may receive equal wages, enjoy the same rights as men in social, economic, and civic liberties, and use her vote for the defense of the Mexican and Spanish American people, and of American democracy.[70]

Emphatically breaking with the older stance against naturalization and upholding one first staked out by American unions, El Congreso called upon Mexicans to become American citizens and vote in American elections. Congress officials were adamant in their description of the Mexican immigrant population as steady contributors to American society who had for decades been unfairly denied access to the fruits of American democracy. They blamed the low rate of naturalization on high fees and educational requirements, citing red tape and the long waiting period which discouraged potential citizens. Most important, they believed that no one should be denied citizenship on the basis of race or political views. El Congreso also urged Mexican American citizens to exercise their right to vote and to energetically engage in political action to combat discrimination. It advised them "to endorse candidates only on the basis of their sincerity and devotion to the fundamental causes of the people, and not on the basis of nationality."[71]

Given their commitment to Mexican American political and social activity, it is not surprising that El Congreso would lead the ethnic fight against fascism. By the time of the entry of the United States into World War II, Germany had violated the Nazi-Soviet pact by invading Russian territory, leading American Communists to decisively align themselves against Nazism. With Mexican Americans already volunteering for overseas duty, El Congreso quickly urged wartime unity against fascism and described the battle against Nazism as the most profound civil rights issue of the day. A call for the fourth state convention of the congress at the downtown Embassy Auditorium in May 1942 made it clear what would happen if Germany and Japan won the war:

Life becomes death. A whisper of protest leads to the concentration camp. Work becomes slavery. Poverty becomes starvation. To our mixed races it would mean complete slavery. For Hitler has said that only the pure Aryan has the right to rule, that all other peoples must be his slaves. . . . So we, the Spanish speaking Americans, must act in concert with the rest of the nation. In factories and fields, in the armed forces, on land, on the sea, in the air . . . . Spanish speaking American men, women, youth—unite and step forward proudly to defend your homes.[72]

Members of El Congreso were anxious to fight for American democracy, because they believed it represented not a society of equality but rather "a chance for change." Continuing to promote activism among its members, the "Call" to their fourth state convention observed that "in a democracy voices can be lifted. Songs can be sung. You can belong to a union, a club, to a church. And through that union, that club, or that church you can ask for a chance. Do you want a chance?" Although acknowledging Franklin Roosevelt's recent decree to end discrimination in defense employment, El Congreso's call for patriotism was predicated on an image of the United States as a land where social activism had the potential for creating equality, not one where equality already existed.[73]

World War II, however, curtailed militant activity by El Congreso, since both Communist party and CIO leaders opted for playing down civil rights activity in order to promote wartime unity. Despite a few rhetorical attempts to continue the campaign in this new context, enthusiasm for the struggle against fascism overran arguments for continued pressure on questions of civil rights. Moreover, El Congreso's membership declined when many key members were inducted into the armed forces. Meanwhile, other organizations competed for the time and commitment of those that remained. Increasingly, Fierro de Bright and congress members battled for the rights of Mexicans in other venues. Luisa Moreno and Frank López, for example, served on the Fair Employment Practices Commission in California, while some were appointed to commissions and boards by California governor Culbert Olsen. Fierro de Bright, Bert Corona, and Eduardo Quevedo became instrumental in setting up the Sleepy Lagoon Defense Committee in late 1942, a group that defended a group of Mexican American youths unjustly accused of murder. Similar issues, such as police violence against young Mexicans, consumed members' time through the late 1940s. Ironically, much of this work would mark the remaining membership of El Congreso with labels of "un-American" and "Communist" as early as 1943.[74]

The upsurge in Chicano political activity that occurred in the 1930s and early 1940s, however, involved at its core an attempt by the children of the immigrant generation and those who had arrived in the United States as youngsters to integrate themselves into American society. Although leaders were drawn from both groups of immigrants and their children, it was the second-generation experience that shaped most profoundly the emergence of Mexican American activism, linking workers' rights to civil rights. From the start, women were central to this effort. Moreover, the task of working on these issues usually involved multiethnic organizing, even if it had a primarily Mexican focus. Tellingly, Communist party members and other Anglo leftists often provided Mexican immigrants and their children with their first exposure to American politics in the New Deal era. Ironically, this labor and political activity often served as the greatest "Americanizing agent" of the 1930s and early 1940s.

For immigrants who were not American citizens, political activity could directly result in harassment by local law enforcement and social service officials. Although a few of the older generation were moved enough by Roosevelt's programs to become naturalized, most did not. Thus, the new interest in politics predominated among the American-born. Yet these adolescents and young adults had a profound impact on their parents, who generally supported the organizing efforts of their children. In certain industries, particularly the female-dominated types of garment and cannery work, immigrant parents often labored side-by-side and engaged in work stoppages alongside sons and daughters. Conversely, the worries and difficulties of their elders still greatly influenced the larger agenda of the community. When local officials encouraged *all* Chicanos to return to Mexico or fired Mexican Americans from jobs with impunity, repatriation made clear that Mexican ethnicity, rather than citizenship status, defined the Chicano experience in Los Angeles. This understanding tied the generations together.

Union activity, however, played a crucial role in changing the perspective of Mexican Americans toward political activity in the United States. CIO locals, in particular, encouraged active electoral support for President Roosevelt and local politicians who championed the union cause. The union framed its appeal in ethnic terms, while New Deal rhetoric bolstered the notion that political coalitions should include ethnic Americans of all nationalities. The local Democratic party at least tacitly validated this approach. In 1938, Eduardo Quevedo, Sr., ran unsuccessfully for city council and became the first Mexican American candidate in the twentieth century to aspire actively to local office. By World War II, Quevedo was chairman of the 40th District Democratic Council and a member of the Los Angeles Democratic Advisory Committee. He even became an instructor of adult citizenship classes at Roosevelt High School.[75] A decade later, Edward Roybal would successfully capture a seat on the city council, drawing strength in his campaign from well-developed community support founded on a base of labor union activity.

The three Roosevelt reelection campaigns produced increasing appeals to this newfound consciousness by connecting improvements in local conditions for Mexican Americans with the national policies of Franklin Delano Roosevelt. Anticipating the Viva Kennedy clubs of 1960, the new Mexican American political leadership of Los Angeles in the late 1930s and early 1940s, including Eduardo Quevedo, Manuel Ruiz, Jr., and Edward Roybal, took these ethnically focused appeals to the city's Mexican American community. One flyer from the campaign of 1944, for example, asked "American workers of Mexican descent" whether they wanted to return to 1932, when agricultural jobs were the only jobs available to them, and Mexicans were forced to pay high rents for overcrowded housing in Belvedere where there was "not enough plumbing for a tenth of the people living there." Claiming that Roosevelt had reduced employment discrimination through the Fair Employment

Practices Committee, protected the rights of workers to join unions, and built public housing for better homes at lower rents, this appeal stressed that the election was *"particularly important* to the 3 million Americans of Mexican birth or descent":

> Of these 3 million citizens of Mexican descent, 1 1/2 million are entitled to vote. But many of them do not take advantage of their right to choose the President of the United States and their representatives in Washington. Less than one-fifth of these citizens have cast votes in previous elections. If you fail to vote for Roosevelt, you are helping Hoover and Dewey. If you fail to vote, you are bringing back the days of 1932, of joblessness, and hard times. Are you going to vote this year?[76]

What fifteen years previously had been antithetical to the interests of the Mexican community of Los Angeles—naturalization, registration, and voting in an American election—now had become the quintessential way to express solidarity with the larger community of Chicanos, be they citizens or non-citizens.

Though many members of the Mexican immigrant population continued to shun American political activity, others became involved in supporting candidates and participating in union rallies, though few made the decisive leap to naturalization and voting. The dangers inherent in political activity for the immigrant generation, however, were made clear by later developments in the lives of the courageous labor leaders of the 1930s. This intense and fruitful period of organizing Mexican workers would be cut short by the onslaught of McCarthyism in the late 1940s. Although Rose Pesotta had returned to being a sewing machine operator as early as 1942, Josefina Fierro de Bright would flee to Mexico to avoid the witch hunt that took shape at the end of the decade.[77] It would be Luisa Moreno, however, who would suffer the most for her activities by being deported to her native Guatemala.[78] Speaking before the 12th Annual Convention of the California CIO Council in 1949, she came out of retirement to take on the Tenney Committee, a statewide precursor to the national McCarthy-led witchhunt of radicals:

> Strange things are happening in this land. Things that are truly alien to traditions and threaten the very existence of those cherished traditions. . . . Yes, tragically, the unmistakable signs are before us—before us, who really love America. And it is we who must sound the alarm, for the workers and the people to hear and take notice. For it seems that today, as the right to organize and strike was fought for and won, as the new labor agreements were fought for and won, as the fight against discrimination is being fought but far from won, so the fight for the very fundamentals of American democracy must again be fought for and re-established.[79]

Ironically, Moreno would include herself and the other Latino labor leaders of the period as individuals who "really love America" even while the U.S. government was trying to define her and others like her as aliens, outsiders to the American tradition. As immigrants, though some had become naturalized American citizens, these individuals faced deportation for the political activities they had engaged in under the rubric of a newfound ethnic "Americanism." Having reshaped the contours of their American identity to include active union participation and political organizing as Mexican Americans, segments of the American government now sought to define this work as "un-American." Yet despite the McCarthyites' successful campaign to deport Luisa Moreno and other immigrant labor leaders, these young activists managed to root a new ethnic identity among the Mexican-origin population in Los Angeles, an identity which combined ethnicity with Americanism. These new "Mexican Americans," steeped in the strong base of working-class experience and Mexican traditions, immediately involved themselves in directions which reformulated the boundaries of Chicano culture and society.

CHAPTER 12

The Rise of
the Second Generation

In May 1943, a Mexican American youngster charged with disturbing the peace in Venice, a local beachside community in southern California, stood before municipal judge Arthur S. Guerin. Alfred Barela and a group of his friends had been picked up by the police on suspicion of being involved in a noisy ruckus. According to Barela, the police pushed them around, ridiculed their dress, grabbed them by the hair, and threatened to shear their heads. Though the disorderly mob was apparently overwhelmingly non-Mexican, the police singled out uninvolved Mexican youths for harassment. Weighing the facts of the case, Judge Guerin dismissed all charges against the boys, but not before lecturing the group to stay out of trouble.[1]

Barela, still fuming over his treatment by the police and the courts, later wrote Guerin an angry, but thoughtful letter. In response to the judge's characterization of the boys as "a disgrace to the Mexican people" and Mexican youth in general as a grave problem for local authorities, Barela stated:

> Ever since I can remember I've been pushed around and called names because I'm a Mexican. I was born in this country. Like you said I have the same rights and privileges of other Americans. . . . Pretty soon I guess I'll be in the army and I'll be glad to go. But I want to be treated like everybody else. We're tired of being pushed around. We're tired of being told we can't go to this show or that dance hall because we're Mexican or that we better not be seen on the beach front, or that we can't wear draped pants or have our hair cut the way we want to. . . . I don't want any more trouble and I don't want anyone saying my people are in disgrace. My people work hard, fight hard in the army and navy of the United States. They're good Americans and they should have justice.[2]

Barela's personal resentment toward harassment of Mexican American youth reflected a widespread anti-Mexican attitude in Los Angeles during the early 1940s which culminated in the "Zoot Suit Riots" of 1943.

Problems with police harassment and conflict with various public institutions reveals much about the adjustment by Mexican American youth to the sociopolitical realities of life in Los Angeles. Barela's first-hand experience with discrimination confirmed in him an understanding of how American institutions were unequal with regard to racial minorities. Like many of his generation, he attempted to cope with the knowledge of being a member of a society where equality and justice did not apply fully to all citizens.

For Mexican Americans in Los Angeles, the issue of cultural adjustment in the second generation was not salient until the 1930s and 1940s.[3] Although immigrants and their children had arrived in the United States throughout the nineteenth century, it was not until the decade of the Mexican Revolution, and the ensuing mass immigration north, that large numbers of Mexicans made Los Angeles home. Increasingly during the second and third decades of the twentieth century, Mexican immigrants settled permanently in communities throughout the southern California basin. The generation of American-born Mexican Americans, therefore, did not emerge as an influential factor in the city's history until their maturation during the Depression decade.

Most historical analyses of this generation have focused on organizations centered outside of California during this period: LULAC, the League of United Latin American Citizens, and the G.I. Forum, both based in Texas, and the Alianza Hispano Americana, an organization founded in Arizona which later spread throughout the Southwest.[4] LULAC, which has garnered the most attention, was organized in 1929 in Corpus Christi by a small but influential group of Mexican American entrepreneurs, small businessmen, and professionals. As articulators of an emerging middle-class philosophy, argues Richard Garcia, LULAC members in San Antonio contrasted their organization with that of the Mexico-focused elites that had until then dominated intellectual life. Eventually the LULAC ideology of duality in ethnic life—"Mexican in culture and social activity, but American in philosophy and politics"— won out over a more nationalistic approach.[5]

In Los Angeles, the patterns of Mexican immigration and residential instability resulted in a different set of attitudes and philosophies which developed within the Mexican community. While immigration to Texas during the 1920s spurred an average annual increase in the size of its Mexican population of 7.6 percent, California experienced a 20.4 percent yearly gain.[6] Los Angeles received the heaviest in-migration, and, consequently, recent immigrants dominated community life, though an elite group of refugees who were centered around the consulate office controlled barrio politics. The only middle-class challenge to this Mexico-focused leadership that developed was the nascent Belvedere merchants' organization. But this group included large numbers of non-Mexican businessmen and, thus, never played as crucial a role as LULAC or its predecessors in defining a separate Mexican American agenda for the community. Furthermore, repatriation had weakened the traditionally ac-knowledged leadership of the immigrant generation, particularly the elite refugees from the turmoil of the Mexican Revolution that had coalesced around the Mexican consulate.

Stepping into this leadership void by the latter half of the 1930s were second-generation Mexican Americans emerging from the large working-class population in Los Angeles. This new young group of lead-ers were individuals born in the United States yet by no means ready to sever themselves from their parents' culture. To understand this transi-tion in community leadership, one must examine the different cultural frames of reference of the children of immigrants as compared with that of their parents.

In the 1930s, three institutions most clearly framed the experience of Mexican American adolescents and young adults in Los Angeles: the family, the school, and the workplace. Each exposed young people to a different set of values and expectations which fostered an understanding of their own ethnic identity. This process was often accompanied by con-siderable internal conflict, although for some it also led to novel interpre-tations of self and one's goals. Though this soul-searching was an individ-ual experience, it also manifested itself in new organizations that encouraged particular approaches to problems of identity. Though they were concerned with the balance between what was "Mexican" in one's past with what was "American" in one's present, they were concerned about the future and what the term "Mexican American" implied.

An important organization that mirrored the conflict and resolution of Mexican American identity during this period is the Mexican Ameri-can Movement (MAM). It emerged from Young Men's Christian Associ-ation (YMCA) clubs in southern California and was composed of second-generation young people. MAM is noteworthy for a variety of reasons. It was evidently the first Chicano organization formed by and for students, functioning in this capacity from 1934 to 1950. It empha-sized the progress of Mexican American people through education. Orig-inally it consisted mostly of high school students, but as the founding members entered postsecondary education, the rank and file were primar-ily college undergraduates. During the 1940s, when MAM members be-gan careers as teachers, social workers, and other professionals, the group tried to stimulate interest among younger students by creating a youth division. Although the organization did not officially break from the YMCA until 1944, and though it remained active until 1950, it was the early development of MAM during the late 1930s and early 1940s that reflected the cultural outlook of a certain segment of Mexican American youth.[7]

In 1934, the YMCA of southern California sponsored an Older Boys' Conference at the San Pedro branch for the benefit of Mexican American youth. This effort evolved from the work of various Protestant demoninations with immigrants from Mexico during the 1920s.[8] The

YMCA attempted to take leaders of Mexican descent and further develop their leadership skills and build fellowship between them outside the barrio. Later called the Mexican Youth Conference, these YMCA meetings continued annually, and participants recruited others who were interested in the problems of their people. From the start, the group was selective: those invited were Mexican American adolescents, many from Protestant backgrounds, who were identified as potential leaders among the boys who congregated at southern California YMCAs and who took part in athletics and other organized peer activities.

It did not take these young men long to realize that their work required more than an annual meeting. Increasingly, the most ambitious of the group began to hold meetings on their own, and at one such meeting in 1938 they decided to publish a newspaper, the *Mexican Voice,* and to use this "inspirational/educational paper" as the organ for the Mexican Youth Conference. MAM's most active members regularly contributed to the newspaper, which became a mouthpiece for some of the most articulate young people in the Chicano barrios of southern California. The first young editor, Félix Gutiérrez, served in this capacity for ten consecutive years.[9]

The establishment of the *Voice* was only a first step in attempts to broaden the group's work. In 1939, leadership training institutes were inaugurated and regional conferences began supplementing the annual meeting. In addition, leaders sponsored a Mexican American Girls' conference in 1940 in San Pedro, an event which increased female participation in all aspects of the group's work. The 1940 annual meeting marked the organization's important transition from a self-help group, geared mostly to boys, to a full-fledged organization of professionals committed to working with Mexican youth. At the San Pedro meeting, a Mexican American Teachers' Association was formed from the ranks of MAM's college graduates—members who had applied MAM's commitment to education to their own careers. MAM also initiated contact with young Mexican American leaders outside California, particularly in Arizona and Texas.

A focus on MAM in the years before World War II reveals the development of ideologies among second-generation youth and demonstrates how they viewed themselves and their surroundings. In general, little scholarly attention has been paid to the history of adolescents and young adults in American society; this is especially true for Mexican American youth. Moreover, more attention has been paid to Chicano intellectual history and political history for the post–World War II era. In fact, there is a tendency in Chicano history to view World War II as a watershed, a period—it is argued—when returning servicemen for the first time fought for their rights as citizens.[10] I argue, rather, that much of the cultural identity and sense of self of the Mexican American second generation was already shaped before the war. Examining an organization such as MAM provides substantial evidence to support this argument.

The conviction that Mexican people in the United States could succeed only through education was central to MAM's philosophy. Each member was a dedicated student who had seen his or her own horizons expand through attending high school or college. Their own experience and success in education led them to simplify a complex problem. For many MAM leaders, only lack of knowledge kept Mexicans from advancing in American society.

MAM members put forth three arguments as to why Mexican youth should continue in school. First, educated Mexicans were less likely to be targets of discrimination and prejudice. A college degree, therefore, held out the possibility of acceptance by the larger society despite one's race. Second, MAM members saw education as key to understanding the world and, thereby, transcending the limited confines of the barrio. Finally, they understood education to be a way of advancing socioeconomic mobility. Social mobility, they argued, promoted not only personal advancement but progress for Mexican American people as a group.

An issue of the *Mexican Voice,* published in 1938, emphasized the importance of staying in school: "Education is the only tool which will raise our influence, command the respect of the rich class, and enable us to mingle in their social, political and religious life." The author, José Rodríguez, asserted that increased competition made a college education an absolute necessity for success. "Education means a complete knowledge of yourself, a good knowledge of your fellowmen and a thorough knowledge of the world in which you live," he explained. *"EDUCA-TION,"* he concluded, "is our *only weapon!"* [11]

Most Chicano students in Los Angeles schools, however, had difficulty heeding such advice. A study conducted in the early 1930s, for example, found that 53.7 percent of Mexican girls and 43.7 percent of Mexican boys dropped out of school between the ages of fourteen and sixteen. Though immediate financial problems at home were identified as the primary reason for leaving school, about 13 percent of students reported that they quit school because they were simply not encouraged to stay. Other studies blamed a curriculum that was not designed with the needs of Mexican children in mind. Some cited racial discrimination on the part of teachers. [12] The dismal drop-out figures only strengthened MAM's resolve to find ways to encourage school attendance.

Disseminated among youth in various Chicano barrios throughout southern California, the MAM newsletter periodically contained short biographies of its leaders who were considered to be role models. MAM leaders justified such articles by reminding readers that an individual's success story benefited others in the community. Mainstream American heroes were also identified. A special cultural hero was Abraham Lincoln. They pictured him as a poor, humble person who, like themselves, had to struggle to overcome many obstacles to success. What made Lincoln's story even more appealing was the fact that he worked in order to study and "was able to save enough money for books which he read by the pale glow of his hearth."[13]

MAM leaders saw themselves as modern-day Abraham Lincolns. One article described MAM members as "New Modern Mexicans"— young men who "threw away the old timeworn, worthless ill-put slogan, 'A Mexican hasn't a chance.' " Praising the accomplishments of four Mexican students at Compton Junior College, the author cited them for proving that Mexicans "are as good a race as any other—artistically, mentally and physically!"[14] José Rodríguez was featured as a person never doubtful of his success, despite his dark skin. Studying at San Bernardino Junior College for an eventual career in foreign trade, he felt that there was "always room at the top."[15] A UCLA graduate, Joe Vidal, was described in January 1940 as a young man who had worked his way through high school and college by tutoring and working in a department store. He became a public school administrator and an authority in Spanish American literature. According to the *Voice,* "the United States is proud to have such men as citizens, and Mexican parents may well be proud of such sons."[16]

Probably the most admired individual in the organization was Stephen A. Reyes. President of the 1937 San Pedro Conference, Reyes's biography appeared in the *Voice* the following year. Despite the fact that he had no left arm, Reyes picked oranges during the summers throughout high school in order to attend junior college. After receiving his associate of arts degree, he entered UCLA in the fall of 1933. Though he obtained a small loan, he could not afford the high rents in Westwood; he commuted, borrowed additional funds, and worked part-time to stay in school. After graduating from UCLA in 1938, Reyes taught night school at a junior college, directed a local playground, and aspired to return to UCLA for work on a master's degree.[17]

The idea of self-help was a central message in these success stories. Although not a new message by any means in the Chicano community or within the society at large, these youth saw their opportunity through educational institutions. However, they also realized that education for Mexican Americans was increasingly segregated and unequal.[18] As early as 1928, for example, 64 schools in eight counties in southern California reported enrollments of between 90 and 100 percent Mexican. In 1931, another survey found that 80 percent of school districts with substantial Mexican American enrollment practiced segregation. In Los Angeles, as in other districts with large Chicano enrollments, attendance zones were manipulated by officials in order to insure white residents that most Mexican American pupils attended separate schools. Segregation was, for the most part, de facto until 1935, when Mexicans (identified as part Indian) were included along with "Chinese, Japanese, Mongolians, and Indians" in a long-standing statute in the state educational code which permitted segregation of these racial minorities.[19]

Segregation in schools increased as the enrollment of Chicano students multiplied and they became a growing concern of local officials. By the late 1920s and early 1930s, Chicano students could hardly be ignored. Over 65,000 Chicano students—nearly 10 percent of the state's public school population—were already enrolled in California schools by 1927. The great majority (88 percent) were concentrated in southern California, with 50 percent in Los Angeles County alone. Despite the loss of thousands during the repatriation period, Los Angeles County in 1941 continued to enroll 36,000 Chicano students in its public schools, a slight increase from 1927, representing nearly 16 percent of the county's 230,700 public school students.[20]

Although Los Angeles city school board officials deliberately segregated black students in the central district and designated certain schools in the San Fernando Valley as "all-Mexican," the segregation of Chicanos resulted from both residential segregation and the placement of Mexican students in separate facilities because, it was argued, of their language deficiencies. In addition, IQ testing in the 1920s and 1930s contributed to the belief held by educators that most Mexican children were "retarded," thus providing additional evidence to separate them from other students. Even when Chicano youngsters attended the same schools as Anglos, they were often separated into distinct classes for those deemed "slow" or those who could aspire to nothing more than vocational training.[21]

The cumulative results of these policies were evident in the racial makeup of schools in East Los Angeles. Among 14 primary schools located in this area in 1926, three contained an estimated 80 percent Spanish surname enrollment and three others had at least a 60 percent Spanish surname enrollment. By 1939, each of these six elementary schools had more than 80 percent Mexican enrollment, while two other schools—which had few Mexican students in 1926—now contained approximately 35 percent.[22]

These educational conditions haunted MAM's leadership. Though they attempted to portray segregation realistically, they still were optimistic about the future. Their willingness to decry prejudice was particularly evident from local councils which operated in more rural or more isolated parts of southern California. A report from the Monrovia club, for example, described attempts to end the practice of restricting Mexicans in the use of the city pool to the one day reserved for blacks. Monrovia's city council refused to consider persons of Mexican descent as "white Americans." The club concluded that before Americans criticized Germany's treatment of the Jews, they should first "look at what Americans are doing to Americans in America."[23] Reports of discrimination in public services and education in other cities increasingly found their way into MAM's publications, along with highlights of local efforts to eradicate the problem.

But even as they recognized the potency of bigotry, MAM members often naïvely portrayed discrimination as an obstacle any individual could

overcome. Descriptions of racial discrimination were usually alongside
encouraging words for those who faced acts of prejudice. An article in
the *Mexican Voice,* for example, instructed members to bolster their self-
confidence so that "we wouldn't attribute our shortcomings and defeats
to segregation and prejudice."[24] Some writers, however, saw some prejudice against Mexicans as understandable given the lack of resolve they
themselves perceived in their people. To illustrate, Mary Martinez, a
young social activist from the "Utah Flats" area of Los Angeles, recalled
that she had "wanted to fight and fight that terrible prejudice against our
people" ever since she could remember, but acknowledged—without going into any detail—that "some of it is well-founded."[25] These sentiments were certainly products of youthful frustration with Mexicans who
were perceived as unwilling to work to improve their condition or unable
to advance educationally or economically. Moreover, they speak to the
pressures exerted by assimilation ideology and the internalization by
some Chicanos of stereotypes and prejudices held by whites.

For the most part, the *Mexican Voice* encouraged Mexican American
youth to overcome the personal effects of discrimination while participating in its social eradication. MAM members avoided any public exploration of the negative psychological and social consequences of racism on
Mexican Americans. Instead, they exhibited exaggerated optimism toward the future. They saw America moving away from discrimination
and inequality and toward greater justice and opportunity for all. They
believed Mexican Americans could be a part of the American democracy
if they applied themselves. Perhaps the most eloquent expression of their
vision appeared in an undated memo distributed to all members:

> Experience reveals that Equality, like its companion, Freedom, exists in
> four modes—
> > the Equality which God gives,
> > the Equality which the State gives,
> > the Equality which a man wins for himself,
> > the Equality which one bestows on another.[26]

MAM focused primarily on the third mode, hoping to instill a sense
of confidence and optimism in its members during the turmoil and pessi-
mism of the Great Depression. Certainly the rosy portrait of the future
painted by Franklin Roosevelt and other New Deal leaders had some
influence on these young people. Like other Angelinos, Mexican Americans were perhaps more attentive to national politics than to local issues
during the 1930s. Much of that attention focused on President Roosevelt. Roosevelt had an astonishing effect on ordinary people. According
to historian Paul Conkin, he "gave millions of Americans a transfusion
of courage . . . . From his confidence, his optimism, they gleaned bits
of hope in times of trouble and confusion."[27] Whether that hope was
centered on real progress or on illusions, it had an important impact on
the Los Angeles population, and particularly on the Mexican American

generation. Juan B. Ruiz, a pharmacist on Main Street and an influential businessman in the Mexican community, commented in a 1937 interview that "I much admire the things that President Roosevelt has done. The Mexicans generally in this country are greatly attracted to him."[28]

Not only were American-born Chicanos affected by Roosevelt; the marked acceleration in initial requests for naturalization papers between 1934 and 1936 suggests that the Mexican-born also felt the changing atmosphere in American politics (see Table 15, above). This increase in naturalization requests was particularly powerful among those who had migrated to the United States as young children, and therefore shared many of the same cultural experiences as the American-born generation.[29] Historian Rodolfo Acuña has argued that "most Chicanos have been nurtured to believe in the Virgin of Guadalupe, the Sacred Heart, and the party of Franklin D. Roosevelt." Beatrice Griffith, who worked with Mexican juveniles in Los Angeles during this period, believed that "Franklin D. Roosevelt's name was the spark that started thousands of Spanish-speaking persons to the polls." While being careful not to upset white conservative support, the Democratic party sent out strong messages to Mexican Americans, much as it did to African Americans, arguing that their votes were crucial to the success of the New Deal coalition.[30]

On the local level, Mexican Americans witnessed an increasing number of influential Anglo Americans grow interested in their plight. Many Anglo educators served on the advisory board of the Mexican American Movement. Moreover, important political figures, such as Carey McWilliams, chief of the California Division of Immigration and Housing, and John Anson Ford, Los Angeles County Supervisor, sent messages of encouragement concerning MAM's work.[31] White supporters of Chicano organizations such as MAM were evident throughout the 1930s. For example, a substantial group of local figures, especially well-known Hollywood actors and actresses with liberal or leftist political views, participated in many fund-raising benefits and celebrity functions in support of the Los Angeles Chicano community. By the early 1940s, when the Sleepy Lagoon case and Zoot Suit Riots dramatized the plight of the Mexican American generation, Hollywood celebrities were conspicuous among the defense committees and investigatory panels created to address the issues raised by these events.[32]

More radical political alternatives also motivated a significant proportion of barrio residents to action. Several historians have suggested that Mexicans in Los Angeles gave widespread support to Upton Sinclair's gubernatorial candidacy in 1934. After two attempts at the governorship as the Socialist party's candidate, Sinclair organized the "End Poverty in California" (EPIC) campaign and received the Democratic nomination for governor of California with the highest vote ever achieved in a primary. Although he lost the general election, Sinclair received nearly 900,000 votes out of 2.3 million cast. Two-thirds of his support came from southern California, where both Mexicans and unemployment were more highly concentrated.[33] After Sinclair's defeat, some Los Angeles Mexicans supported Dr. Francis Townsend, a resident of nearby Long Beach, who pushed a plan to aid the elderly.[34] Other Mexican residents turned to the U.S. Communist party, which was active in the Los Angeles barrio during the Depression. About 10 percent of the Communist party's recruits in 1936–37 were Spanish and Mexican workers.[35] Whether Chicanos in Los Angeles were attracted to the Democratic party, EPIC, or the Communist party, they began to participate in American politics as they had never done before. Many individuals established organizational connections with groups that were completely or primarily "American" in focus.

This change, however, evolved slowly and did not involve the entire community. A major proportion of barrio residents were still not American citizens and did not have the opportunities for political expression available to citizens of the United States. Not surprisingly, second-generation Mexican Americans tended to participate in American electoral politics more than their parents. By the late 1930s, when many of the children of immigrants from Mexico came of voting age, their political opportunities were often very different from those of their non-citizen parents.

Intergenerational tensions, whether reflected in political attitudes or national identity, were evident by the late 1930s. The Mexican American Movement and the writing of its members give ample expression to these tensions. Most MAM members felt it imperative to acknowledge that they had no divided loyalties between Mexico and the United States. They considered themselves fully American citizens, with all the rights and responsibilities that citizenship implied. As Félix Gutiérrez, editor of the *Mexican Voice,* put it: "Very few of us pin our future in Mexico. Our future is here."[36] Rebecca Muñoz of Tempe, Arizona, went even further, displaying annoyance with those in her community who refused to participate politically in the United States because of the pull of Mexico. She blamed Mexicans in the United States for "sadly lacking the desire to make the best of these opportunities . . . partly due to a certain indifference to anything that may happen here, since they prefer to live in their memories of the old country."[37] Muñoz clearly expressed the frustrations of the American-born generation with the political focus of the immigrant generation. Although few MAM members were as harsh on the immigrant generation as Muñoz, most believed that it was time to focus attention on issues related to Mexican Americans north of the border.

MAM members were forced to confront both the differences and similarities they shared with their parents. In particular, these sons and daughters came to realize that they had developed very different attitudes toward *la madre patria*. Although MAM members emphasized that they shared much of Mexican culture with their parents, they had much less allegiance to the Mexican state. Paul Coronel, MAM's president in 1938 and 1939, criticized Mexico's inefficient government for not creating an educated citizenry and failing to provide working conditions that gave Mexicans a chance to get ahead. Since Mexico had put so many obstacles in the path of ambition, Coronel called on Mexican Americans to focus on individual advancement above and beyond loyalty to Mexico. His opinion was typical of descendants of immigrants who believed that the mother country had provided little opportunity: "We're not awakening ourselves for Mexico nor the United States, but for ourselves."[38]

Besides intergenerational differences in national identity, there were other sources of tension within Mexican American families resulting from the emergence of a new ethnic identity among the young. A series of articles written by young women and directed at Mexican immigrant parents illustrates some of these tensions. (These were the first articles to appear in the *Voice* in Spanish since its inception.) Dora Ibañez, a graduate of La Verne College and a local music teacher, voiced the frustrations she and others of the second generation faced in balancing the desires of their immigrant parents and the reality of their American existence. In an August 1939 article, she acknowledged the disparity between the social and moral world of youth and their parents. Showing the utmost respect for the parents' perspective, she praised their sacrifices, their self-respect, and their caring attitude, all of which she credited to Mexican tradition. Then she asked the questions most on the minds of these parents:

> What is happening with our children? Why do they reject our behavior? Why don't they respond harmoniously with our way of thinking? Don't they feel the warmth of our traditions and customs like we do? Many of you don't get answers to these questions and see that your son or daughter doesn't find satisfaction in themselves, nor in the home, nor in the the community nor in their own people in general.

Ibañez's response made clear to parents that their children could never feel as close to Mexican tradition as they, because she and her peers were born in the United States. Although they admired the beauty and tradition of Mexico, they were Americans. Ibañez urged parents to accept their children's new identity since, after all, they themselves had left Mexico to advance their family's socioeconomic and educational opportunities. She reminded parents that their children hoped to dignify what is "Mexican" in this country by elevating themselves, their communities, and their people. Ibañez, furthermore, called on parents to begin a "parallel march" with their children and to guide them in their advancement. In this way, she felt, the generation gap could be bridged.[39]

Ibañez argued that Mexican American children were living out their parents' dream by making the best of opportunities provided in the United States. Though their Mexican heritage encouraged and inspired them, she argued, it was not a useful guide for success in America. Ibañez

hoped that such gentle criticism explained why children needed to be left
to find their own path. Yet in spite of such sentiments, she made it clear
that members of MAM were not about to reject their ethnicity; it connected them to their families, communities, and to the other young Mexicans they wished to reach.

The strategy of balancing the cultural connection between oneself
and one's Mexican heritage together with one's identity as an American
was repeated by other members of MAM. In an article entitled "Are We
Proud of Being Mexican?," Manuel Ceja, a student at Compton Junior
College who aspired to become a lawyer, criticized Mexican American
youth who denied their Mexican nationality and who claimed to be
"Spanish." Ceja asserted that the dual background of Mexican American
youth was a distinct advantage in the United States, especially in a society increasingly conscious about Latin America. For him, assimilation
did not mean abandonment of Mexican culture:

> The Mexican Youth in the United States is, indeed, a very fortunate person.
> Why? Where else in one country do you have two cultures and civilizations
> of the highest type that have been developed come together to form into
> one? The Mexican Youth comes from a background of the highest type of
> Aztec and Spanish cultures, and now is living in a country whose standard
> of life is one of the highest and where there are the best opportunities for
> success. Take the best of our background, and the best of the present one
> we are now living under, and we shall have something that cannot be
> equaled culturally.[40]

In fact, condemning Mexicans who denied their heritage was a constant theme among members of the organization. Paul Coronel instructed MAM members not to "pretend to be Spanish or a Californian
when you know very well that you are a Mexican." He denounced Mexican American professionals who, instead of becoming leaders for progress among their people, "become ashamed of being Mexicans because of
the common notion that we Mexicans are a bunch of stupid field laborers."[41] MAM hoped to promote a new type of middle-class leadership in
Chicano communities that would not deny its Mexican and working-
class origins.

In the late 1930s, however, Chicano working-class youth in Los
Angeles became increasingly estranged from a society unable to provide
adequate jobs or education. MAM members found it ever more difficult
to claim that they represented the vast majority of Chicanos in their late
teens and early twenties. Drop-out rates from high school and junior
highs in East Los Angeles remained extremely high as Chicanos of working age were forced to contribute to a family income still recovering
from the effects of the Depression. Moreover, job possibilities remained
stagnant, often holding out little hope for advancement. De facto segregation in schools and pernicious discrimination in public facilities contra-
dicted the messages that MAM promoted.

Within a few years, the most visible emblem of this estrangement became a highly stylized form of male dress—the zoot suit. With a broad-brimmed hat, long jacket, and draped trousers tapered at the ankles, the zoot suit was adopted by thousands of working-class youth in urban centers during the early 1940s. In Los Angeles, both African American and Mexican American young men were wearing what one commentator has referred to as "a refusal: a subcultural gesture that refused to concede to the manners of subservience."[42] Although the origins of the zoot suit are unclear, this form of dress emerged out of identification with the motion pictures and jazz culture of the late 1930s. As the United States entered World War II, however, this form of self-expression increasingly was frowned upon by American officials concerned with conformity. In March 1942, the War Production Board drew up regulations for the wartime manufacture of streamlined suits which required a minimum of fabric. The demand for zoot suits, however, continued to grow, with bootleg tailors supplying the outlawed suits.[43] Though it was estimated by law enforcement officials that as many as two-thirds of the Mexican working-class young men in Los Angeles wore the zoot suit in the early 1940s, the number was probably much lower.[44]

Since their personal appearance was more strictly supervised by parents, young Mexican women were often the first to conflict with their families over fashion. Swimwear, bloomers, and short skirts were major points of friction between parents and their daughters. Increased use of cosmetics in the late 1930s and early 1940s resulted both from attempts to emulate Hollywood starlets and from the growing number of barrio beauty pageants.[45] Adolescent women often tested the boundaries of tradition and their own personal freedom through experimentation and innovation in dress and image. Others, however, stuck closer to the prescribed behavior of their parents. As was indicated by a Chicana who described fashion in the early 1940s:

> We actually started seeing the drape pants around that time, but we didn't approve of it and we didn't dress that way. . . . And we never went for that excessive makeup or the tight skirts. . . . We figured it's an exaggerated style of dress that's going to come and go away, never expecting them to take over as much as they did eventually.[46]

The growth of disagreement over proper modes of attire was only one manifestation of the independence second-generation youth demanded from their immigrant parents. The war only exacerbated this division by providing greater independence and expanded employment opportunities for young American-born individuals, while increasing discriminatory practices against those born outside the country. Parents had to rely more on the income of their children to sustain the family. This economic dependence affected relationships of authority in the family and fostered greater independence among those adolescents old enough to secure jobs. At some level, this process was a continuation of the demise of the immigrant generation's influence, a change set in motion by the massive repatriation which had occurred a decade before. Yet the turn of events during the war brought about even greater recognition that the problems of the second generation had come to dominate the Chicano community of Los Angeles.

No other event best symbolizes the ascendancy of second-generation Mexican Americans to public consciousness than the arrest and trial of a group of youths for a single murder which occurred in East Los Angeles. On the morning on August 2, 1942, José Díaz was found on a dirt road near the so-called Sleepy Lagoon, a water-filled gravel pit used for recreational purposes. Apparently, on the previous evening a clash had occurred between members of the 38th Street Club and guests at a party being held at the nearby Williams Ranch. Díaz had attended the party, and after police discovered his body, they jailed the entire gang and charged twenty-two of them with criminal conspiracy. Trial publicity was passionately anti-Mexican. While the press portrayed the Sleepy Lagoon defendants as Mexican hoodlums, the police submitted a report to the Grand Jury that declared Mexicans inherently criminal and biologically prone to violence. The trial itself was full of irregularities, including the refusal by Judge Charles W. Fricke to allow the defendants to cut their hair or change their zoot-suit–style clothes, despite months in jail. On January 12, 1943, seventeen of the young defendants were found guilty of crimes ranging from assault to first-degree murder—the largest mass conviction in California history.[47]

The young defendants, ranging in age from sixteen to twenty-two, represented an element of Mexican American youth of the second-generation who stood in profound contrast to the success stories of MAM members. Almost all had dropped out of high school and had worked at a variety of industrial jobs in Los Angeles, particularly in the furniture industry. Most resided in or near the growing industrial sector of Vernon, a district in Los Angeles south of downtown. Most of their parents had been born in Mexico, although three fathers had taken out first papers for naturalization and one had already been naturalized. Almost a quarter of the families had experienced disruption, more often by death of a spouse than by divorce or desertion. Three of the defendants had married at an early age and already had children. Most had been in trouble with the law before, although many of these past incidents had been relatively minor. At least three mothers had found it necessary to request help in controlling their sons: one had called the Catholic Welfare Bureau, the other two had appealed to the police.[48]

The 38th Street Club was one of at least thirty-five youth "gangs" in Los Angeles during the early 1940s. These groups differed from the complex east-side gangs that have operated in Los Angeles in recent decades. One contemporary report, for example, issued by a Citizens Committee formed by Governor Earl Warren to investigate these neighborhood clubs, found that "many of them are wholesome social groups, meeting in public schools, in sub-police stations, and under the supervision of responsible officials."[49] Some studies traced the origins of these groups in the Chicano community to the 1920s when Protestant organizations promoted youth groups as a means of encouraging leadership among the Mexican population. Only in the late 1940s and early 1950s, when drugs and recurring violence became part of the gang subculture, did a few of these clubs eventually become well-organized gangs.[50]

However benign their activities during the World War II–era, Chicano youth were increasingly viewed as a threat to the stability of Los Angeles by a large proportion of the Anglo American population. In 1942 and 1943, a series of newspaper articles published in the Los Angeles *Times* presented lurid accounts of isolated acts of alleged crime and dramatized these activities so as to launch a campaign against the "zoot suiter menace." Accounts of Mexican juvenile delinquency either replaced or were printed alongside stories of supposed disloyalty among interned Japanese Americans. Chicano youth were increasingly depicted as the "enemy within." The war catalyzed the growing antagonism of the Anglo American community toward diversity and difference. Mounting tensions finally culminated in early June 1943, when ten days of violent clashes between Mexican American youth and Anglo servicemen, joined quickly by civilians, erupted into the so-called "Zoot Suit Riots" of Los Angeles.[51]

Perhaps the nature of these encounters can best be understood by looking at the experiences of Pedro García, a graduating senior at Roosevelt High School in Boyle Heights. The American-born son of Mexican immigrants, García considered himself American. On June 7, 1943, he went to see a movie at the RKO theater on Hill Street, as he and many other Mexican American young people like him had done on countless Saturday nights. He had taken an aisle seat and was enjoying the picture when a group of Anglo American servicemen burst into the theater looking for Mexican "zoot suiters." They grabbed Pedro, dragging him outside the theater and into the street. As a large group of Anglo American bystanders looked on in amusement, the sailors ripped off his clothes, kicked and beat him, and left him bleeding and unconscious. Nearby policemen witnessed the spectacle, but did not make a move.[52]

This incident and other incidents of the Zoot Suit Riots made clear to many second-generation Chicanos that much of their optimism about the future had been misguided. As the riots became an international incident, the problems of the second generation in Los Angeles took centerstage. Local officials searched for the causes of the disorder in the perceived degradation of the Mexican community, but MAM and other Mexican American organizations knew that the occurrences symbolized a central American dilemma, the Anglo American cultural intolerance of racial and cultural differences, and the special difficulties of a generation of youth suspended between two cultures.

The Mexican writer Octavio Paz—who spent two years in Los Angeles during the mid-1940s—also noted the peculiar position of Mexican American youth. In the opening chapter of his classic book on life and thought in Mexico, *The Labyrinth of Solitude,* his own perceptions on what it meant to be Mexican were jarred by the presence of "pachucos"—the adolescent "zoot suiters" who had only recently been the object of such controversy. For him the Chicanos he saw on the streets of Los Angeles were certainly Mexican, but an extreme caricature of his cousins south of the border:

> They have lived in the city for many years, wearing the same clothes and speaking the same language as the other inhabitants, and they feel ashamed of their origin, yet no one would mistake them for authentic North Americans. . . . What distinguishes them, I think, is their furtive, restless air; they act like persons who are wearing disguises, who are afraid of a stranger's look because it could strip them and leave them stark naked.[53]

For Paz, the adolescent pachuco symbolized the suspension between cultures of the Mexican American. "He does not want to become Mexican again," he wrote, "at the same time he does not want to blend into the life of North America." As a member of the elite in Mexico, Paz tended to see the emerging Chicano culture in negative terms, as "a tangle of contradictions, an enigma." The desire to remain different appeared to be simple rebelliousness, the "empty gesture" of an eccentric. Paz viewed the pachuco as self-defeating in his efforts to assimilate. "The pachuco is the prey of society, but instead of hiding he adorns himself to attract the hunter's attention."[54] Struggling to understand the cultural identity of those Mexicans and their offspring who had decided to stay in the United States, he could see little but pathology.

Lost in all the attention focused on the problems of second-generation youth was the legacy of the immigrant generation. In fact, the only mention in the Los Angeles *Times* of the parents of any "zoot suiter" during those days in early June 1943 was a short article entitled "Mother Tears Up Zoot Suit of Boy Wounded in Clash." The *Times* reported that a young fifteen-year-old boy shot in the leg during a "zoot suit riot" had recently been visited in the hospital by his mother. In an obvious attempt to show the disapproval of responsible Mexican adults, the *Times* photographed his mother ceremoniously tearing "his natty peg-top black trousers," before her son's very eyes.[55] But this scene, probably staged by the reporter, rarely appeared in the public discussion accompanying the riots. By and large, the "zoot suiters" were depicted as without parental guidance or control, almost "orphans" in the history of Mexican immigration to Los Angeles.

Little more than a decade had passed since the Mexican immigrant had cornered the attention of city and county officials during repatriation. The child of the immigrant had replaced the newcomer, yet American society was inhospitable to their generation as well. Too often lost amid the dramatic conflicts or court trials were the creative adaptations of Mexican American youth to a society that did not make them welcome. Their cultural adaptability was an important development in the process of Chicano cultural expression in the United States. It was a legacy that was shared among generations of Chicanos in Los Angeles and throughout the Southwest.

Conclusion

When I arrived in the United States, I lived for a while in Los Angeles, a city inhabited by over a million persons of Mexican origin. At first sight, the visitor is surprised not only by the purity of the sky and the ugliness of the dispersed and ostentatious buildings, but also by the city's vaguely Mexican atmosphere, which cannot be captured in words or concepts. This Mexicanism—delight in decoration, carelessness and pomp, negligence, passion and reserve—floats in the air. I say "floats" because it never mixes or unites with the other world, the North American world based on precision and efficiency. It floats, without offering any opposition; it hovers, blown here and there by the wind, sometimes breaking up like a cloud, sometimes standing erect like a rising skyrocket. It creeps, it wrinkles, expands and contracts; it sleeps or dreams; it is ragged or beautiful. It floats, never quite existing, never quite vanishing.

—Octavio Paz from the *Labyrinth of Solitude* (1950)[1]

Stand in the lobby of the transnational terminal at Los Angeles International Airport the week before Christmas. Witness the transformation of immigrant adaptation to American society. Here in this airport, which has become the single largest port of entry for immigrants to this country, one sees continuous movement back to native countries during the holiday season. At the separate terminal which handles Mexico-bound flights, thousands of individuals carry huge packages full of the latest toys, games, and other American consumer items bound for their friends and relatives back home. Venture south to Tijuana at the border and witness a less conspicuous, but equally important movement to the Mexican interior aboard thousands of buses carrying less prosperous migrants to their loved ones. After the holiday season, having reconnected themselves with the lives they lived before migration, most of these individuals will return to their jobs and homes in Los Angeles.

The modern conveniences of the late twentieth century—air travel, the telephone, television—have made the world a smaller place and have altered the nature of immigrant adjustment. Communication with lands left behind is now possible, and for immigrants dealing with new surroundings in this country this contact is crucial. In many ways, the close

271

relationship that Mexican immigrants maintained with their home communities across the border in the early part of this century can be replicated now by almost all newcomers to American society. No longer are time and space the barriers to this communication, but rather financial resources and, for some, the imposed restrictions on immigration due to legal status. Christmas 1988, for example, saw a huge increase in movement across the border when amnesty provisions in the Immigration Reform and Control Act of 1986 legalized the status of hundreds of thousands living in this country, reducing fears of being discovered by the authorities.[2]

Analyzing the cultural adaptation of Mexican immigrants in the early part of the century can help us understand the broader implications of acculturation and ethnicity in late twentieth-century American society. Rarely did migration to the United States uproot all vestiges of one's native culture, but neither did Mexican culture remain unchanged in the United States. Rather, cultural adaptation occurred gradually, particularly among those who made conscious decisions to remain north of the border. Though changes were evident in the values and practices of the immigrant generation, a more profound adaptation usually occurred among their children.

Despite the subtle pace of cultural change exhibited by Mexican immigrants, change most definitely occurred. By the time that Nobel Prize-winning author Octavio Paz spent a few years in Los Angeles as a young man in the mid-1940s, he would have difficulty pinpointing the city's "vaguely Mexican atmosphere." The million Mexican-origin residents of Los Angeles would appear to him to be no longer Mexican and yet not quite American either, but suspended between two cultures. To this elite Mexican intellectual, the mostly working-class Mexican American population appeared unable to ever truly be at home in their new homeland. Yet Mexican Americans themselves did not necessarily concur. Most had no difficulty seeing themselves as both Mexican and American. They knew that they had become cultural bridges between two lands; in fact, they had created a new borderlands in the east-side barrios in which cultural revival and re-creation were ever-present.

The back-and-forth nature of Mexican migration throughout the twentieth century—with the exception of the 1930s—insured the constant infusion of Mexican culture into Chicano communities in the United States. Moreover, it also guaranteed that American culture would be brought deep into Mexico by returning migrants. The sharp divisions between American and Mexican cultures were blunted by the migrants themselves, as they carried products, values, and memories across the border. The forced cessation of Mexican immigration during the Great Depression, however, created a unique context for the development of the Mexican American generation in Los Angeles. Being witness to the repatriation of thousands of Mexicans early in the decade, Mexican American adolescents struggled to find their identity on American soil without benefit of recent newcomers. As Americans, they attempted to take a middle ground—searching for ways to reconcile their Mexican heritage with a new role as citizens of the United States. All of this occurred within a society that too often denied them equal opportunity.

For both the immigrant generation and their children, most adaptation to American society occurred within the confines of the working class. During the 1920s, this meant that Mexicans learned about American life in the ethnically mixed neighborhoods in central and east Los Angeles. Beginning with Belvedere, communities east of the river began to take on more of the characteristics of barrios isolated from other communities in the city. During most of the period from 1900 to 1945, however, Mexicans were integrated into American working-class life, living among other ethnics also coming to terms with what it meant to be American.

Different families, of course, adopted different cultural strategies. Intermarriage with Anglo Americans, or with American-born Chicanos, often led to a particular familial pattern. Even among families formed by two Mexican immigrants, variation was marked. Individuals who did not marry were often able to find a niche for themselves among the ethnically integrated neighborhoods west of the river. In East Los Angeles, however, life was increasingly family-focused.

Certain patterns did emerge which seem to characterize Chicano culture in Los Angeles. Catholicism, which had played such an important role in village life in Mexico, was challenged by Protestant denominations and particularly by secular culture. Often, religious practice became the domain of women in the family, and was displayed most emphatically within the home. Other aspects of Mexican culture were transformed as American consumerism contributed to assimilation. Spanish language music, for example, became commercialized and streamlined for a more technologically sophisticated audience, one growing accustomed to the radio and the phonograph. Even American manufacturers became aware of the potential in packaging one's product in an immigrant's native tongue.

It is clear that the power of governmental bodies to encourage or dictate particular forms of cultural adaptation was minimal. Neither Americanization nor Mexicanization programs succeeded in eliciting intended responses among immigrants in Los Angeles. At best, these programs encouraged the creation of an identity as ethnic Americans among the Mexican immigrant population. Economic, social, and cultural forces in the city and relations between the two nations had more influence on motivating particular behavior or attitudes than organized governmental efforts. Indeed, Communist organizers and New Deal labor activists may have played a more important role in "Americanizing" the Mexican working-class population than government simply by characterizing the labor union and ethnic political organizing of the 1930s as a quintessential American activity. Moreover, immigrants themselves usually negotiated the most critical decisions regarding their cultural future, in spite of how and in what ways their options were delimited by socioeconomic realities in the Chicano community.

Indeed, it is possible to argue that Los Angeles provided Mexican immigrants more latitude than any other community in the Southwest in shaping a Mexican American identity. Far enough away from the border to encourage experimentation with new cultural influences, newcomers there were still close enough to the population centers of Mexico to receive constant imput from newly arrived immigrants. A city where Mexicans were only the most numerous of many ethnic groups, diversity marked even the barrios that most immigrants called home. Today, as in the past, Mexican American culture in Los Angeles is the product of international influences and an adaptive process among the individuals, families, and communities that constitute the ever-changing Chicano society in the City of Angels.

Even Zeferino Ramírez, long considered a leader in this "Mexico de afuera" community, would eventually alter his perspective. Although he never considered officially exchanging his Mexican citizenship to become an American, a trip to the interior of Mexico in the late 1930s did convince him never to return permanently to live in his native country. He would stay the rest of his life in Los Angeles, seeing his children mature as Mexican Americans, content in providing for them the resources which allowed them to flourish. But Zeferino himself would also contribute to the creation of a Mexican American identity in East Los Angeles, by remaining active in the cultural world now dominated by his children's generation.

In 1945, at the conclusion of World War II, one of his daughters, Julia, prevailed upon him to become involved in the construction of a memorial to the Mexican American soldiers who had died in the war. After forming a committee that persuaded the Los Angeles city council of the worthiness of the project, Ramírez himself contributed the final $4000 in construction costs. Still standing at the triangle formed by Lorena, Brooklyn, and Indiana streets, this monument also marks the dividing line between the city and county of Los Angeles and the entrance to Belvedere, Zeferino Ramírez's adopted home.[3] More emphatically for the purposes of this story, it also marks the transition from a Mexico-centered leadership to one focused on political and social advancement in American society.

On Sources

This study makes extensive use of naturalization records obtained from the regional branch of the National Archives, located in Laguna Niguel, California. Every Mexican who applied for naturalization in the Federal District Court of Los Angeles before January 1, 1940, was included in a computerized data set. These naturalization records offer extensive personal information about each of the 2,238 individuals who applied for naturalization and his or her family, including information regarding an applicant's migration pattern into the country. I utilized this information to identify the communities from which Mexican immigrants had migrated to Los Angeles, and the manner by which immigrants reached the city. Beginning in Chapter 6, I limited my sample to 1,740 families who listed an address in the city of Los Angeles or its surrounding adjacent unincorporated areas (especially Belvedere and Watts), in order to discuss developments in the city proper.

Many studies have concluded that Mexicans had one of the lowest rates of naturalization for any group in the United States during the twentieth century and, therefore, that the usefulness of naturalization records for studying the Mexican immigrant population may be questioned.[1] My research, however, indicates that these materials are much more valuable than previously believed. Previous studies of Chicanos in Los Angeles during this period have provided an understanding of the total population at any given time by utilizing census data, city directories, and marriage records.[2] I have tried to separate those Mexican immigrants who were in Los Angeles temporarily from those who at some point in their lives made a decision to remain permanently in the United States, in order to better understand the process of cultural adaptation. Naturalization records, therefore, seem a more appropriate vehicle for studying this subset of the total Chicano population than materials which survey the entire population. Since even most immigrants who remained in Los Angeles permanently did not choose to naturalize, however, my conclusions from these data should be viewed as tentative and subject to further refinement.

The data are useful for other reasons as well. I have found that 45 percent (1,001 individuals) of all Mexican citizens who applied for naturalization in Los Angeles did not complete the process. Thus, early studies that have concluded that the low total number of Mexicans naturalized indicates a lack of willingness to consider changing citizenship fail to account for the much larger figures of Mexicans who applied but were unable to finish the process of naturalization.[3] The second step was much more rigorous than the first, involving a two-year hiatus from extended visits to Mexico, a facility with the English language, and a knowledge of the fundamentals of American history and government. My data, however, includes records of all persons who initiated the first application, along with the much smaller group that completed the second. Not only does this provide a larger sample, but it is also a broader sample of those who considered changing citizenship status.

Like any other data, the sample is not without its biases. It severely underrepresents the female immigrant population, since male immigrants were much more likely to apply for naturalization than women. It also tends to favor migrants who came to the United States as children and/ or as unmarried persons, as opposed to older migrants who arrived with their families. I attempt to deal with some of these problems within the context of my discussion of the pertinent groups (see Chapters 3 and 6, for example). On the other hand, these data do allow me to describe the overall process of cultural adaptation, because they link together information from the person's past, most notably his or her migration experience, with details of life in Los Angeles, including occupation, marital status, and address. In sum, this information is critical for understanding the cultural adaptation of Mexican immigrants who chose to reside permanently in Los Angeles.

Notes

Introduction

1. Studies of Mexican-origin people in the United States are full of debates over appropriate labeling. I will add little toward a resolution of this problem. Since my study involves the process by which Mexican immigrants adapt to American society, I have chosen to simplify for the sake of clarity a rather complex and politically charged issue.

Those born in Mexico who reside temporarily in the United States are called "Mexican," "Mexicano," and "Mexican immigrant" interchangeably. "Mexican American" denotes both those born in the United States and those who change their citizenship status. I also use "Chicano" as an umbrella term for both groups, although I am aware that most of the individuals described in this study would not have used this term to describe themselves. "Latino" is used to describe the entire population of immigrants from Latin America and their descendants.

In this study, I will also use the terms "white," "Anglo American," and "Euro-American" interchangeably, although I also acknowledge the great diversity existing among those so designated. When appropriate, I refer to the national origins of subgroups within this "Anglo American" population.

2. *La Opinión,* 10–13 June 1927.

3. Interview with Zeferino Ramírez, "Biographies and Case Histories" II folder, Z-R5, Manuel Gamio collection, Bancroft Library, University of California, Berkeley; interview with Beatrice Palomares, 3 Jan. 1991, conducted by George J. Sánchez.

4. Interview with Zeferino Ramírez, 2, 5–7, Gamio collection; interview with Beatrice Palomares.

5. See *La Opinión* editorial, 11 June 1927.

6. "The Quest for National Character," in *The Reconstruction of American History,* John Higham, ed. (New York: Hutchinson, 1962), 197–98.

7. *The Uprooted: The Epic Story of the Great Migrations That Made the American People,* 2nd ed., enlarged (1951; Boston: Little, Brown, 1973), 3. Handlin extended his argument in the early 1970s to include non-whites in the United States in this enlarged 2nd edition. See the combative, if rambling, added Chap-

ter 13. See also *The Newcomers: Negroes and Puerto Ricans in a Changing Metropolis* (Cambridge, Mass.: Harvard Univ. Press, 1959) and *The American People in the Twentieth Century* (Cambridge, Mass.: Harvard Univ. Press, 1954).

8. The first historian directly to take on Handlin's work was Rudolph J. Vecoli in *"Contadini* in Chicago: A Critique of *The Uprooted," Journal of American History* 51 (1964), 404–17. Since then, many critiques of Handlin's work have followed. The most notable are Virginia Yans-McLaughlin, *Family and Community: Italian Immigrants in Buffalo, 1880–1930* (Ithaca: Cornell Univ. Press, 1977), 18, 26, 57, 62, 181; John W. Briggs, *An Italian Passage: Immigrants to Three American Cities, 1890–1930* (New Haven: Yale Univ. Press, 1978), 118; and Josef J. Barton, *Peasants and Strangers: Italians, Rumanians, and Slovaks in an American City, 1890–1950* (Cambridge, Mass.: Harvard Univ. Press, 1975), 2.

9. Howard N. Rabinowitz, "Race, Ethnicity, and Cultural Pluralism in American History," in *Ordinary People and Everyday Life: Perspectives on the New Social History,* James B. Gardner and George Rollie Adams, eds. (Nashville: American Association for State and Local History, 1983), 28.

10. Werner Sollors, *Beyond Ethnicity: Consent and Descent in American Culture* (New York: Oxford Univ. Press, 1986), 33.

11. John Bodnar, *The Transplanted: A History of Immigrants in Urban America* (Bloomington: Indiana Univ. Press, 1985). One result of the stress on the continuity of culture among European immigrants was the strengthening of the argument that the immigrant cultural experience could be equated with that of racial minority groups, thereby discounting the effect of unequal economic opportunities. For the best example of the reduction of issues of race and class to ones of "ethnicity" in sociological writings, see the essays in Nathan Glazer and Daniel P. Moynihan, eds., *Ethnicity: Theory and Experience* (Cambridge, Mass.: Harvard Univ. Press, 1975). For a more contemporary discussion in literature from the same perspective, see Sollors, *Beyond Ethnicity.* For two devastating critiques of this trend, see Stephen Steinberg, *The Ethnic Myth: Race, Ethnicity, and Class in America* (Boston: Beacon Press, 1982), and Michael Omi and Howard Winant, *Racial Formation in the United States: From the 1960s to the 1980s* (New York: Routledge & Kegan Paul, 1986).

12. Octavio Romano was the first to criticize the stereotyped version of Chicano culture in Anglo American social science writings. See "The Anthropology and Sociology of the Mexican-Americans: The Distortion of Mexican-American History," *El Grito* 2:1 (Fall 1968), 13–26. See Alex M. Saragoza, "Recent Chicano Historiography: An Interpretive Essay," *Aztlán* 19:1 (Spring 1988–90), 1–77, for a fuller discussion of the emergence of Chicano history as a field.

13. Juan Gómez-Quiñones, "On Culture," *Revista Chicano-Riquena* 5:2 (1977), 29.

14. Ibid., 33, 35, 39. For a more recent discussion of these issues by the author, see *Chicano Politics: Reality and Promise, 1940–1990* (Albuquerque: Univ. of New Mexico Press, 1990). For confirmation of Gómez-Quiñones's role in shaping the field of Chicano history in the 1970s, see Richard Griswold del Castillo, "Southern California Chicano History: Regional Origins and National Critique," *Aztlán* 19:1 (Spring 1988–90), 112–13.

15. The pull toward depicting cultural worlds as bipolar opposites is a general feature of literature and history, particularly when comparing a largely urban

setting, as in Los Angeles, with an agrarian country like Mexico. See Raymond Williams, *The Country and the City* (New York: Oxford Univ. Press, 1973), for a full exploration of this tendency in writings on Great Britain.

16. Richard Griswold del Castillo, "La Raza Hispano Americana," 255; Albert Camarillo, *Chicanos in a Changing Society: From Mexican Pueblos to American Barrios in Santa Barbara and Southern California, 1848–1930* (Cambridge, Mass.: Harvard Univ. Press, 1979), 154. See also Rabinowitz, "Race, Ethnicity, and Cultural Pluralism," 39.

17. See George P. Rawick, *From Sundown to Sunup: The Making of the Black Community* (Westport, Conn.: Greenwood, 1972); John W. Blassingame, *The Slave Community: Plantation Life in the Antebellum South* (New York: Oxford Univ. Press, 1972); Eugene D. Genovese, *Roll, Jordan, Roll: The World the Slaves Made* (New York: Pantheon, 1974); Herbert G. Gutman, *The Black Family in Slavery and Freedom, 1750–1925* (New York: Pantheon, 1976); Lawrence W. Levine, *Black Culture and Black Consciousness: Afro-American Folk Thought from Slavery to Freedom* (New York: Oxford Univ. Press, 1977). The counterpart to Handlin in the historiography of slavery in shaping a generation of scholarship was Stanley M. Elkins, *Slavery: A Problem in American Institutional and Intellectual Life* (Chicago: Univ. of Chicago Press, 1959).

18. Gordon H. Chang, "Asian Americans and the Writing of Their History," *Radical History Review* 53 (1992), 108. For the most recent examples of this perspective, see Ronald Takaki, *Strangers from a Different Shore: A History of Asian Americans* (New York: Penguin Books, 1990), and Sucheng Chan, *Asian Americans: An Interpretive History* (Boston: Twayne, 1991).

19. See the essays in Ronald Takaki, ed., *From Different Shores: Perspectives on Race and Ethnicity in America* (New York: Oxford Univ. Press, 1987), for a full understanding of this debate.

20. See Camarillo, *Chicanos in a Changing Society*; Rodolfo Acuña, *Occupied America: The Chicano's Struggle toward Liberation* (San Francisco: Canfield, 1972); Mario Barrera, *Race and Class in the Southwest: A Theory of Racial Inequality* (Notre Dame: Univ. of Notre Dame Press, 1979); Mario T. García, *Desert Immigrants: The Mexicans of El Paso, 1880–1920* (New Haven: Yale Univ. Press, 1981); David Montejano, *Anglos and Mexicans in the Making of Texas, 1836–1986* (Austin: Univ. of Texas Press, 1987).

21. For an insightful essay on the impact of Chicano-related writings on the historiography of the American West, see Richard White, "Race Relations in the American West," *American Quarterly* 38 (1986), 396–416.

22. Mario T. García, *Desert Immigrants: The Mexicans of El Paso, 1880–1920* (New Haven: Yale Univ. Press, 1981), 231. See also García, "La Frontera: The Border as Symbol and Reality in Mexican-American Thought," *Mexican Studies/ Estudios Mexicanos*: 1:2 (Summer 1985), 195–225.

23. Arnoldo De León, *The Tejano Community, 1836–1900* (Albuquerque: Univ. of New Mexico Press, 1982), 206.

24. See Roger Rouse, "Mexican Migration and the Social Space of Postmodernism," *Diaspora* 1:1 (Spring 1991), 11.

25. Renato Rosaldo, *Culture and Truth: The Remaking of Social Analysis* (Boston: Beacon Press, 1989), 28–29.

26. For the best summary of these trends, see Richard Johnson, "What Is Cultural Studies Anyway?," *Social Text* 16 (1986–87), 38–80.

27. For two critical, yet different, appraisals of the relationship between

cultural studies and Chicano/a studies, see José David Saldivar, "The Limits of Cultural Studies," *American Literary History* 2:2 (Summer 1990), 251–66; and George Lipsitz, "*Con Safos:* Can Cultural Studies Read the Writings on the Walls?" (unpublished paper, Jan. 1991). For feminist critiques of the nationalist position, see Norma Alarcón, "Chicano Feminism: In the Tracks of 'the' Native Women," *Cultural Studies* 4:3 (Oct. 1990), 248–56.

28. For one of the best discussions of the evolution of ideas about "culture" in anthropology, see Clifford Geertz, *The Interpretation of Cultures* (New York: Basic Books, 1973), 3–30. See also Rosaldo, *Culture and Truth,* pp. 25–45, for a more recent description of the changes in the field of anthropology.

29. James Clifford, "Introduction: Partial Truths," in *Writing Culture: The Poetics and Politics of Ethnography,* James Clifford and George E. Marcus, eds. (Berkeley: Univ. of California Press, 1986), 18–19.

30. George E. Marcus, "Contemporary Problems of Ethnography in the Modern World System," in Clifford and Marcus, eds., *Writing Culture,* 177–78.

31. Rouse, "Mexican Migration," 8, 18–19.

32. The term "trans-creation" comes from Juan Flores and George Yudice, "Living Borders/Buscando America: Languages of Latino Self-Formation," *Social Text* 8:2 (1990), 74.

33. Gloria Anzaldúa, *Borderlands/La Frontera: The New Mestiza* (San Francisco: Spinsters/Aunt Lute, 1987), 79. See also Rosaldo, *Culture and Truth,* 215–17.

34. See Benedict Anderson, *Imagined Communities: Reflections on the Origin and Spread of Nationalism,* rev. ed. (London: Verso, 1983, 1991), for an analysis of the problem of nationalism as a worldwide phenomena. For the Mexican case, see Ilene V. O'Malley, *The Myth of the Revolution: Hero Cults and the Institutionalization of the Mexican States, 1920–1940* (New York: Greenwood Press, 1986).

35. Eric Hobsbawm, "Introduction: Inventing Traditions," in *The Invention of Tradition,* Eric Hobsbawm and Terence Ranger, eds. (Cambridge, Eng.: Cambridge Univ. Press, 1983), 4–5.

36. These are fully noted in Kevin Starr's collected works, including *Americans and the California Dream: 1850–1915* (New York: Oxford Univ. Press, 1973), *Inventing the Dream: California through the Progressive Era* (New York: Oxford Univ. Press, 1985), and *Material Dreams: Southern California through the 1920s* (New York: Oxford Univ. Press, 1990).

37. Etienne Balibar, "The Nation Form: History and Ideology," in Balibar and Immanuel Wallerstein, eds., *Race, Nation, Class: Ambiguous Identities* (London: Verso, 1991), 96. For another discussion of the link between "peoplehood" and nation-building, see Immanuel Wallerstein, "The Construction of Peoplehood: Racism, Nationalism, Ethnicity," in *Race, Nation, Class,* 71–85.

38. Rosaldo, *Culture and Truth,* 209.

39. Michael M. J. Fischer, "Ethnicity and the Post-modern Arts of Memory," in Clifford and Marcus, eds., *Writing Culture,* 195.

40. George Lipsitz, *Time Passages: Collective Memory and American Popular Culture* (Minneapolis: Univ. of Minnesota Press, 1990), 16–17.

41. Stuart Hall, "Cultural Identity and Diaspora," in *Identity: Community, Culture, Difference,* Jonathan Rutherford, ed. (London: Lawrence & Wishart, 1990), 225.

42. This study would have been impossible were it not for the work of earlier Chicano historians who have focused on Los Angeles. See especially Ri-

cardo Romo, *East Los Angeles: History of a Barrio* (Austin: Univ. of Texas Press, 1983); Richard Griswold del Castillo, *The Los Angeles Barrio, 1850–1890: A Social History* (Berkeley: University of California Press, 1979); Pedro G. Castillo, "The Making of a Mexican Barrio: Los Angeles, 1890–1920" (Ph.D. diss., University of California, Los Angeles, 1979); and Douglas G. Monroy, "Mexicanos in Los Angeles, 1930–1941: An Ethnic Group in Relation to Class Forces" (Ph.D. diss., University of California, Los Angeles, 1978).

43. Not only have 90 percent of Mexican immigrants settled in California, Texas, New Mexico, Arizona, or Colorado, but also the majority of immigrants from Asian countries first landed in the West Coast. Furthermore, a diverse group of Indian tribes make up significant parts of the populations of Arizona and New Mexico, while southerners brought black slaves into Texas and cast its history very early on in a southern mold. The Anglo American heritage of the region, of course, is a product of migrants or descendants of migrants to this area, most of whom arrived in the twentieth century. Finally, European immigration, though never reaching the levels of East Coast cities, has also been significant.

44. Lipsitz, *Time Passages,* 21.

Part One
Crossing Borders

1. James Clifford, "Introduction: Partial Truths," in *Writing Culture: The Poetics and Politics of Ethnography,* James Clifford and George E. Marcus, eds. (Berkeley: Univ. of California Press, 1986), 10.

2. Chas. T. Connell and R. W. Burton, Report to James J. Davis, Secretary of Labor, "Survey of Labor Conditions on the Mexican Border and Contiguous Territory as Relates to Unskilled Labor," 15 Nov. 1922, File 165/223, p. 15, U.S. Mediation & Conciliation Service, RG 280, National Archives, Washington, D.C. Hereafter referred to as Connell and Burton Survey.

3. See Phillip Sonnichsen, Notes to "Texas-Mexican Border Music, Vols 2 & 3, Corridos Parts 1 & 2" (Berkeley: Arhoolie Records and Chris Strachwitz, 1975), 6–7; and Nellie Foster, "The *Corrido:* A Mexican Culture Trait Persisting in Southern California" (Master's thesis, University of Southern California, 1939), 180–82.

Chapter 1
Farewell Homeland

1. Manuel Gamio, *The Life Story of the Mexican Immigrant* (1931; rpt., New York: Dover, 1971), 87–91.

2. Ricardo Romo, "The Urbanization of Southwestern Chicanos in Early Twentieth Century," *New Scholar* 6 (1977), 194. Paul S. Taylor explains the difficulties inherent in producing accurate statistical figures for Mexican immigration in this period from American and Mexican government sources in *Mexican Labor in the United States: Migration Statistics,* Univ. of California Publications in Economics, 6, No. 3 (Berkeley: Univ. of California Press, 1930), I, 237–55.

During the 1920s, Mexicans accounted for over 11 percent of the total legal immigration to the United States. Since 1820, Mexico ranks eighth among the nations of origin of legal immigrants to the United States. See Alejandro

Portes and Robert L. Bach, *Latin Journey: Cuban and Mexican Immigrants in the United States* (Berkeley: Univ. of California Press, 1985), 79; and Michael C. LeMay, *From Open Door to Dutch Door: An Analysis of U.S. Immigration Policy Since 1820* (New York: Praeger, 1987), 2–3.

3. For some of the best work from a socio-economic perspective, see Mark Reisler, *By the Sweat of Their Brow: Mexican Immigrant Labor in the United States, 1900–1940* (Westport, Conn.: Greenwood, 1976); Lawrence A. Cardoso, *Mexican Emigration to the United States, 1897–1931: Socio-economic Patterns* (Tucson: Univ. of Arizona Press, 1980). More recent immigration has been analyzed by Portes and Bach, *Latin Journey;* Douglas Massey, Rafael Alarcón, Jorge Durand, and Humberto González, *Return to Aztlán: The Social Process of International Migration from Western Mexico* (Berkeley: Univ. of California Press, 1987).

4. Arthur F. Corwin, "Early Mexican Labor Migration: A Frontier Sketch, 1848–1900," in *Immigrants—and Immigrants: Perspectives on Mexican Labor Migration to the United States,* Arthur F. Corwin, ed. (Westport, Conn.: Greenwood, 1979), 25–37; Oscar J. Martinez gives higher estimates of the Mexican population during this period primarily based on an estimated 40 percent margin of error in official statistics. See "On the Size of the Chicano Population: New Estimates, 1850–1900," *Aztlán: International Journal of Chicano Studies Research* 6 (1975), 43–67.

5. Reisler, *Sweat,* 3–8; Cardoso, *Emigration,* 18–27.

6. William S. Bernard, "A History of U.S. Immigration Policy," in *Immigration: Dimensions of Ethnicity,* Reed Ueda, ed. (Cambridge, Mass.: Belknap-Harvard Univ. Press, 1982), 87–99.

7. Reisler, *Sweat,* 8–13, 24–42.

8. Ibid., 15–16; Cardoso, *Emigration,* 38–44.

9. Cardoso, *Emigration,* 13–17.

10. David M. Pletcher, *Rails, Mines, and Progress: Seven American Promoters in Mexico, 1867–1911* (Ithaca: Cornell Univ. Press, 1958). British, French, and German capital were also crucial in the development of the Mexican railway system.

11. Ibid., 3; Charles S. Aiken, "The Land of Tomorrow," *Sunset Magazine* 22 (1909), 509–10.

12. John H. Coatsworth, *Growth against Development: The Economic Impact of Railroads in Porfirian Mexico* (Dekalb: Northern Illinois Univ. Press, 1981), 149–74.

13. Consul General, Mexico City, Letter to U.S. Secretary of State, 12 Sept. 1912, File No. 812.4051, Records of the Department of State Relating to the Internal Affairs of Mexico, 1910–29, RG 59, National Archives, Washington, D.C.; Lorena May Parlee, "Porfirio Díaz, Railroads, and Development in Northern Mexico: A Study of Government Policy toward the Central and Nacional Railroads, 1876–1910" (Ph.D. diss., University of California, San Diego, 1981), 145, 148–52; William H. Beezley, *Judas at the Jockey Club and Other Episodes of Porfirian Mexico* (Lincoln: Univ. of Nebraska, 1987), 4–5, 16, 19–26.

14. Parlee, "Díaz," 113–14; Arthur P. Schmidt, Jr., "The Social and Economic Effect of the Railroad in Puebla and Veracruz, Mexico, 1867–1911" (Ph.D. diss., Indiana University, 1973), 191; Robert Redfield, *Tepoztlán, A Mexican Village: A Study in Folk Life* (Chicago: Univ. of Chicago Press, 1930), 86–87.

15. Redfield, *Tepoztlán*, 39; Ernesto Galarza, *Barrio Boy* (Notre Dame: Univ. of Notre Dame Press, 1971), 37.

16. Massey et al., *Return to Aztlán*, 63–65.

17. Redfield, *Tepoztlán*; Manuel Gamio, *La población del Valle de Teotihuacan* (Mexico City: Talleres Gráficas de la Nación, 1922); Luis González, *San José de Gracia: Mexican Village in Transition*, trans. John Upton (Austin: Univ. of Texas Press, 1972); Carlos B. Gil, *Life in Provincial Mexico: National and Regional History Seen from Mascota, Jalisco, 1867–1972*, UCLA Latin American Studies, 53 (Los Angeles: UCLA Latin American Center Publications, 1983); Erich Fromm and Michael Maccoby, *Social Character in a Mexican Village: A Sociopsychoanalytic Study* (Englewood Cliffs: Prentice-Hall, 1970); George M. Foster, *Tzintzuntzan: Mexican Peasants in a Changing World*, rev. ed. (New York: Elsevier, 1979).

18. In this manner, they share much in common with the accounts of Anglo American travelers south of the border. See Beezley, *Judas*, 78–83.

19. Cynthia Hewitt de Alcántara, *Anthropological Perspectives on Rural Mexico* (London: Routledge and Kegan Paul, 1984), particularly chap. 1. Redfield, however, would later be criticized by Oscar Lewis for romanticizing Mexican village life. See Susan M. Rigdon, *The Culture Facade: Art, Science, and Politics in the Work of Oscar Lewis* (Urbana: Univ. of Illinois Press, 1988), 40–44.

20. Sumner W. Cushing, "The Distribution of Population in Mexico," *Geographical Review* 11 (1921), 227, 233. See also Beezley, *Judas*, 12, for a description of the capital, with 330,000 residents in 1890, as "one of the world's largest villages."

21. Redfield, *Tepoztlán*, 58–59, 88–90, 218–21; Gil, *Mascota*, 38.

22. Galarza, *Barrio Boy*, 41–45. In a similar environment, Sarah Deutsch argues that women's activities in the traditional communal villages of northern New Mexico were vital in sustaining the community, particularly once men migrated from the village for work; see "Women and Intercultural Relations: The Case of Hispanic New Mexico and Colorado," *Signs: Journal of Women in Culture and Society* 12 (1987), 719–39.

23. See Sarah Deutsch, *No Separate Refuge: Culture, Class, and Gender on an Anglo-Hispanic Frontier in the American Southwest, 1880–1940* (New York: Oxford Univ. Press, 1987), 41–62.

24. Paul Friedrich, *Agrarian Revolt in a Mexican Village*, 2nd ed. (Chicago: Univ. of Chicago Press, 1970, 1977), 15, 22, 45–49.

25. González, *San José*, 87, 110; Gil, *Mascota*, 37–38.

26. González, *San José*, 96, 110; Gil, *Mascota*, 111–14; Mary Kay Vaughn, "Primary Education and Literacy in Nineteenth-Century Mexico: Research Trends, 1968–1988," *Latin American Research Review* 25:1 (1990), 46.

27. Vaughn, "Primary Education," 42–43; Gil, *Mascota*, 181–82; *Estadísticas Históricas de México: Tomo I*, Segunda Edición (Instituto Nacional de Estadística, Geografía e Informática, 1990), 95.

28. Vaughn, "Primary Education," 43–45; *Estadísticas Históricas*, 95–108.

29. González, *San José*, 87; Cynthia Nelson, *The Waiting Village: Social Change in Rural Mexico* (Boston: Little, Brown, 1971), 34–35. It is probable that priests exerted much less influence in the north, where the shortage of clerics severely limited their ability to integrate themselves into more than a handful of communities.

30. This equates to a transition from hamlet status to a community deemed important enough to house an alderman.

31. González, *San José*, 104–6; Redfield, *Tepoztlán*, 66–67. For more recent examples of this distrust of government in Mexican villages see Foster, *Tzintzuntzan*, 167–83; and Lola Romanucci-Ross, *Conflict, Violence, and Morality in a Mexican Village* (Palo Alto: National Press Books, 1973), 116–30.

32. Redfield, *Tepoztlán*, 30, 108, 191; González, *San José*, 101.

33. González, *San José*, 112.

34. Michael C. Meyer and William L. Sherman, *The Course of Mexican History*, 2nd ed. (New York: Oxford Univ. Press, 1983), 457; Alan Knight, "Racism, Revolution, and *Indigenismo:* Mexico, 1910–1940," in *The Idea of Race in Latin America, 1870–1940*, Richard Graham, ed. (Austin: Univ. of Texas Press, 1990), 73–76. As for mestizos forming the bulk of migrants to the United States, in addition to Table 1, see Manuel Gamio, *Mexican Immigration to the United States: A Study of Human Migration and Adjustment* (1930; rpt. New York: Dover, 1971), 57–58.

35. The quote is from Paul S. Taylor, *A Spanish-Mexican Peasant Community: Arandas in Jalisco, Mexico*, Ibero-Americana, 4 (Berkeley: Univ. of California Press, 1933), 18–20; Knight, "Racism," 78–102.

36. The "tradition" of parental control over a Mexican child's choice of spouse was itself a product of the growth of state power over that of the Catholic Church dating from a 1771 law instituted by the Fourth Provincial Council of the Mexican Catholic Church and the Spanish Royal Pragmatic on Marriage in 1776. See Ramón Arturo Gutiérrez, "Marriage, Sex and the Family: Social Change in Colonial New Mexico, 1690–1846" (Ph.D. diss., University of Wisconsin-Madison, 1980), 237–69.

37. A total of 133 separate marriages were identified as occurring in Mexico, with 38 of them between individuals born in the same locale. Only 53 marriages (or 39.8%) were between individuals born in the same state (Petitions for Naturalization, National Archives, Laguna Niguel, Calif.). For more discussion of the selection of marriage partners, see Chapter 6.

38. Gutiérrez, "Marriage, Sex and the Family," 408. See Friedrich, *Agrarian Revolt*, 49, for this process in the early twentieth-century Mexican village.

39. Foster, *Tzintzuntzan*, 67–74. González found the *robo* in San José de Gracia as early as 1901. See *San José*, 89. This practice may have been an adaptation reversing the order of marriage and *robo* in a new cultural world. Ramón Gutiérrez notes the existence in colonial New Mexico of an ancient ritual of the supposed "kidnapping" of the bride from the groom by the bride's kin following the marriage in order to further symbolize the comingling of two clans. See Gutiérrez, "Marriage, Sex and the Family," 198.

40. Gutiérrez, "Marriage, Sex and the Family," 171–74.

41. Gil, *Mascota*, 94–95.

42. Ibid., 95–98. See also Friedrich, *Agrarian Revolt*, 18–22.

43. For a description of studies on *machismo* within the Mexican and Chicano family, see Alfredo Mirandé and Evangelina Enríquez, *La Chicana: The Mexican-American Woman* (Chicago: Univ. of Chicago Press, 1979), 108–17.

44. González, *San José*, 111; Fromm and Macoby, *Social Character*, chaps. 7 and 8.

45. Gutiérrez, "Marriage, Sex and the Family," 31–52; Richard Griswold del Castillo, "Patriarchy and the Status of Women in the Late Nineteenth-

Century Southwest," in *The Mexican and Mexican American Experience in the 19th Century,* Jaime E. Rodríguez O., ed. (Tempe: Bilingual Press, 1989), 85; Richard Griswold del Castillo, *La Familia: Chicano Families in the Urban Southwest, 1848 to the Present* (Notre Dame: Univ. of Notre Dame Press, 1984), 26–29.

46. Gil, *Mascota,* 81–92, 97–99.

47. Oscar Lewis, *A Death in the Sánchez Family* (New York: Random House, 1969), xiv.

48. This percentage amounts to 1,145 migrants out of 2,622 in the sample where both birth date and date of first arrival into the United States were recorded. Naturalization Records, National Archives, Laguna Niguel, Calif.

49. Gamio, *Immigration,* 13–32.

50. Taylor, *Arandas,* 43.

51. The one notable exception to this pattern was among Irish immigrants, where single women formed a large percentage of those coming to the United States in the nineteenth century. See Hasia Diner, *Erin's Daughters in America: Irish Immigrant Women in the Nineteenth Century* (Baltimore: John Hopkins Univ. Press, 1983).

52. *Resumen del censo general de habitantes de 30 de noviembre de 1921* (México: Talleres Gráficos de la Nación, 1928). In recent decades, female migration from Mexico has increased dramatically, including both single women and female-headed households. See Wayne Cornelius, "Immigration, Mexican Development Policy, and the future of U.S.-Mexican Relations," in *Mexico and the United States,* R. McBride, ed. (Englewood Cliffs: Prentice-Hall, 1981), 104–27; and Sylvia Guendelman and Auristela Perez-Itriago, "Double Lives: The Changing Role of Women in Seasonal Migration," *Women's Studies* 13 (1987), 249–71.

53. Taylor, *Arandas,* 47, 53.

54. Some European immigrants, of course, also had this vision of their emigration, most notably Italians and Poles. For a description of this return migration, see Dino Cinel, *From Italy to San Francisco: The Immigrant Experience* (Stanford: Stanford Univ. Press, 1982), 1–2, 46–59; John J. Bukowczyk, *And My Children Did Not Know Me: A History of the Polish Americans* (Bloomington: Indiana Univ. Press, 1987), 12–14, 21, 34–35.

55. Taylor, *Arandas,* 29, 32–34.

56. Ibid., 13.

57. Fromm and Maccoby, *Social Character,* 37.

Chapter 2
Across the Dividing Line

1. For a vivid description of attempts to define the border over time, see Patricia Nelson Limerick, *Legacy of Conquest: The Unbroken Past of the American West* (New York: W.W. Norton, 1987), chap. 7.

2. Supervising Inspector F. W. Berkshire, Immigration Service, El Paso, Texas, Letter to Commissioner-General of Immigration, 30 June 1910, File 52546/31B, p. 2, Immigration and Naturalization Service (INS), RG 85, National Archives, Washington, D.C.

3. Immigrant Inspector Frank R. Stone, Letter to Supervising Inspector F. W. Berkshire, Immigration Service, El Paso, Texas, 23 June 1910, File 52546/31B, pp. 4–5, INS, National Archives. Hereafter referred to as Stone Report.

4. Ibid., 7.

5. Berkshire, Letter to Commissioner-General of Immigration, 30 June 1910, pp. 2–3, INS, National Archives; Chas. T. Connell and R. W. Burton, Report to James J. Davis, Secretary of Labor, "Survey of Labor Conditions on the Mexican Border and Contiguous Territory as Relates to Unskilled Labor," 15 Nov. 1922, File 165/223, p. 24, U.S. Mediation and Conciliation Service, RG 280, National Archives, Washington, D.C. Hereafter referred to as Connell and Burton Survey.

6. Stone Report, p. 16.

7. Paul S. Taylor, *A Spanish-Mexican Peasant Community: Arandas in Jalisco, Mexico,* Ibero-Americana, 4 (Berkeley: Univ. of California Press, 1933), 47–55.

8. For a description of widespread European return migration, see John Bodnar, *The Transplanted: A History of Immigrants in Urban America* (Bloomington: Indiana Univ. Press, 1985), 53–54.

9. Stone Report, Table "A." Women, however, made up 1,003 (25%) of the 4,017 "statistical," or permanent, immigrants admitted, further evidence that women largely arrived at the border in family groups seeking permanent settlement in the United States.

10. My sample contains eight nuns who sought naturalization and four older female students, although many others apparently originally crossed as students. In comparison, only 83 other single women who had crossed the border above the age of fifteen are found in my sample, and many of these were probably grown children who migrated with their entire families. A total of 365 Mexican women sought naturalization in this period, but the majority arrived in the United States as children.

11. *Estadísticas Históricas de México,* Tomo I (Mexico City: Instituto Nacional de Estadística, Geografía e Informática, 1985), 24, 27; Robert V. Kemper and Anya Peterson Royce, "Mexican Urbanization since 1821: A Macro-historical Approach," *Urban Anthropology* 8 (1979), 270–72, 277–78.

12. Stone Report, 7, 19.

13. Jean-Pierre Bastian, "Metodismo y clase obrera durante el Porfiriato," *Historia Mexicana* 33 (1983), 39–71.

14."Julian Ruiz," Racial Minorities Survey, San Diego Project, Federal Writers Project collection, MS 306, Department of Special Collections, University of California, Los Angeles.

15. Oscar J. Martínez, *Border Boom Town: Ciudad Juárez since 1848* (Austin: Univ. of Texas Press, 1975), 23.

16. Ibid., 20–21.

17. Ibid., 22–25.

18. Ibid., 21.

19. The research undertaken by Manuel Gamio in the 1920s to identify the backgrounds of Mexican immigrants in the United States was framed in the assumption that these immigrants were in the U.S. temporarily and would eventually return to Mexico. Consequently, he identified the states of origin of these migrants by utilizing money orders sent home to relatives by Mexican workers. More permanent settlers who had severed financial ties with their families or brought their families north were therefore overlooked. *Mexican Immigration to the United States: A Study of Human Migration and Adjustment* (1930; rpt., New York: Dover, 1971), chap. 2.

20. Gamio, *Immigration,* 25–30.

21. Friedrich Katz, "Labor Conditions on Haciendas in Porfirian Mexico: Some Trends and Tendencies," *Hispanic American Historical Review* 54 (1974), 31–37.

22. Martínez, *Boom Town,* 20.

23. Clifford Alan Perkins, *Border Patrol: With the U.S. Immigration Service on the Mexican Boundary, 1910–54* (El Paso: Texas Western, 1978), 54.

24. Stone Report, 15–16.

25. Ibid., Exhibit 16, Appendix.

26. Seraphic report on conditions on the Mexican border, 1906–7, File 51423/1, p. 16, INS, National Archives. Hereafter referred to as the Seraphic Report.

27. Ibid., 16; interview with Mike Romo, 8 and 14 Oct. 1975, conducted by Oscar J. Martínez, University of Texas at El Paso, Institute of Oral History, No. 215, 8, 24.

28. Interview with Charles Armijo, 30 Jan. 1973, conducted by Leon C. Metz, University of Texas at El Paso, Institute of Oral History, No. 106, pp. 2–3; interview with Cleofas Calleros, 14 Sept. 1972, conducted by Oscar J. Martínez, No. 157, p. 9; interview with Conrado Mendoza, 4 Dec. 1976, conducted by Mike Acosta, No. 252, p. 3; interview with Señor Pedro González, 3 Sept. 1976, conducted by Oscar J. Martínez, No. 313, pp. 10–11.

29. F.W. Berkshire, Supervising Inspector, Letter to the Commissioner-General of Immigration, 1 Nov. 1909, File 52546/31, p. 9, INS, National Archives.

30. Ibid., 8; Connell and Burton Survey, 5; *Mexicans in California,* Report of Governor C.C. Young's Mexican Fact-Finding Committee (1930; rpt. San Francisco: R and E Research Associates, 1970), 31.

31. Berkshire to Commissioner-General, 13 Nov. 1909, 8.

32. Perkins, *Border Patrol,* p. 54.

33. F. W. Berkshire, Letter to Commissioner-General, 24 May 1912, File 52546/31E, p. 2; Berkshire, Letter to Commissioner-General, 1 Nov. 1909, File 52546/31, p. 4, INS, National Archives.

34. Berkshire to Commissioner-General, 1 Nov. 1909, p. 5; Berkshire to Commissioner-General, 24 June 1911, File 52546/31C, p. 7; Attorneys Caldwell and Sweeney, Letter to Commissioner-General of Immigration, 11 Jan. 1912, File 52546/31E, pp. 1–3, 6; Berkshire to Commissioner-General, 24 May 1912, File 52546/31E, p. 7, INS, National Archives.

35. F. W. Berkshire, Letter to Commissioner-General, 24 June 1911, File 52546/31C, p. 2, INS, National Archives.

36. Immigrant Inspector Richard H. Taylor, Letter to Commissioner-General, 9 July 1912, File 52546/31E, p. 2; Assistant Solicitor Edward E. Quigley, Letter to the Secretary of Commerce and Labor, 3 Oct. 1912, File 52546/31E, pp. 15–16, INS, National Archives.

37. Solicitor Charles Earl, Letter to the Secretary of Commerce & Labor, 10 Oct. 1912, File 52546/31E; W. W. Husband, Letter to the Commissioner-General, 12 Feb. 1913, File 52546/31F, p. 1, INS, National Archives.

38. Harry C. Carr, Letter to Senator Works, El Paso, Texas, 1 July 1916, File 53912/25, p. 1, INS, National Archives.

39. Ibid., 1–2.

40. Ibid., 2.

41. Interview with J.C. Machuca, 9 May 1975, conducted by Oscar J. Martínez, University of Texas at El Paso, Institute of Oral History, No. 152, p. 8.

42. Inspector in Charge Will E. Soult, Letter to Supervisor, 13 Dec. 1923, File 52903/29, p. 1, INS, National Archives.

43. Ibid., 2.

44. Ibid., 2–5.

45. Interview with Señora X (Anonymous), 6 Dec. 1979, conducted by María Nuckolls, University of Texas at El Paso, Institute of Oral History, No. 722, pp. 24–26; interview with José Cruz Burciaga, 16 Feb. 1974, conducted by Oscar J. Martínez, No. 143, pp. 20–21.

46. Inspector in Charge Walter F. Miller, Letter to Supervisor, 8 Dec. 1923, File 52903/29, p. 1, INS, National Archives.

47. Soult to Supervisor, 13 Dec. 1923, p. 3.

48. Ibid.; Miller to Supervisor, 8 Dec. 1923, p. 2.

49. Mark Reisler, *By the Sweat of Their Brow: Mexican Immigrant Labor in the United States, 1900–1940* (Westport, Conn.: Greenwood, 1976), 12, 59, 265; U.S. Department of Labor, *Annual Report of the Commissioner-General of Immigration* (1918), 316; Soult to Supervisor, 13 Dec. 1923, p. 2.

50. One can see the extensive transcripts of these "Boards of Special Inquiry" in the Records of the Immigration and Naturalization Service, RG 85, Box 250, National Archives, Washington, D.C.

51. Immigrant Inspector J. H. Bradford, Letter to the Commissioner-General, 3 June 1913, File 52546/31H, 9, INS, National Archives.

52. Connell and Burton Survey, 17; Department of Labor, *Annual Report of the Commissioner-General of Immigration* (1920), 433. See also Perkins, *Border Patrol,* 17–19, 44, for precedence of railroad-hopping among illegal Chinese entrants.

53. Perkins, *Border Patrol,* chaps. 3, 5–8.

54. Connell and Burton Survey, 14.

55. Reisler, *Sweat,* 265; Perkins, *Border Patrol,* 94.

56. Interview with Wesley E. Stiles, Jan. 1986, conducted by Wesley C. Shaw, University of Texas at El Paso, Institute of Oral History, No. 756, p. 4.

57. Interview with Edwin M. Reeves, 25 June 1974, conducted by Robert H. Novak, University of Texas at El Paso, Institute of Oral History, No. 135, p. 3.

58. Interview with J. C. Machuca, 9 May 1975, conducted by Oscar J. Martínez, University of Texas at El Paso, Institute of Oral History, No. 152, pp. 10–12, 21.

59. Perkins, *Border Patrol,* pp. 54–55; interview with Cleofas Calleros, 10. More on this subject will be presented in the following chapter.

60. Interview with José Cruz Burciaga, 16 Feb. 1974, conducted by Oscar J. Martínez, University of Texas at El Paso, Institute of Oral History, No. 143, pp. 25–29.

61. Interview with Jesús Pérez, 8 May 1976, conducted by Magdaleno Cisneros, University of Texas at El Paso, Institute of Oral History, No. 249, pp. 20–21.

62. Interview with Catalina Aranda, 30 Nov. 1976, conducted by María del Carmen González, University of Texas at El Paso, Institute of Oral History, No. 269, pp. 3–6.

63. Interview with Angel Oaxaca, 26 Feb. 1977, conducted by Oscar J.

Martínez, University of Texas at El Paso, Institute of Oral History, No. 418, pp. 34–35.

64. Interview with Epitacio Armendáriz, 14 Oct. 1979, Conducted by Virgilio H. Sánchez, University of Texas at El Paso, Institute of Oral History, No. 551, pp. 30–32.

65. Attorney Caldwell, Memorandum to the Office of the Solicitor General, regarding charges against Supervising Inspector F.W. Berkshire, 11 Jan. 1912, File 52546/31E, p. 5, INS, National Archives.

Chapter 3
Newcomers in the City of the Angles

1. "Vidas (Ia. Zeferino Velázquez)," Documento 32, No. 94, Interview by Luis Felipe Recinos, Los Angeles, California, 20 April 1927, "Biographies & Case Histories" I folder, Z-R5, Manuel Gamio collection, Bancroft Library, University of California, Berkeley.

2. Naturalization Records, RG 21, National Archives, Laguna Niguel, California. From this chapter forward, all information taken from naturalization records is limited to those individuals whose residence was in the city of Los Angeles or the adjacent communities of Belvedere and Watts. Previously all Mexican immigrants living in southern California who took out their naturalization forms in Los Angeles were included in my figures.

3. See Table 5, Chapter 2.

4. *Mexicans in California*, Report of Governor C.C. Young's Mexican Fact-Finding Committee (1930; rpt., San Francisco: R and E Research Associates, 1970), 33–35.

5. Mario T. García, *Desert Immigrants: The Mexicans of El Paso, 1880–1920* (New Haven: Yale Univ. Press, 1981), 17–24, 65–74; Ricardo Romo, "The Urbanization of Southwestern Chicanos in the Early Twentieth Century," *New Scholar* 6 (1977), 185–86; interview with Inspector Hudson, U.S. Immigration Service, Eagle Pass, Texas, "American Government Officials" folder, pp. 130–33, 74/187c, Paul S. Taylor collection, Bancroft Library, University of California.

6. Romo, "Urbanization," 185–87; David Montejano, *Anglos and Mexicans in the Making of Texas, 1836–1986* (Austin: Univ. of Texas Press, 1987), 106–10, 167–69, 197–219.

7. Romo, "Urbanization," 196.

8. Mario Barrera, *Race and Class in the Southwest: A Theory of Racial Inequality* (Notre Dame: Univ. of Notre Dame Press, 1979), 72, 83–84.

9. Barrera, *Race and Class,* 60, 82–83; Camille Guerin-Gonzáles, "Cycles of Immigration and Repatriation: Mexican Farm Workers in California Industrial Agriculture, 1900–1940" (Ph.D. diss., University of California, Riverside, 1985), 60–61, 76; Lloyd Fellows, "Economic Aspects of the Mexican Rural Population in California with Special Emphasis on the Need for Mexican Labor in California" (Master's thesis, University of Southern California, 1929), 18–23. For a broader discussion of the forces which determined the growth of the Imperial Valley in the early twentieth century, see Donald Worster, *Rivers of Empire: Water, Aridity, and the Growth of the American West* (New York: Pantheon Books, 1985), 194–212, 221–22; and Kevin Starr, *Material Dreams: Southern California through the 1920s* (New York: Oxford Univ. Press, 1990), 20–44.

10. Interview with Benito Rodríguez, at Consulate General, El Paso, Texas,

undated (but probably 1929), Nos. 61–66, 113–17, "Mexican Officials" folder, 74/187c, Taylor collection; interview with Mr. Goeldner, Asst. to General Manager, Atchinson, Topeka & Santa Fe, Topeka, Kansas, 16 Aug. 1928, p. 1, "Railroads" folder, Taylor collection; interview with Inspector A.J. Milliken, U.S. Immigration Service, in charge of Santa Fe Bridge, El Paso, Texas, undated (but probably 1929), "American Government Officials" folder, Taylor collection.

11. Interview with J. R. Silva, employment agent for Mexican agricultural laborers, El Paso, Texas, undated (but probably 1929), 106–10, "Labor Contractors and Agencies" folder, 74/187c, Taylor collection; interview with Benito Rodríguez, 113–17, Taylor collection.

12. Interview with Mexican in office of J. R. Silva, 317–488, Taylor collection; interview with Manager and Assistant Manager, Gunn Supply Company, Ogden, Utah, 24 April 1928, No. 51–851, Taylor collection; Interview with G.J. Gonzales, Apex Employment Agency, Salt Lake City, Utah, undated (but probably 1928), No. 58–858, Taylor collection.

13. Douglas G. Monroy, "Mexicanos in Los Angeles, 1930–1941: An Ethnic Group in Relation to Class Forces" (Ph.D. diss., University of California, Los Angeles, 1978), 65–69.

14. Franklin Walker, "Pacific Electric," in *Los Angeles: Biography of a City,* John and LaRee Caughey, eds. (Berkeley: Univ. of California Press, 1976), 218–22; Ashleigh Brilliant, *The Great Car Craze: How Southern California Collided with the Automobile in the 1920s* (Santa Barbara: Woodbridge Press, 1989), 147–48.

15. Interview with Mr. Johnson, L. H. Manning Company, El Paso, Texas, undated (but probably 1929), "Labor Contractors and Agencies" folder, Taylor collection.

16. Interview with Mr. Morrill, Holmes Supply Company, Los Angeles, California, undated (but probably 1929), No. 1–173, "Labor Contractors and Agencies" folder, Taylor collection.

17. Richard Griswold del Castillo, *The Los Angeles Barrio, 1850–1890: A Social History* (Berkeley: Univ. of California Press, 1979), 36–39.

18. Pedro G. Castillo, "The Making of a Mexican Barrio: Los Angeles, 1890–1920" (Ph.D. diss., University of California, Santa Barbara, 1979), 116–19; Ricardo Romo, *East Los Angeles: History of a Barrio* (Austin: Univ. of Texas Press, 1983), 124–28; Stephan Thernstrom, *The Other Bostonians: Poverty and Progress in the American Metropolis, 1880–1970* (Cambridge: Harvard Univ. Press, 1973), 222, 226.

19. For estimates of the overall Chicano population in Los Angeles, see Albert Camarillo, *Chicanos in a Changing Society: From Mexican Pueblos to American Barrios in Santa Barbara and Southern California, 1848–1930* (Cambridge, Mass.: Harvard Univ. Press, 1979), 200. For figures on foreign-born residents, see *Mexicans in California,* 57; *Fifteenth Census of the United States: 1930, Population: Special Report on Foreign-Born White Families by Country of Birth of Head with an Appendix Giving Statistics for Mexican, Indian, Chinese and Japanese Families* (Washington, D.C.: GPO, 1933), Table 40, p. 213.

20. Geographic analysis of "Pioneers" mailing list, Manuel Ruiz papers, M295, Box 4, Folder 5, Department of Special Collections, Stanford University; Leo Carrillo, *The California I Love* (Englewood Cliffs: Prentice-Hall, 1961); Leonard Pitt, *The Decline of the Californios: A Social History of the Spanish-Speaking Californians, 1846–1890* (Berkeley: Univ. of California Press, 1966), 262–66.

21. Pitt, *Decline*, 286–90; Kevin Starr, *Inventing the Dream: California through the Progressive Era* (New York: Oxford Univ. Press, 1985), 75–89; Carey McWilliams, *Southern California: An Island on the Land* (New York: Duell, Sloan and Pearce, 1946), 64–67; Douglas Monroy, *Thrown among Strangers: The Making of Mexican Culture in Frontier California* (Berkeley: Univ. of California Press, 1990), 258–71.

22. David Lavender, *California: Land of New Beginnings* (Lincoln: Univ. of Nebraska Press, 1972), 313, 346, 372.

23. See Theodore E. Treutlein, "Los Angeles, California: The Question of the City's Original Spanish Name," *Southern California Quarterly* LV (Spring 1973), 1–7, for a full detailing of the original naming of the pueblo.

24. W. W. Robinson, *Los Angeles from the Days of the Pueblo: A Brief History and Guide to the Plaza Area* (San Francisco: California Historical Society, 1981), chaps. 1–5.

25. Olive P. Kirschner, "The Italian in Los Angeles" (Master's thesis, University of Southern California, 1920), 7; Romo, *East Los Angeles,* 70–71.

26. Romo, *East Los Angeles,* 63–67, 70–72, 77; Castillo, "Mexican Barrio," 87–91, 105–14; Gilbert G. González, "Factors Relating to Property Ownership of Chicanos in Lincoln Heights Los Angeles," *Aztlán* 2 (1971), 115–18; Proposal from Church of All Nations to Centennial Fund of the Board of Home Missions, Container #59, Bounded Volume: Parish Records 2, G. Bromley Oxnam papers, Manuscript Division, Library of Congress, Washington, D.C.

27. G. Bromley Oxnam, "Los Angeles: A City of Many Nations," *California Christian Advocate,* 18 Feb. 1926, p. 14.

28. Romo, *East Los Angeles,* 81–82; *Mexicans in California* Report, 177; Romo, "Urbanization," 185.

29. John E. Kienle, "Housing Conditions among the Mexican Population of Los Angeles" (Master's thesis, University of Southern California, 1912), 5.

30. Castillo, "Mexican Barrio," 106–07.

31. I disagree here with Ricardo Romo's assessment of residential clustering in *East Los Angeles,* 79. For further discussion of the operation of racially restrictive residential practices, see Albert Camarillo, *Chicanos in California: A History of Mexican Americans in California* (San Francisco: Boyd & Fraser, 1984), 35–38; John Modell, *The Economics and Politics of Racial Accommodation: The Japanese of Los Angeles, 1900–1942* (Urbana: Univ. of Illinois Press, 1977), 55–66.

32. Richard Griswold del Castillo, "La Raza Hispano Americana: The Emergence of an Urban Culture among the Spanish Speaking of Los Angeles, 1850–1880" (Ph.D. diss., University of California, Los Angeles, 1974), 276; see also Castillo, "Mexican Barrio," 72–74.

33. Griswold del Castillo, *Barrio,* 146.

34. For blacks in the Boyle Heights area, see Lawrence B. De Graff, "The City of Black Angels: Emergence of the Los Angeles Ghetto, 1890–1930," *Pacific Historical Review* 39 (1970), 333–36; Emory J. Tolbert, *The UNIA and Black Los Angeles: Ideology and Community in the American Garvey Movement* (Los Angeles: Center for Afro-American Studies, Univ. of California, 1980), 27–29, 31–33; and J. Max Bond, "The Negro in Los Angeles" (Ph.D. diss., University of Southern California, 1936), 68, 76, 83–84. For Japanese in Boyle Heights, see Modell, *Racial Accommodation,* 56–59, 72–75. For Russian Molokans in Boyle Heights, see Pauline V. Young, "The Russian Molokan Community in Los Angeles," *American Journal of Sociology* 35 (1929), 393–402. For Jews in

Boyle Heights and Brooklyn Heights, see Max Vorspan and Lloyd P. Gartner, *History of the Jews of Los Angeles* (San Marino: Huntington Library, 1970), 118–19. For Italians in Lincoln Heights, see González, "Property Ownership," 111, 116–18.

35. For a description of the continuance of this control in the post–World War II era, see Rodolfo F. Acuña, *A Community under Siege: A Chronicle of Chicanos East of the Los Angeles River, 1945–1975* (Los Angeles: Chicano Studies Research Center Publications, Univ. of California, 1984).

36. William W. McEuen, "A Survey of the Mexicans in Los Angeles" (Master's thesis, University of Southern California, 1914), 42; Neil M. Locke, "The Lodging Houses of Los Angeles" (Master's thesis, University of Southern California, 1913–14).

37. Castillo, "Mexican Barrio," 91–98; Kienle, "Housing Conditions," 8, Tables 5 and 7, 41–42; McEuen, "Survey," 32, 37; G. Bromley Oxnam, *The Mexican in Los Angeles* (Los Angeles: Interchurch World Movement of North America, 1920), p. 6–8.

38. McEuen, "Survey," 62–66.

39. Ibid., 23, 32; Kienle, "Housing Conditions," 7–8, 11; Castillo, "Mexican Barrio," 98; Romo, *East Los Angeles,* 71–72.

40. *Mexicans in California* Report, 177–78; Kienle, "Housing Conditions," 8; Oxnam, *Mexican in Los Angeles,* 7; Romo, *East Los Angeles,* 81.

41. Emory S. Bogardus, *The Mexican in the United States* (Los Angeles: Univ. of Southern California Press, 1934), 19, 31; *Mexicans in California* Report, 177; González, "Property Ownership," 122–23. For a discussion of the phenomenon of home ownership among other ethnic groups in another American city, see Olivier Zunz, *The Changing Face of Inequality: Urbanization, Industrial Development, and Immigrants in Detroit, 1880–1920* (Chicago: Univ. of Chicago Press, 1982), 152–61.

42. Bogardus, *Mexican in the United States,* 27.

43. Kienle, "Housing Conditions," 2; Municipal League of Los Angeles *Bulletin,* 2:7, 16 Feb. 1925, p. 5.

44. Kienle, "Housing Conditions," 6–7; Castillo, "Mexican Barrio," 104.

45. Bogardus, *Mexican in the United States,* 33.

46. *Mexicans in California* Report, 182–87, 193; Gladys Patric, *A Study of the Housing and Social Conditions in the Ann Street District of Los Angeles, California* (Los Angeles, 1918). See also Mary Lanigan, "Second Generation Mexicans in Belvedere" (Master's thesis, University of Southern California, 1932), 36–38.

47. Oxnam, *Mexicans in Los Angeles,* 10; *Mexicans in California* Report, 183–84.

48. Oxnam, *Mexicans in Los Angeles,* 6, 15–16.

49. For a fuller description of this WASPish boosterism, see Kevin Starr, *Material Dreams: Southern California through the 1920s* (New York: Oxford Univ. Press, 1990), particularly part 2.

50. Timothy C. Turner, "Where Folks Are Folks," *Los Angeles Times Illustrated Magazine,* 14 Sept. 1924.

51. "L.A. Tenement Problem and the Bubonic-Pneumonic Plagues," Municipal League of Los Angeles *Bulletin,* 2:7, 16 Feb. 1925, pp. 2–3.

52. Ibid., 5.

Part Two
Divided Loyalties

1. Ellwood P. Cubberly, *Changing Conceptions of Education* (Boston: Houghton Mifflin, 1909), 15–16.

2. Interview with Cleofas Calleros, 14 Sept. 1972, conducted by Oscar J. Martínez, University of Texas at El Paso, Institute of Oral History, No. 157, p. 14.

3. Original draft of *Mexican Immigration to the United States,* 79, Manuel Gamio collection, Bancroft Library, University of California, Berkeley. This particular paragraph did not make it to the final printed version.

4. "Educacion Mexicana en Estados Unidos," *La Opinión* editorial, 21 June 1930.

Chapter 4
Americanization and the Mexican Immigrant

1. Carey McWilliams, *Southern California: An Island in the Land* (New York: Duell, Sloan and Pearce, 1946), 165; Robert M. Fogelson, *The Fragmented Metropolis: Los Angeles, 1850–1930* (Cambridge, Mass.: Harvard Univ. Press, 1967), 80, Table 7; see also Moses Rischin, "Immigration, Migration, and Minorities in California: A Reassessment," *Pacific Historical Review* 41 (1972), 71–90, for a description of California's high proportion of newcomers.

2. Fogelson, *Fragmented Metropolis,* 77–81.

3. Ibid., 108–34, 223–28; Frank L. Beach, "The Effects of the Westward Movement on California's Growth and Development, 1900–1920," *International Migration Review* 3 (1969), 25–26.

4. As a result, consumption shot up from next to nothing in 1885 to forty oranges per American per year by 1914. George F. Mowry, *The California Progressives* (Berkeley: Univ. of California Press, 1951), 7; Kevin Starr, *Inventing the Dream: California through the Progressive Era* (New York: Oxford Univ. Press, 1985), 161–64.

5. Quoted in Fogelson, *Fragmented Metropolis,* 73–74. See also Bruce Henstell, *Sunshine and Wealth: Los Angeles in the Twenties and Thirties* (San Francisco: Chronicle Books, 1984), 19–29.

6. U.S. Census figures, although used by most scholars, are generally regarded as low estimates of ethnic populations because of notoriously widespread undercounting. For differing estimates of the population of these groups using census figures as a base, see Fogelson, *Fragmented Metropolis,* 75–84; Albert Camarillo, *Chicanos in a Changing Society: From Mexican Pueblos to American Barrios in Santa Barbara and Southern California, 1848–1930* (Cambridge: Harvard Univ. Press, 1979), 200; and John Modell, *The Economics and Politics of Racial Accommodation: The Japanese of Los Angeles, 1900–1942* (Urbana: Univ. of Illinois Press, 1977), 23.

7. Lewis Atherton, *Main Street on the Middle Border* (Bloomington: Indiana Univ. Press, 1984), 75.

8. Richard J. Jensen, *The Winning of the Midwest: Social and Political Conflict, 1888–96* (Chicago: Univ. of Chicago Press, 1971), 89–148, 232–37, 277–79; see also Paul Kleppner, *The Cross of Culture: A Social Analysis of Midwestern Politics, 1850–1900* (New York: Free Press, 1970).

9. Beach, "Effects of the Westward Movement," 27.

10. Fogelson, *Fragmented Metropolis,* 72.

11. Timothy C. Turner, "Where Folks Are Folks," *Los Angeles Times Illustrated Magazine,* 14 Sept. 1924, p. 7.

12. *Los Angeles Examiner,* 5 June 1909; Fogelson, *Fragmented Metropolis,* 191–92.

13. Fogelson, *Fragmented Metropolis,* 144–47; see also Starr, *Inventing the Dream,* 246–48, for a more developed analysis of this southern California ideal.

14. Fogelson, *Fragmented Metropolis,* 191–97; McWilliams, *Southern California,* 166–75.

15. Elaine Tyler May, *Great Expectations: Marriage and Divorce in Post-Victorian America* (Chicago: Univ. of Chicago Press, 1980), 25.

16. Kenneth D. Rose, "'Dry' Los Angeles and Its Liquor Problems in 1924," *Southern California Quarterly* 69 (1987), 55–70.

17. Robert H. Wiebe, *The Search for Order, 1877–1920* (New York: Hill and Wang, 1967).

18. See Mowry, *California Progressives,* 119–20, 128–29, 131–32, for a description of the largely one-issue campaign run by Johnson; Michael Rogin and John L. Shover, *Political Change in California: Critical Elections and Social Movements, 1890–1966* (Westport, Conn.: Greenwood, 1970), 41–43, 52–54; Scott L. Bottles, *Los Angeles and the Automobile: The Making of the Modern City* (Berkeley: Univ. of California Press, 1987), 29–41. For a recent chronicle of Henry Huntington's role in shaping early twentieth-century Los Angeles, see William B. Friedricks, *Henry E. Huntington and the Creation of Southern California* (Columbus: Ohio State Univ. Press, 1992).

19. Spencer C. Olin, *California's Prodigal Sons: Hiram Johnson and the Progressives, 1911–1917* (Berkeley: Univ. of California Press, 1968), 76–80; see David G. Herman, "Neighbors on the Golden Mountain: The Americanization of Immigrants in California. Public Instruction as a Agency of Ethnic Assimilation, 1850 to 1933" (Ph.D. diss., University of California, Berkeley, 1981), chap. 4, for a complete discussion of the CIH's role in statewide Americanization efforts.

20. Herman, "Neighbors," 313–14, 318–21, 333–34; see John Higham, *Strangers in the Land: Patterns of American Nativism, 1860–1925* (New York: Atheneum, 1965), 116–23, and Allen F. Davis, *Spearheads for Reform: The Social Settlements and the Progressive Movement, 1890–1914* (New York: Oxford Univ. Press, 1967), 84–102, for a fuller discussion of the treatment of immigrants by social settlements.

21. Higham, *Strangers,* 234–63; Edwin Layton, "The Better America Federation: A Case Study of Superpatriotism," *Pacific Historical Review* 30 (1961), 137–47.

22. Mowry, *California Progressives,* 139–54; Olin, *Prodigal Sons,* 34–49; Rogin and Shover, *Political Change,* 54–56, 62–85.

23. Herman, "Neighbors," 381–83, 481–85, 546–49.

24. Ellwood P. Cubberly, *Changing Conceptions of Education* (Boston: Houghton Mifflin, 1909), 15–16.

25. Most Californians believed either that Mexicans would remain an insignificant factor in the state's future or that immigration from Mexico was largely temporary. See Mark Reisler, *By the Sweat of Their Brow: Mexican Immigrant Labor in the United States, 1900–1940* (Westport, Conn.: Greenwood, 1976), 24–42, for a discussion of the frustrations of those advocating Mexican immigra-

tion restrictions in the 1910s. Also see Olin, *Prodigal Sons,* 80–90, and Roger Daniels, *The Politics of Prejudice: The Anti-Japanese Movement in California and the Struggle for Japanese Exclusion* (Berkeley: Univ. of California Press, 1962), 46–64, for anti-Asian sentiment among progressives.

26. Reisler, *Sweat,* pp. 151–69.

27. "The Docile Mexican," *Saturday Evening Post,* 10 March 1928, 43.

28. Reisler, *Sweat,* 169; Ricardo Romo, "Responses to Mexican Immigration, 1910–1930," *Aztlán* 6 (1975), 187.

29. U.S. Congress, House, Committee on Immigration and Naturalization, *Hearings on Temporary Admission of Illiterate Mexican Laborers,* 69th Congress, 1st session (1926), 191, and *Hearings on Seasonal Agricultural Laborers from Mexico,* 69th Congress, 1st session (1926), 46.

30. Herman, "Neighbors," 244, 651–652.

31. See Douglas Monroy, "Like Swallows at the Old Mission: Mexicans and the Racial Politics of Growth in Los Angeles in the Interwar Period," *Western Historical Quarterly* 14 (1983), 448–52. See also Mario T. García, "Americanization and the Mexican Immigrant, 1880–1930," *Journal of Ethnic Studies* 6 (1978), 19–34, for another description of efforts directed at Mexicans, largely focused on El Paso.

32. Emory S. Bogardus Collection, University Archives, University of Southern California, including "Ph.D.s Awarded in Sociology"; M. Lorena Chambers, "Urban Reform: The Effect of Social Work on Mexicans in Los Angeles, 1910–1936" (Senior honors thesis in history, University of California, Los Angeles, 1989). See Emory S. Bogardus, *The Mexican in the United States* (Los Angeles: Univ. of Southern California Press, 1934), for his most complete work on the subject.

33. California, Commission of Immigration and Housing (hereafter CIH), *Report of an Experiment Made in Los Angeles in the Summer of 1917 for the Americanization of Foreign Born Women* (Sacramento: State Printing Office, 1917), 11.

34. Helen Walker, "Mexican Immigrants as Laborers," *Sociology and Social Research* 13 (1928), 56.

35. Ibid., 57.

36. Alfred White, "The Apperceptive Mass of Foreigners as Applied to Americanization: The Mexican Group" (Master's thesis, University of California, 1923), 34–35.

37. CIH, *The Home Teacher, Immigrant Education Leaflet No. 5* (San Francisco: State Printing Office, 1916), 8.

38. Pearl Idelia Ellis, *Americanization through Homemaking* (Los Angeles: Wetzel, 1929), 31.

39. Herbert C. Gutman, *Work, Culture, and Society in Industrializing America* (New York: Vintage-Random House, 1977), 13.

40. See Linda Kerber, *Women of the Republic: Intellect & Ideology in Revolutionary America* (Chapel Hill: Univ. of North Carolina Press, 1980), for the origins of this ideology.

41. Ellis, *Americanization through Homemaking,* 65.

42. Interestingly, the clash between domestic duties and work outside the home became a much-addressed, yet unresolved, issue in the 1920s among middle-class, college-educated Anglo American women—the very group recruited to become Americanization teachers. See Carl Degler, *At Odds: Women and the Family in America from the Revolution to the Present* (New York: Oxford

Univ. Press, 1980), 411–13 and Lois Scharf, *To Work and to Wed: Female Employment, Feminism, and the Great Depression* (Westport, Conn.: Greenwood, 1980), 21–43.

43. Victor S. Clark, "Mexican Labor in the United States," U.S. Bureau of Labor *Bulletin,* No. 78 (Washington, D.C.: GPO, 1908), 496.

44. Mario T. García, "The Chicana in American History: The Mexican Women of El Paso, 1880–1920—A Case Study," *Pacific Historical Review* 49 (1980), 325–28.

45. CIH, *A Discussion of Methods of Teaching English to Adult Foreigners, with a Report on Los Angeles County* (Sacramento: State Printing Office, 1917), 21.

46. CIH, *Primer for Foreign-Speaking Women, Part II,* compiled under Mrs. Amanda Matthews Chase (Sacramento: State Printing Office, 1918), 3.

47. Ibid., 5.

48. Jay S. Stowell, *The Near Side of the Mexican Question* (New York: George H. Doran, 1921), 102.

49. Bogardus, *Mexican in the United States,* 81.

50. White, "Apperceptive Mass," 34–35.

51. CIH, *A Discussion of Methods,* 12–14.

52. CIH, *Primer,* 9.

53. Ellis, *Americanization through Homemaking,* 19, 21, 29.

54. See Laura Shapiro, *Perfection Salad: Women and Cooking at the Turn of the Century* (New York: Farrar, Straus and Giroux, 1986), particularly chap. 6, for a thorough description of this movement.

55. Ellis, *Americanization through Homemaking,* 26.

56. Ibid., 47, 64; Bogardus, *Mexican in the United States,* 33.

57. Higham, *Strangers,* 147–48; Reisler, *Sweat,* 154–56.

58. Bogardus, *Mexican in the United States,* 33, 25; Ellis, *Americanization through Homemaking,* 61–62.

59. Mario T. García, "La Familia: The Mexican Immigrant Family, 1900–1930," in *Work, Family, Sex Roles, Language: The National Association of Chicano Studies, Selected Papers 1979,* Mario Barrera, Alberto Camarillo, and Francisco Hernández, eds. (Berkeley: Tonatiuh-Quinto Sol International, 1980), 124–27.

60. For an excellent discussion of occupational sex segregation in this period, see Ruth Milkman, "Women's Work and the Economic Crisis: Some Lessons from the Great Depression," *Review of Radical Political Economics* 8 (Spring 1976), 75–78.

61. Ellis, *Americanization through Homemaking,* 15, 13.

62. Ibid., 35; Bogardus, *Mexican in the United States,* 43.

63. White, "Apperceptive Mass," 43; Ellis, *Americanization through Homemaking,* 43.

64. *Mexicans in California,* Report of Governor C. C. Young's Mexican Fact-Finding Committee (1930; rpt., San Francisco: R and E Research Associates, 1970), 61–74.

65. CIH, *Annual Report* (Sacramento, 1927), 8.

66. James William Cameron, "The History of Mexican Public Education in Los Angeles, 1910–1930" (Ph.D. diss., University of Southern California, 1976), 110–16, 125–36; Gilbert G. González, "Racism, Education, and the Mexican Community in Los Angeles, 1920–30," *Societas* 4 (1974), 287–301; Harvey Kantor, "Choosing a Vocation: The Origins and Transformation of Vocational Guidance in California, 1910–1930," *History of Education Quarterly* 26 (1986), 351–75.

67. White, "Apperceptive Mass," 3.

68. Jackson Putnam, "The Persistence of Progressivism in the 1920's: The Case of California," *Pacific Historical Review* 35 (1966), 398.

69. Romo, *East Los Angeles,* 124–28.

70. CIH, *The Home Teacher,* 3.

71. Abraham Hoffman, "Mexican Repatriation Statistics: Some Suggested Alternatives to Carey McWilliams," *Western Historical Quarterly* 3 (Oct. 1972), 391–404; Abraham Hoffman, *Unwanted Mexican Americans in the Great Depression: Repatriation Pressures, 1929–1939* (Tucson: Univ. of Arizona Press, 1974). Also see Chapter 10.

Chapter 5
The "New Nationalism," Mexican Style

1. *El Heraldo de México,* 15, 20, and 29 Sept. 1921.

2. Los Angeles *Times,* 7 Sept. 1921. For a lengthy discussion of the various activities of the Alianza Hispano Americana, see Jose Amaro Hernandez, *Mutual Aid for Survival: The Case of the Mexican American* (Malabar, Fla.: Robert E. Krieger, 1983), 30–59.

3. Los Angeles *Times,* 7 Sept. 1921; *El Heraldo de México,* 15 Sept. 1921. The amount of coverage and publicity given the independence celebrations was extraordinary in *El Heraldo:* a front-page story on the planning and fruition of the festivities appeared every day through the month of Sept..

4. F. Arturo Rosales, "Shifting Self Perceptions and Ethnic Consciousness among Mexicans in Houston, 1908–1946," *Aztlán* 16:1–2 (1987), 71. See also Mario T. Garcia, "La Frontera: The Border as Symbol and Reality in Mexican-American Thought," *Mexican Estudios/Estudios Mexicanos* 1:2 (Summer 1985), 196–205. In "Prensa y patria: The Spanish-Language Press and the Biculturation of the Tejano Middle Class, 1920–1940," *Western Historical Quarterly* (Nov. 1991), 451–72, Roberto R. Trevino more carefully delineates a complicated promotion of ethnic identity in the Spanish-language press as being rooted in the complex questions of identity among the middle class.

5. Francisco E. Balderrama, in the only extended study of a Mexican consulate available, casts this involvement as largely reactive to developments in the United States in defending Mexican immigrants. See *In Defense of La Raza: The Los Angeles Mexican Consulate and the Mexican Community, 1929 to 1936* (Tucson: Univ. of Arizona Press, 1982).

6. Juan Gómez-Quiñones, "Piedras contra la Luna, México en Aztlán y Aztlán en México: Chicano-Mexican Relations and the Mexican Consulates, 1900–1920," in *Contemporary Mexico: Papers of the IV International Congress of Mexican History* (1973, Santa Monica, California), James W. Wilkie, Michael C. Meyer, and Edna Monzón de Wilkie, eds. (Berkeley: Univ. of California Press, 1976), 500–505.

7. Gómez-Quiñones, "Piedras," 505–09; Juan Gómez-Quiñones, *Sembradores: Ricardo Flores Magón y El Partido Liberal Mexicano: A Eulogy and Critique* (UCLA Chicano Studies Center Publications, 1973), 29, 36–37, 51–53; Edward Joseph Escobar, "Chicano Protest and the Law: Law Enforcement Responses to Chicano Activism in Los Angeles, 1850–1936" (Ph.D diss. University of California, Riverside, 1983), 135–38.

8. Gómez-Quiñones, "Piedras," 509–18.

9. Ibid., 516–17, 520–21.

10. Ramón Eduardo Ruíz, *The Great Rebellion: Mexico, 1905–1924* (New York: Norton, 1980), 179–81; Dr. Atl, "Obregón y el principio de la renovación social," in *Obregón, Aspectos de su Vida* (Mexico: Editorial "Cultura," 1935), 71.

11. Alan Knight, "Racism, Revolution, and *Indigenismo:* Mexico, 1910–1940," in *The Idea of Race in Latin America, 1870–1940* (Austin: Univ. of Texas Press, 1990); Henry C. Schmidt, *The Roots of Lo Mexico: Self and Society in Mexican Thought, 1900–1934* (College Station: Texas A&M Univ. Press, 1978), 77–84, 97–123; Ilene V. O'Malley, *The Myth of the Revolution: Hero Cults and the Institutionalization of the Mexican State, 1920–1940* (Westport, Conn.: Greenwood, 1986), particularly chap. 3; *La Opinión,* 26 Nov. 1926.

12. Rafael de la Colina, *Una vida de hechos* (Mexico City: Secretaría de Relaciones Exteriores, Archivo Histórico Diplomático Mexicano: Serie Testimonios 1, 1989), 27.

13. Michael C. Meyer and William L. Sherman, *The Course of Mexican History,* 2nd ed. (New York: Oxford Univ. Press, 1983), 542. See also Peter H. Smith, "La política dentro de la Revolución: El congreso constituyente de 1916–1917," *Historia Mexicana* 22 (1973), 363–95.

14. This is a social structure quite unlike that of San Antonio, Texas, described by Richard A. Garcia in *Rise of the Mexican American Middle Class: San Antonio, 1929–1941* (College Station: Texas A&M Univ. Press, 1991). Garcia's San Antonio had two well-developed upper classes, an exiled Mexico-focused group of elites and a dual-focused Mexican-American middle class, competing for the leadership of the large working-class population in the 1920s and 1930s.

15. See Alfonso Fábila, *El problema de la emigración de obreros y campesinos mexicanos* (Mexico City: Talleres Gráficos de la Nación, 1929), for the most extreme anti-emigration sentiments by a Mexican government official of the period.

16. Lawrence A. Cardoso, *Mexican Emigration to the United States, 1897–1931: Socio-economic Patterns* (Tucson: Univ. of Arizona Press, 1980), chaps. 4 and 6.

17. Ibid., 106, 116.

18. See George J. Sanchez, "Becoming Mexican American: Ethnicity and Acculturation in Chicano Los Angeles, 1990–1943," (Ph.D. diss., Stanford University, 1989), 266–67, for a description of the social backgrounds of the Mexican immigrant elite in Los Angeles during the 1920s.

19. Gómez-Quiñones, "Piedras," 494–527.

20. Interview with Zeferino Ramírez, "Biographies and Case Histories," II folder, Z-R5, Manuel Gamio collection, Bancroft Library, University of California, Berkeley; Zeferino Ramírez, "Origins of the Ramírez Family, from What I Remember" (unpublished manuscript, 1983); interview with Beatrice R. Palomarez, Pico Rivera, California, 3 Jan. 1991.

21. Eustace L. Williams, "Racial Minorities Survey—Mexican: Juan B. Ruiz," 6 May 1937, and Walton D. Fore, "American Guide: Cultural Centers-LA-265.2: Mexican," 287–88, Federal Writers Program collection, MS 306, Department of Special Collections, University Research Library, University of California, Los Angeles.

22. Fore, "American Guide," 285–89.

23. *La Opinión,* 26 Nov. 1926. It is important to remember that Spanish-language newspapers published in the United States did not necessarily report adequately on the barrios in which they were published. See Francine Medeiros,

"*La Opinión,* a Mexican Exile Newspaper: A Content Analysis of Its First Years, 1926–1929," *Aztlán* 11:1 (1980), 68–75.

24. *La Opinión,* 17 Feb. 1927.

25. *La Opinión,* 17 Feb. 1927 and 1 Sept. 1929.

26. *La Opinión,* 10 May 1928.

27. Mary Kay Vaughan, *The State, Education, and Social Class in Mexico, 1880–1928* (DeKalb: Northern Illinois Univ. Press, 1982), 215–238.

28. Beatrice R. Palomarez, "Escuela Mexico," unpublished paper in possession of the author.

29. Ibid. See Vaughan, *Education,* 174–78.

30. Palomarez, "Escuela Mexico."

31. *La Opinión,* 17 Feb. and 31 July, 1927.

32. *La Opinión,* 31 July 1927 and 30 Nov. 1929.

33. *La Opinión,* 12 Nov. 1927, 1 Sept. 1929, 18 Feb. and 12 Oct. 1930.

34. *La Opinión,* 17 Feb. and 12 Nov. 1927, 1 Sept. 1929, and 21 June 1930.

35. *La Opinión,* 12 Oct. 1930.

36. Vaughn, *The State, Education, and Social Class,* 171.

37. Ibid., 179. To place the discussion of education in a broader framework for viewing cultural developments in Mexico, see Enrique Krauze, *Caudillos culturales en la revolucion mexicana* (Mexico: Siglo veintiuno editores, 1976).

38. Vaughn, *The State, Education, and Social Class,* 184–85.

39. See Knight, "Race, Revolution, and *Indigenismo,*" 72–78, for a thorough exploration of this issue.

40. See David A. Brading, "Manuel Gamio and Official Indigenismo in Mexico," *Bulletin of Latin America Research* 7:1 (1988), 75–89. Manuel Gamio's most important early works include *Forjando patria,* 2nd ed. (Mexico: Porrua, 1960) and *Introduccion, sintesis y conclusion a la poblacion del Valle de Teotihuacan* (Mexico: Secretaria de Educacion Publica, 1924).

41. Manuel Gamio, *Mexican Immigration to the United States. A Study of Human Migration and Adjustment* (1930; rpt., New York: Dover, 1971) and *The Life Story of the Mexican Immigrant: Autobiographical Documents Collected by Manuel Gamio* (1931; rpt., New York: Dover, 1971).

42. Brading, "Official Indigenismo," 77.

43. Manuel Gamio, "Empiricism of Latin-American Governments and the Empiricism of Their Relations with the United States," *Mexican Review* 3:5 (Aug. 1919), 6–7.

44. Ibid., 7.

45. Ibid., 7–8.

46. Gamio, *Mexican Immigration,* 64.

47. Ibid., 65.

48. Ibid., 175–84.

49. John Skirius, "Vasconcelos and *México de Afuera,*" *Aztlán* 7:3 (Fall 1976), 486–87, 492.

50. Gamio, *Mexican Immigration,* 184.

51. Original draft of *Mexican Immigration to the United States,* 79, Manuel Gamio collection, Bancroft Library, University of California Berkeley. This particular paragraph did not make it to the final printed version.

52. "Educacion Mexicana en Estados Unidos," *La Opinion* editorial, 21 June 1930.

53. Balderrama, *In Defense,* 15–32; Abraham Hoffman, *Unwanted Mexican Americans in the Great Depression: Repatriation Pressures, 1929–1939* (Tucson: Univ. of Arizona Press, 1974), 38–115.

54. *La Opinión,* 26 Nov. 1926.

Part Three
Shifting Homelands

1. *The Tidings,* 14 Dec. 1934.

2. "Vida de Estanislao Gómez," 1, interview by Luis Felipe Recinos, 2 April 1927; Biographical & Case Studies II folder, Manuel Gamio collection, Bancroft Library, University of California, Berkeley.

3. Stuart Hall, "Cultural Identity and Diaspora," in *Identity: Community, Culture, Difference,* Jonathan Rutherford, ed. (London: Lawrence & Wishart, 1990), 224.

4. Quoted in Nellie Foster, "The *Corrido:* A Mexican Culture Trait Persisting in Southern California" (Master's thesis, University of Southern California, 1939), 166–68.

Chapter 6
Family Life and the Search for Stability

1. Marita Hernández and Robert Montemayor, "Mexico to U.S.—a Cultural Odyssey," *Los Angeles Times,* 24 July 1983, pp. 1, 18.

2. Oscar Handlin, *The Uprooted: The Epic Story of the Great Migrations That Made the American People,* 2nd ed., enlarged (1951; Boston: Little, Brown, 1973), 210. Virginia Yans-McLaughlin refers to the body of work portraying immigrant family life in this fashion as a "disorganization" school in *Family and Community: Italian Immigrants in Buffalo, 1880–1930* (Ithaca: Cornell Univ. Press, 1971), 19.

3. Hernández and Montemayor, "Odyssey," 1, 18–21.

4. Yans-McLaughlin, *Family and Community,* 61, 64. For a historical account of the Mexican immigrant family from this perspective, see Mario T. García, "La Familia: The Mexican Immigrant Family, 1900–1930," in *Work, Family, Sex Roles and Language,* Mario Barrera, Alberto Camarillo, and Francisco Hernández, eds. (Berkeley: Tonatiuh-Quinto Sol, 1980), 117–39.

5. David Alvirez and Frank D. Bean, "The Mexican American Family," in *Ethnic Families in America,* Charles H. Mindel and Robert W. Haberstein, eds. (New York: Elsevier, 1976); Miguel Montiel, "The Social Science Myth of the Mexican American Family," *El Grito: A Journal of Contemporary Mexican American Thought* 3 (1970), 56–63; Nathan Murillo, "The Mexican American Family," in *Chicanos: Social and Psychological Perspectives,* Nathaniel N. Wagner and Marsha J. Haug, eds. (St. Louis: C.V. Mosby, 1971), 97–108; Octavio Ignacio V. Romano, "The Anthropology and Sociology of the Mexican-Americans: The Distortion of Mexican-American History," *El Grito* 2:1 (Fall 1968), 13–26. For an overview of this literature, see Alfredo Mirandé, "The Chicano Family: A Reanalysis of Conflicting Views," *Journal of Marriage and the Family* 39 (1977), 750–51. An excellent summary of various theoretical approaches to family history and their possible application to Chicano history is provided by Richard Griswold

del Castillo in chapter 1 of *La Familia: Chicano Families in the Urban Southwest, 1848 to the Present* (Notre Dame: Univ. of Notre Dame Press, 1984), 1–9.

6. Mirandé, "The Chicano Family," 748–51.

7. For a description of widespread European return migration, see John Bodnar, *The Transplanted: A History of Immigrants in Urban America* (Bloomington: Indiana Univ. Press, 1985), 53–54.

8. The following account is derived from transcripts of interviews conducted by a Board of Special Inquiry at Nogales, Arizona, from 25 Aug. 1917 to 1 July 1918, Records of the Immigration and Naturalization Service, RG 85, Box 250, Folder 54281/36B, National Archives, Washington, D.C. These interviews were conducted beginning in 1917 when entrance to the United States was requested by children under the age of sixteen unaccompanied by their parents. Their purpose was to determine whether the child was likely to become a public charge, therefore detailed information regarding their family situation in Mexico and the United States was obtained.

9. Dino Cinel, *From Italy to San Francisco: The Immigrant Experience* (Stanford: Stanford Univ. Press, 1982), 168; Naturalization Records, RG 21, National Archives, Laguna Niguel, California.

10. "Interview with Arturo Morales," No. 11, p. 1, interviewed by Luis Felipe Recinos, 8 April 1927, Biographies & Case Histories III folder, Z-R5, Manuel Gamio collection, Bancroft Library, University of California, Berkeley.

11. "Vida de Estanislao Gómez," 1, interview by Luis Felipe Recinos, 2 April 1927; Biographical & Case Studies II folder, Gamio collection.

12. See "Vida de Leova López," 1, interview by M. Robles, 19 April 1927, Biographies & Case Histories II folder; and "Interview with Juana Martínez," No. 102, p. 1, interview by Luis Felipe Recinos, 6 April 1927, Biographies & Case Histories I folder, Gamio collection.

13. For descriptions of the growth of the central business district and nearby factories, see Robert M. Fogelson, *The Fragmented Metropolis: Los Angeles, 1850–1930* (Cambridge, Mass.: Harvard Univ. Press, 1967), 147–51; Scott L. Bottles, *Los Angeles and the Automobile: The Making of the Modern City* (Berkeley: Univ. of California Press, 1987), 200–201; and Howard J. Nelson, "The Vernon Area, California—A Study of the Political Factor in Urban Geography," *Annals of the Association of American Geographers* 42 (1952), 177–91. For the continuation of these patterns as late as 1940, see Eshref Shevky and Molly Lewin, *Your Neighborhood: A Social Profile of Los Angeles* (Los Angeles: Haynes Foundation, 1949), 24–26.

14. "The Arce (Galván) family," 8, interview by Luis Felipe Recinos, 8 April 1927, Biographies & Case Histories III folder, Gamio collection.

15. See "Interview with Gloria Navas," No. 56, p. 4, interview by Luis Felipe Recinos, 16 April 1927, Biographies & Case Histories I folder; and "Interview with Elisa Morales," No. 53, pp. 3, 5, interview by Luis Felipe Recinos, 16 April 1927, Biographies & Case Histories II folder, Gamio collection. Joanne Meyerowitz describes similar responses to urban environments by these "women adrift" in "Women and Migration: Autonomous Female Migrants to Chicago, 1880–1930," *Journal of Urban History* 13 (1987), 147–68.

16. Analysis of Naturalization documents, National Archives, Laguna Niguel, California.

17. Constantine Panunzio found a 17 percent rate of exogamous marriages among Mexicans in "Intermarriage in Los Angeles, 1924–1933," *American Jour-*

nal of Sociology 48 (1942), 698, 701. Griswold del Castillo finds similar figures in *Familia,* 106–7. See Edward Murguia, *Chicano Intermarriage: A Theoretical and Empirical Study* (San Antonio: Trinity Univ. Press, 1982), for comparisons with other regions over time.

18. This figure is skewed by the fact that the number of Mexican immigrant women who naturalized was small—involving only 111 marriages—and that Mexican women who married Mexican men were probably less likely to naturalize than those who married Anglo American men.

19. The pattern is similar for women, although there are many fewer in the sample.

20. "Interview with Sra. Maria Rovitz Ramos," No. 7, pp. 1, 4, interview by Luis Felipe Recinos, 2 April 1927, Biographies & Case Histories I folder, Gamio collection.

21. From survey of Anglo American spouses in naturalization documents. The argument that ethnic groups tend to intermarry within religious groupings has been made by Milton Gordon in *Assimilation in American Life: The Role of Race, Religion, and National Origins* (New York: Oxford Univ. Press, 1964). Edward Murguia makes a similar argument specifically for Chicanos in *Chicano Intermarriage,* 35. According to at least one local official, Jewish-Mexican intermarriage was rare, despite substantial interaction, because Russian Jews felt that Mexicans were "filthy and godless and without sex morality because of their public courtship." See "Interview with Dr. Miriam Van Waters, Referee of Juvenile Court, Hall of Justice, Los Angeles, California," 2–512, American Officials folder, 74/187c, Paul S. Taylor collection, Bancroft Library, University of California, Berkeley.

22. For examples of lighter-skinned Mexicans intermarrying, see "José Robles," "Sra. Ruhe López," and "Sr. Campos" in Manuel Gamio, *The Life Story of the Mexican Immigrant* (1931; rpt., New York: Dover, 1971), 226–37.

23. "Interview with the Santaella family," No. 45, pp. 2–3, interview by Luis Felipe Recinos, 15 April 1927, Biographies & Case Histories II folder, Gamio collection. "Interview with Maria Rovitz Ramos," 6. See also Richard Romo, *East Los Angeles: History of a Barrio* (Austin: Univ. of Texas Press, 1983), 85. One quantitative analysis of the 1940 U.S. census suggests that significant rates of intermarriage between Mexican women and non-Mexican men occurred in the Southwest during this period, often leading to relatively higher economic standing for these women. Brian Gratton, F. Arturo Rosales, and Hans DeBano, "A Sample of the Mexican-American Population in 1940," *Historical Methods* 21 (1988), 83–85.

24. Quoted in Mary Lanigan, "Second Generation Mexicans in Belvedere" (Master's thesis, University of Southern California, 1932), 20.

25. Ibid., 18–21.

26. Knowledge, let alone use, of effective birth control methods besides abstinence does not appear to have been widespread in the Mexican immigrant community. In addition to strict cultural sanctions against birth control emanating from the Catholic Church, legal restrictions against dispensing birth control information still existed. Mexican women were unlikely to have access to the occasional private physician who might be willing to give such advice. The one exception to this generalization seems to have been the small group of Mexican women who regularly worked as dance hall girls in the Plaza area. See "Interview

with Gloria Navas," No. 56, p. 4, interview by Luis Felipe Recinos, 16 April 1927, Biographies & Case Histories I folder; and "Interview with Elisa Morales," No. 53, pp. 3, 5, interview by Luis Felipe Recinos, 16 April 1927, Biographies & Case Histories II folder, Gamio collection.

27. Analysis of birth and marriage dates in Naturalization documents, National Archives, Laguna Niguel, California. To place marriage patterns in Los Angeles in a larger historical context, see John D'Emilio and Estelle B. Freedman, *Intimate Matters: A History of Sexuality in America* (New York: Harper and Row, 1988), chap. 5, esp. 89–90.

28. The vast majority of the married couples sampled do not list any information for wives because that question was not asked of husbands applying for naturalization until the second phase of the process. The figures do reflect, however, the labor status of the 142 married women who took out first papers, along with wives of men who got to the second level of the process. This percentage therefore is tentative, as it is based only on 187 total couples in the overall sample of 1,249 (15.0%). In addition, this group would tend to represent those most settled in the United States.

29. Albert Camarillo, *Chicanos in a Changing Society: From Mexican Pueblos to American Barrios in Santa Barbara and Southern California, 1848–1930* (Cambridge: Harvard Univ. Press, 1979), 220; Douglas Guy Monroy, "Mexicanos in Los Angeles, 1930–1941: An Ethnic Group in Relation to Class Forces" (Ph.D. diss., University of California, Los Angeles, 1978), 77; Mario T. García, *Desert Immigrants: The Mexicans of El Paso, 1880–1920* (New Haven: Yale Univ. Press, 1981), 75, 200.

30. Carl Degler, *At Odds: Women and the Family in America from the Revolution to the Present* (New York: Oxford Univ. Press, 1980), 384. See also Yans-McLaughlin, *Family and Community*, 173; and Bodnar, *Translated*, 78–80.

31. Griswold del Castillo, *Familia*, 63.

32. See Paul S. Taylor, "Mexican Women in Los Angeles Industry in 1928," *Aztlán* 11 (1980), 104–5.

33. See "The Arce (Galván) family," Gamio collection.

34. "Interview with Santaella family," 4, Gamio collection.

35. Analysis of Naturalization documents, National Archives, Laguna Niguel, California. See also Taylor, "Industry," 106–8.

36. See Alex M. Saragoza, "The Conceptualization of the History of the Chicano Family," in *The State of Chicano Research in Family, Labor and Migration Studies,* Armando Valdez, Albert Camarillo and Tomás Almaguer, eds. (Stanford: Stanford Center for Chicano Research, 1983), 119–20.

37. Naturalization documents, National Archives, Laguna Niguel, California. Other areas of the Southwest also produced early marriages among more stable populations. Sarah Deutsch reports that from 1860 to 1910, Chicanas married in New Mexico between the ages of 15 and 21, while men tended to marry between age 19 and 26. See "Culture, Class, and Gender: Chicanas and Chicanos in Colorado and New Mexico, 1900–1940" (Ph.D. diss., Yale University, 1985), 83.

38. See figures for Mascota, Jalisco, in Carlos B. Gil, *Life in Provincial Mexico: National and Regional History Seen from Mascota, Jalisco, 1867–1972* (Los Angeles: UCLA Latin American Studies Center Publications, 1983), 95–98.

39. Lanigan, "Second Generation," 25–26.

40. Ibid., 26.

41. "Interview with Juana Martínez," 3–4; "Interview with Gloria Navas," 7, Gamio collection.

42. "Interview with Ignacio Sandoval," 5; "Interview with Luis Aguiñaga," 1–2, Gamio collection.

43. Lanigan, "Second Generation," 34.

44. Ibid., 33.

45. Cinel, *From Italy,* 177.

46. In contrast, see ibid., 177–78, or Yans-McLaughlin, *Family and Community,* 256–57, for Italians. See Abraham Cahan, *The Rise of David Levinsky* (1917; rpt., New York: Harper and Row, 1960), 106, for Jews.

47. This terminology was first used to describe the southern Italian family by two anthropologists. See Leonard W. Moss and Walter H. Thomson, "The South Italian Family: Literature and Observation," *Human Organization* 18 (1959), 38.

48. See Deutsch, "Culture," 90, for an analysis which stresses the flexibility of the sexual division of labor within the Chicano family. For studies of the contemporary Chicano family which stress flexibility, see Griswold del Castillo, *Familia,* 118–19; Maxine Baca Zinn, "Marital Roles, Marital Power and Ethnicity: A Study of Changing Families" (Ph.D. diss., University of Oregon, 1978); and Lea Ybarra, "Conjugal Role Relationships in the Chicano Family" (Ph.D. diss., University of California, Berkeley, 1977).

49. Cinel, *From Italy,* 188.

50. Analysis of marital status information from Naturalization documents, National Archives, Laguna Niguel, California. See Elaine Tyler May, *Great Expectations: Marriage and Divorce in Post-Victorian America* (Chicago: Univ. of Chicago Press, 1980), 9, for comparative data for the rest of Los Angeles.

51. "Manuela 'Minnie' Ortiz," interview by J. Isaac Aceves, 14 June 1937, p. 330, Field Continuity, Mexican Population, San Diego Project, Federal Writers Project collection, Department of Special Collections, University of California, Los Angeles.

52. "Interview with Dr. Miriam Van Waters, Referee of Juvenile Court, Hall of Justice, Los Angeles, California," 1–511, Taylor collection.

53. See Mirandé, "The Chicano Family," 751.

54. See Susan E. Keefe and Amado Padilla, *Chicano Ethnicity* (Albuquerque: Univ. of New Mexico Press, 1987), especially chaps. 9 and 10, for a modern account of this process in southern California.

Chapter 7
Religious Adaptations

1. *Associated Catholic Charities Report, 1922* (Los Angeles, 1923), 23.

2. For further information on the work of the Plaza Methodist Episcopal Church, see Samuel M. Ortegon, "Mexican Religious Population of Los Angeles" (Master's thesis, University of Southern California, 1932), 19–24.

3. See Timothy L. Smith, "Religion and Ethnicity in America," *American Historical Review* 83:5 (Dec. 1978), 1155–185, for a provocative exploration of this topic as it relates to European ethnics. When Smith tries to expand his analysis to Hispanics, however (see 1169, 1173–74), he often misses the mark.

4. See Chapter 1 for a thorough discussion of this situation.

5. Robert E. Quirk, *The Mexican Revolution and the Catholic Church, 1910–1929* (Bloomington: Indiana Univ. Press, 1973), 11–12; Deborah J. Baldwin, *Protestants and the Mexican Revolution: Missionaries, Ministers, and Social Change* (Urbana: Univ. of Illinois Press, 1990), 11–13, 30–32.

6. Quirk, *The Mexican Revolution,* 15–16.

7. Jean-Pierre Bastian, "Metodismo y clase obrera durante el Porfiriato," *Historia Mexicana* 33 (1983), 40–47, 50–55; Baldwin, *Protestants and the Mexican Revolution* 30–59; F. LaMond Tullis, *Mormons in Mexico: The Dynamics of Faith and Culture* (Logan: Utah State Univ. Press, 1987), 13–84. See also Lawrence A. Cardoso, "Protestant Missionaries and the Mexican: An Environmentalist Image of Cultural Differences," *New Scholar* 7 (1978), 223–36.

8. Manuel Gamio, *The Life Story of the Mexican Immigrant* (1931; rpt., New York: Dover, 1971), 201–5.

9. Writings of Reverend Francisco Quintanilla, in Clara Gertrude Smith, "The Development of the Mexican People in the Community of Watts, California" (Master's thesis, University of Southern California, 1933), 71–72.

10. Bastian, "Metodismo," 64–67.

11. Roberto Gonzáles, Sunnyvale, Calif., interview, 2 Feb. 1976. Quoted in Antonio R. Soto, "The Church in California and the Chicano: A Sociological Analysis," *Grito del Sol* 4 (1979), 54.

12. Gamio, *Life Story,* 48–49.

13. Gregory H. Singleton, *Religion in the City of Angels: American Protestant Culture and Urbanization, Los Angeles, 1850–1930* (Ann Arbor: UMI Research Press, 1979), 36–41.

14. Ibid., 2, 15, 90–91.

15. Ibid., 91.

16. See ibid., 95–98, 124–26; and Frederic Cople Jaher, *The Urban Establishment: Upper Strata in Boston, New York, Charleston, Chicago, and Los Angeles* (Urbana: Univ. of Illinois Press, 1982), 612–17, 630–33, 639–54.

17. Singleton, *Religion in the City,* 105, 135.

18. Linna E. Bresette, *Mexicans in the United States: A Report of a Brief Survey* (Washington, D.C.: National Catholic Welfare Conference, 1929), 36.

19. R. Douglas Brackenridge and Francisco O. García-Treto, *Iglesia Presbiteriana: A History of Presbyterians and Mexican Americans in the Southwest* (San Antonio: Trinity Univ. Press, 1974), 138; Interchurch World Movement of North America, *The Mexican in Los Angeles: Los Angeles City Survey* (1920; rpt., San Francisco: R and E Research Associates, 1970), 18–19.

20. Brackenridge and García-Treto, *Iglesia Presbiteriana,* 142.

21. Ibid., 142–48.

22. *Associated Catholic Charities Report, 1919* (Los Angeles, 1920), 11.

23. *Associated Catholic Charities Report, 1923* (Los Angeles, 1924), 23.

24. Lawrence J. Mosqueda, *Chicanos, Catholicism and Political Ideology* (Lanham, Md.: Univ. Press of America, 1986), 45–49.

25. Soto, "The Church in California," 53.

26. "El Hogar Feliz—The Happy Home," in *The Tidings,* 7th annual ed. (1910), 80.

27. Francis J. Weber, "Irish-Born Champion of the Mexican-Americans," *California Historical Society Quarterly* 49 (1970), 235.

28. Stephan A. Privett, S.J., "Robert E. Lucey: Evangelization and Cathesis among Hispanic Catholics" (Ph.D. diss., Catholic University of America, 1985),

9–15; Weber, "Irish-Born Champion," 237. See also Saul E. Bronder, *Social Justice & Church Authority: The Public Life of Archbishop Robert E. Lucey* (Philadelphia: Temple Univ. Press, 1982), especially chap. 2.

29. *Associated Catholic Charities Report, 1920* (Los Angeles, 1921), 6; Weber, "Irish-Born Champion," 237.

30. Leo Grebler, Joan W. Moore, and Ralph C. Guzman, *The Mexican American People* (New York: Free Press, 1970), 457; *Associated Catholic Charities Report, 1919* (Los Angeles, 1920), 11; *Associated Catholic Charities Report, 1920* (Los Angeles, 1921), 21.

31. *Associated Catholic Charities Report, 1920* (Los Angeles, 1921), 21–22.

32. Ibid., 22.

33. *Associated Catholic Charities Report, 1921* (Los Angeles, 1922), 10.

34. *Associated Catholic Charities Report, 1919* (Los Angeles, 1920), 11.

35. Weber, "Irish-Born Champion," 236; Privett, "Robert E. Lucey," 17; Rev. Dennis J. Burke, "The History of the Confraternity of Christian Doctrine in the Diocese of Los Angeles, 1922–1936" (Master's thesis, Catholic University of America, 1965), 19–23.

36. *Associated Catholic Charities Report, 1921* (Los Angeles, 1922), 10.

37. Eighth Annual Report of the Catholic Welfare Bureau (Los Angeles, 1926), 28.

38. Burke, "History of the Confraternity," 26.

39. Weber, "Irish-Born Champion," 236; Privett, "Robert E. Lucey," 17.

40. Weber, "Irish-Born Champion," 236–37.

41. *Associated Catholic Charities Report, 1922* (Los Angeles, 1923), 23–24; *Associated Catholic Charities Report, 1923* (Los Angeles, 1924), 5.

42. *Associated Catholic Charities Report, 1923* (Los Angeles, 1924), 23.

43. Ibid., 23.

44. Ibid.

45. Mosqueda, *Catholicism and Political Ideology,* 79.

46. Bresette, *Mexicans in the United States,* 28.

47. Weber, "Irish-Born Champion," 234.

48. See Annual Catholic Welfare Reports from 1929 to 1932.

49. "The Business of the Church," *The Tidings,* 22 Oct. 1937, 8.

50. Bresette, *Mexicans in the United States,* 26.

51. Mosqueda, *Catholicism and Political Ideology,* 58; Clifton L. Holland, *The Religious Dimension in Hispanic Los Angeles, a Protestant Case Study* (South Pasadena: William Carey Library, 1974), 200; Grebler, Moore, and Guzman, *Mexican American People,* 487; Antonio R. Soto, *The Chicano and the Church* (San Francisco: Marfel Associates, 1973), 11–12. The only exception to this 5 to 10 percent estimate is the optimistic Methodist Bishop G. Bromley Oxnam who calculates a very high 30 to 50 percent estimate of Protestant Chicanos in "The Mexican in Los Angeles from the Standpoint of the Religious Forces of the City," *Annals of the American Academy* 93 (1921), 130–33.

52. Interview with Zeferino Ramírez, 21 April 1927, p. 4, "Biographies and Case Histories" II folder, Z-R5, Manuel Gamio collection, Bancroft Library, University of California, Berkeley.

53. Interview with Dr. Vernon M. McCombs, in Smith, "Watts, California," 67–68.

54. Ibid., 68–73.

55. Gamio, *Life Story,* 63.

56. Interview with Julio Cortina, 346, San Diego Project, Racial Minorities Survey, Federal Writers Project collection, RG 306, Department of Special Collections, University of California, Los Angeles.

57. Alice Bessie Culp, "A Case Study of the Living Conditions of Thirty-five Mexican Families of Los Angeles with Special Reference to Mexican Children" (Master's thesis, University of Southern California, 1921), 50–51.

58. Bresette, *Mexicans in the United States,* 43.

59. Interview with Ramón Ramírez, 75, Racial Minorities Survey, Federal Writers Project collection, University of California, Los Angeles.

60. Interview with Arthur Morales, 8 April 1927, p. 4, "Biographies and Case Histories" II folder, Gamio collection.

61. Interview with the Galván Family, 8 April 1927, p. 9, "Biographies and Case Histories" III folder, Gamio collection.

62. Interview with Sr. Francisco Uribe, 19 April 1927, p. 5, "Biographies and Case Histories" I folder, Gamio collection.

63. Interview with Sr. Miguel Alonso, 18 April 1927, p. 5, "Biographies and Case Histories" I folder, Gamio collection.

64. Interview with Sra. María Rocha, 8 April 1927, p. 2, "Biographies and Case Histories" I folder, Gamio collection.

65. Ortegon, "Religious Population," 10–11.

66. Quoted in Mary Lanigan, "Second Generation Mexicans in Belvedere" (Master's thesis, University of Southern California, 1932), 28.

67. Gamio, *Life Story,* 86.

68. Interview with Sr. Tomás de la Cruz, 19 April 1927, p. 3, "Biographies and Case Histories" III folder, Gamio collection.

69. Interview with Julio Cortina, 346, Racial Minorities Survey, Federal Writers Project collection, University of California, Los Angeles.

70. Lanigan, "Second Generation," 30.

71. Mike Davis, *City of Quartz: Excavating the Future in Los Angeles* (London: Verso, 1990), 331. For another, slightly different account of these events, see Francisco E. Balderrama, *In Defense of La Raza: The Los Angeles Mexican Consulate and the Mexican Community, 1929 to 1936* (Tucson: Univ. of Arizona Press, 1982), 73–87.

72. Balderrama, *Defense,* 76–77; *The Tidings,* 30 Nov. 1934.

73. See *The Tidings,* 30 Nov. and 7 Dec. 1934, for this very analysis.

74. *The Tidings,* 30 Nov. and 7 Dec. 1934.

75. *The Tidings,* 21 Dec. 1934.

76. *The Tidings,* 14 Dec. 1934.

77. Ibid.

78. Ibid.

79. *The Tidings,* 15 Feb. 1935.

80. Balderrama, *Defense,* 86.

Chapter 8
Music and the Growth of Mass Culture

1. The Heller Committee for Research in Social Economics of the University of California and Constantine Panunzio, *How Mexicans Earn and Live: A Study of the Incomes and Expenditures of One Hundred Mexican Families in San*

Diego, California, Cost of Living Studies, 5 (Berkeley: Univ. of California Press, 1933), 49–52, 63–64.

2. "Observaciones—Los Salones de Baile," Observations of Luis Felipe Recinos, Los Angeles, Calif., 15 April 1927, Manuel Gamio collection, Bancroft Library, University of California, Berkeley.

3. U.S. Department of Labor, Bureau of Labor Statistics, "Mexican Families in Los Angeles," *Money Disbursements of Wage Earners and Clerical Workers in Five Cities in the Pacific Region, 1934–36,* Bulletin 639 (Washington, D.C.: GPO, 1939), 108–9, 231. For similar figures from San Diego, see Heller Committee and Panunzio, *Earn and Live,* 50–52.

4. See *Ballad of an Unsung Hero,* Cinewest/KPBS, 1983.

5. *La Opinión,* 10 May 1929.

6. *La Opinión,* 3 May 1927.

7. U.S. Department of Labor, *Money Disbursements,* 231; Heller Committee and Panunzio, *Earn and Live,* 50, 71.

8. See Lary May, *Screening Out the Past: The Birth of Mass Culture and the Motion Picture Industry* (Chicago: Univ. of Chicago Press, 1980), and Kathy Peiss, *Cheap Amusements: Working Women and Leisure in Turn-of-the-Century New York* (Philadelphia: Temple Univ. Press, 1986), 139–62.

9. May, *Screening,* 152–58.

10. Larry Carr, *More Fabulous Faces: The Evolution and Metamorphosis of Dolores Del Río, Myrna Loy, Carole Lombard, Bette Davis, and Katharine Hepburn* (Garden City, N.Y.: Doubleday, 1979), 1–51; De Witt Bodeen, "Dolores Del Río: Was the First Mexican of Family to Act in Hollywood," *Films in Review* 18 (1967), 266–83; José Gómez-Sicre, "Dolores Del Río," *Américas* 19 (1967), 8–17.

11. From content analysis of advertisements in Spanish-language newspapers in Los Angeles from 1916 to 1935. *El Heraldo de México,* 3 May 1916, 2 May 1920, and 13 Sept. 1925, and *La Opinión,* 6 April 1930 and 5 May 1935.

12. *El Heraldo de México,* 3 May 1916.

13. See "Juan B. Ruiz," interview by Eustace L. Williams, 6 May 1937, Racial Minorities Survey, Federal Writers Project collection, University of California, Los Angeles.

14. *La Opinión,* 6 April 1930 and 5 May 1935.

15. See Regina Markell Morantz-Sanchez, *Sympathy and Science: Women Physicians in American Medicine* (New York: Oxford Univ. Press, 1985); Mary Roth Walsh, *"Doctors Wanted: No Women Need Apply": Sexual Barriers in the Medical Profession, 1835–1975* (New Haven: Yale Univ. Press, 1977); and Gloria Melnick Moldow, *Women Doctors in Gilded-Age Washington: Race, Gender, and Professionalization* (Urbana: Univ. of Illinois Press, 1987).

16. From Naturalization documents, National Archives, Laguna Niguel, California. The occupational structure of this Mexican immigrant sample will be discussed in full in the next chapter.

17. Take for example the words of conjunto musician Narciso Martínez, who explained: "No puedo ser americano, porque mi papá y mamá eran mexicanos" ("I cannot consider myself American, because my father and mother were Mexicans"). Manuel Peña, *The Texas-Mexican Conjunto: History of a Working-Class Music* (Austin: Univ. of Texas Press, 1985), 54–55.

18. Peña, *Conjunto,* 146–48; Tim Patterson, "Notes on the Historical Application of Marxist Cultural Theory," *Science & Society* 39 (1975), 257–91. The

notion of "organic intellectuals" comes from Antonio Gramsci's theory of cultural hegemony, which can be found in *Selections from the Prison Notebooks,* Quintin Hoare and Geoffrey Newell Smith, eds. and trans. (New York: International, 1971).

19. Analysis of Naturalization documents, National Archives, Laguna Niguel, California.

20. Daniel Castañeda, "La música y la revolución mexicana." *Boletín Latín-Americano de Música* 5 (1941), 447–48; Claes af Geijerstam, *Popular Music in Mexico* (Albuquerque: Univ. of New Mexico Press, 1976), 88–91.

21. Merle E. Simmons, *The Mexican Corrido as a Source for Interpretive Study of Modern Mexico (1870–1950)* (Bloomington: Indiana Univ. Press, 1957), 7, 34. See also Vicente T. Mendoza, *El corrido mexicano* (México, D.F.: Fondo de Cultura Económica, 1974), xv, who calls the years between 1910 and 1929 the "Golden Age" of the corrido.

22. Pedro J. González, *Translated Transcripts* (Cinewest/KPBS, 1983), 75–77.

23. Geijerstam, *Popular Music,* 49–58. For traditional songs from one particular region in central Mexico, see Juan Diego Razo Oliva, *Rebeldes populares del Bajío (Hazanas, tragedias y corridos, 1910–1927)* (Mexico City: Editorial Katun, S.A., 1983).

24. Geijerstam, *Popular Music,* 68–69.

25. Ibid., 83–88; Américo Paredes, "The Ancestry of Mexico's Corridos: A Matter of Definitions," *Journal of American Folklore* 76 (1963), 233. For a collection of folk songs composed by Mexicans in conflict with Anglo Texans, see Américo Paredes, *A Texas-Mexican Cancionero: Folksongs of the Lower Border* (Urbana: Univ. of Illinois Press, 1976), particularly part II.

26. John Holmes McDowell, "The Corrido of Greater Mexico as Discourse, Music and Event," in *"And Other Neighborly Names": Social Process and Cultural Image in Texas Folklore,* Richard Bauman and Roger D. Abrahams, eds. (Austin: Univ. of Texas Press, 1981), 47, 73; Geijerstam, 52–53, 56; Merle E. Simmons, "The Ancestry of Mexico Corridos," *Journal of American Folklore* 76 (1963), 3.

27. Quoted in Nellie Foster, "The *Corrido:* A Mexican Culture Trait Persisting in Southern California" (Master's thesis, University of Southern California, 1939), 7.

28. Peña, *Conjunto,* 40, according to Library of Congress informant Richard Spottswood; Pedro J. González, videocassette tape interviews for *Ballad of an Unsung Hero* (San Diego: Cinewest/KPBS, 1983), 2:12 to 2:19. There is some dispute over whether this recording was indeed the first, since Pedro González remembers recording sessions he participated in as early as 1924. See González, *Transcripts,* 72–74. There is also a dispute over whether, indeed, Gonzalez was the original composer. In 1939, a recording of "El Lavaplatos" was discovered in Los Angeles which claimed that the music and words had been written by Jesus Ororio. See Foster, *"Corrido,"* 24.

29. "El Lavaplatos" (The Dishwasher) from Philip Sonnichsen, *Texas-Mexican Border Music, Vols. 2 & 3: Corridos Parts 1 & 2* (Arhoolie Records, 1975), 9. For more lyrics, see the beginning of Chapter 7.

30. González, *Transcripts,* 82–85, 96–98, 217–19, 227–30.

31. Foster, *"Corrido,"* 30.

32. Naturalization documents, National Archives, Laguna Niguel, California.

33. Peña, *Conjunto,* 58.

34. See ibid., esp. chap. 1.

35. J. Xavier Mondragón, "El Desarrollo de la Canción Mexicana en Estados Unidos," *La Opinión,* 27 March 1932.

36. Nicolás Kanellos, "An Overview of Hispanic Theatre in the United States," in *Hispanic Theatre in the United States,* Nicolás Kanellos, ed. (Houston: Arte Público Press, 1984), 9; Nicolás Kanellos, "Two Centuries of Hispanic Theatre in the Southwest," *Revista Chicano-Riqueña* 11 (1983), 27–35; Tomás Ybarra-Frausto, "I Can Still Hear the Applause: La Farándula Chicana: Carpas y Tandas de Variedad," in *Hispanic Theatre,* 56.

37. González, *Transcripts,* 221.

38. Ibid., 246–49.

39. Foster, "*Corrido,*" 26.

40. Pedro J. González remembers being asked to play for the studios in *Transcripts,* 96–98.

41. In this fashion, Chicano musicians in Los Angeles resembled their counterparts in Texas. See Peña, *Conjunto,* 146–48.

42. Ibid., 49–50.

43. Foster, *Corrido,* 19.

44. Quoted in Stephen Joseph Loza, "The Musical Life of the Mexican/Chicano People in Los Angeles, 1945–1985: A Study in Maintenance, Change, and Adaptation" (Ph.D. diss., University of California, Los Angeles, 1985), 110–11.

45. Analysis of Naturalization documents, National Archives, Laguna Niguel, California.

46. González, *Transcripts,* 124–25.

47. Peña, *Conjunto,* 39–41.

48. Ibid., 42; exceedingly rare was González's claim that he made $50 a recording and began receiving royalties in the 1930s. Since he was considered the top Mexican artist by 1930, this claim is understandable. González, *Transcript,* 118, 234, 240.

49. According to Lalo Guerrero, in Loza, "Musical Life," 118.

50. González, *Transcript,* 74; Foster, "*Corrido,*" 22–23. The only other musical outlet I have discovered that advertised in the Spanish-language press was La Platt Music Company. For go-betweens in San Antonio, see Peña, *Conjunto,* 40.

51. Loza, "Musical Life," 105, 117–20.

52. Roland Marchand, *Advertising the American Dream: Making Way for Modernity, 1920–1940* (Berkeley: Univ. of California Press, 1985), 89–94.

53. González, *Transcript,* 99; Félix F. Gutiérrez and Jorge Reina Schement, *Spanish-Language Radio in the Southwestern United States* (Austin: Center for Mexican American Studies, Univ. of Texas, 1979), 5.

54. González, *Transcript,* 102–6.

55. Gutiérrez and Schement, *Spanish-Language Radio,* 5–7; Gonzalez, *Transcript,* 98–99.

56. Gutiérrez and Schement, *Spanish-Language Radio,* 7.

57. Foster, "*Corrido,*" 20–22; González, *Transcript,* 114–15; see also 249–51 for a description of rural jestering with pig.

58. González, *Transcript,* 101–9, 119–37; Gene Fowler and Bill Crawford, *Border Radio* (Austin: Texas Monthly Press, 1987), 207–8.

59. See Fowler and Crawford, *Border Radio,* esp. 162, 247.

60. Kanellos, "Overview," 10, and "Two Centuries," 36.

61. See Carl J. Mora, *Mexican Cinema: Reflections of a Society, 1896–1980* (Berkeley: Univ. of California Press, 1982), 25–51, for a description of this period of transition in the Mexican film industry.

62. Loza, "Musical Life," 102–4.

63. Mary Lanigan, "Second Generation Mexicans in Belvedere" (Master's thesis, University of Southern California, 1932), 58.

64. Ibid., 62.

65. Ibid., 103–5.

66. Ibid., 22.

Chapter 9
Workers and Consumers: A Community Emerges

1. Quoted in Philip Sonnichsen, *Texas-Mexican Border Music, Vols. 2 & 3: Corridos Parts 1 & 2* (Arhoolie Records, 1975), 9; Pedro J. González, videocassette tape interviews for "Ballad of an Unsung Hero" (San Diego: Cinewest/KPBS, 1983), 2:12 to 2:19, 4:19 to 4:25.

2. See Albert Camarillo, *Chicanos in a Changing Society: From Mexican Pueblos to American Barrios in Santa Barbara and Southern California, 1848–1930* (Cambridge: Harvard Univ. Press, 1979), 210–13, 217–25; Ricardo Romo, *East Los Angeles: History of a Barrio* (Austin: Univ. of Texas Press, 1983), 112–28; Pedro G. Castillo, "The Making of a Mexican Barrio: Los Angeles, 1890–1920" (Ph.D. diss., University of California, Santa Barbara, 1979), 139–73; and Douglas G. Monroy, "Mexicanos in Los Angeles, 1930–1941: An Ethnic Group in Relation to Class Forces" (Ph.D. diss., University of California, Los Angeles, 1978), 76–78.

3. Herbert G. Gutman, *Work, Culture and Society in Industrializing America: Essays in American Working-Class and Social History* (New York: Vintage–Random House, 1977), 8–9; original song printed in Paul Taylor, *Mexican Labor in the United States: Chicago and the Calumet Region* (Berkeley: Univ. of California Press, 1932), vi–vii.

4. The 1930 Census, for example, reported only 23 percent of Mexican laborers employed in manufacturing in the five southwestern states, as compared with 41 percent employed in agriculture. See Mario Barrera, *Race and Class in the Southwest: A Theory of Racial Inequality* (Notre Dame: Univ. of Notre Dame Press, 1979), 95. An exception to this pattern existed in the Midwest, where Chicanos did move into industrial occupations in significant numbers. See Zaragoza Vargas, "Mexican Auto Workers at Ford Motor Company, 1918–1933" (Ph.D. diss., University of Michigan, 1984), 10; and Francisco Arturo Rosales and Daniel T. Simon, "Mexican Immigrant Experience in the Urban Midwest: East Chicago, Indiana, 1919–1945," in *Forging a Community: The Latino Experience in Northwest Indiana, 1919–1975,* James B. Lane and Edward J. Escobar, eds. (Chicago: Cattails Press, 1987), 138–39.

5. Interview with Mr. Hill, A.T.& S.F., Kerchoff Building, Los Angeles, Calif., p. 1, 74/187, Paul S. Taylor collection, Bancroft Library, University of California, Berkeley.

6. Interview with Mr. Martínez, Clovis, Calif., 5 Sept. 1928, 1, Taylor collection.

7. Ibid.

8. Mark Reisler, *By the Sweat of Their Brow: Mexican Immigrant Labor in the United States, 1900–1940* (Westport, Conn.: Greenwood, 1976), 128–31.

9. Note by Dr. George Clements, Clements Papers, Box 72, UCLA Department of Special Collections.

10. Reisler, *Sweat,* especially chaps. 7 and 8.

11. Romo, *East Los Angeles,* 119.

12. Castillo, "Mexican Barrio," 166; Camarillo, *Changing,* 218.

13. Douglas Monroy, "Like Swallows at the Old Mission: Mexicans and the Racial Politics of Growth in Los Angeles in the Interwar Period," *Western Historical Quarterly* 14 (1983), 438–39.

14. Problems with this designation for musicians will be discussed later in this chapter.

15. See Monroy, "Mexicanos," especially chap. 3, and Luis Leobardo Arroyo, "Industrial Unionism and the Los Angeles Furniture Industry, 1918–1954" (Ph.D. diss., University of California at Los Angeles, 1979).

16. See Thomas Kessner, *The Golden Door: Italian and Jewish Immigrant Mobility in New York City, 1880–1915* (New York: Oxford Univ. Press, 1977), 33, 35, and 52, for a similar placement of Italian immigrant musicians.

17. Romo, *East Los Angeles,* 123.

18. Interview with Mr. Martínez, 5, Taylor collection.

19. "The Success of Mexican Women in Industrial Work," 3, Taylor collection.

20. See the All Nations Foundation, *A Friend by the Side of the Road* (Los Angeles: All Nations Boys Club, 1939), Container 58, G. Bromley Oxnam collection, Manuscript Division, Library of Congress, Washington, D.C.

21. Ricardo Romo, "The Urbanization of Southwestern Chicanos in the Early Twentieth Century," *New Scholar* 6 (1977), 185.

22. *Fifteenth Census of the United States: 1930, Population: Special Report on Foreign-Born White Families by Country of Birth of Head with an Appendix Giving Statistics for Mexican, Indian, Chinese and Japanese Families* (Washington, D.C.: GPO, 1933), 211–12.

23. The Heller Committee for Research in Social Economics of the University of California and Constantine Panunzio, *How Mexicans Earn and Live: A Study of the Incomes and Expenditures of One Hundred Mexican Families in San Diego, California,* Cost of Living Studies, 5 (Berkeley: Univ. of California Press, 1933), 29, 38–40.

24. Stephan Thernstrom, *Poverty and Progress: Social Mobility in a Nineteenth Century City* (Cambridge, Mass.: Harvard Univ. Press, 1964), 116–22.

25. Thernstrom, *Poverty,* 155–57; Stephan Thernstrom, *The Other Bostonians: Poverty and Progress in the American Metropolis, 1880–1970* (Cambridge, Mass.: Harvard Univ. Press, 1973), 101.

26. Daniel Luria, "Wealth, Capital, and Power: The Social Meaning of Home Ownership," *Journal of Interdisciplinary History* 7 (1976), 267–69.

27. Olivier Zunz, *The Changing Face of Inequality: Urbanization, Industrial Development, and Immigrants in Detroit, 1880–1920* (Chicago: Univ. of Chicago Press, 1982), 153.

28. The man north of Mr. Martínez, under the trees at Clovis, 5 Sept. 1928, 1, Taylor collection.

29. Heller Committee and Panunzio, *Earn and Live,* 46–47.

30. Ibid., 43, 46.

31. Zunz, *Inequality,* 170–76.

32. Richard Griswold del Castillo, *La Familia: Chicano Families in the Urban Southwest, 1848 to the Present* (Notre Dame: Univ. of Notre Dame Press, 1984), 63, 103.

33. Heller Committee and Panunzio, *Earn and Live,* 11; Paul S. Taylor, "The Success of Mexican Women in Industrial Work," unpublished manuscript, p. 1, Taylor collection.

34. See Vicki Ruiz, *Cannery Women/Cannery Lives: Mexican Women, Unionization, and the California Food Processing Industry, 1930–1950* (Albuquerque: Univ. of New Mexico Press, 1987), 17–19, 34–36.

35. *Mexicans in California,* Report of Governor C. C. Young's Mexican Fact-Finding Committee (1930; rpt., San Francisco: R and E Research Associates, 1970), 213–14; Heller Committee and Panunzio, *Earn and Live,* 13, 15–19; "The Success of Mexican Women in Industrial Work," 2, Taylor collection. Louise A. Tilly and Joan W. Scott describe a similar pattern for French married women and their children in *Women, Work, & Family* (New York: Holt, Rinehart and Winston, 1978), 134–36, 200–202.

36. Heller Committee and Panunzio, *Earn and Live,* 14–19; *Mexicans in California,* 211–12. For comparison with European immigrant and native-born American families, see Virginia Yans-McLaughlin, *Family and Community: Italian Immigrants in Buffalo, 1880–1930* (Urbana: Univ. of Illinois Press, 1971), 165, 173–74; and John Modell and Tamara K. Haraven, "Urbanization and the Malleable Household: An Examination of Boarding and Lodging in American Families," in *The American Family in Social-Historical Perspective,* 2nd ed., Michael Gordon, ed. (New York: St. Martin's Press, 1978), 51–68.

37. Heller Committee and Panunzio, *Earn and Live,* 29, 49.

38. Ibid., 28–29; Robert Redfield, *Tepoztlán: A Mexican Village: A Study of Folk Life* (Chicago: Univ. of Chicago Press, 1930), 85–87.

39. From content analysis of advertisements in Spanish-language newspapers in Los Angeles from 1916 to 1935. *El Heraldo de México,* 3 May 1916, 2 May 1920, and 13 Sept. 1925, and *La Opinión,* 6 April 1930 and 5 May 1935.

40. Heller Committee and Panunzio, *Earn and Live,* 30–35.

41. Ibid., 36; Manuel Gamio, *Mexican Immigration to the United States: A Study of Human Migration and Adjustment* (1930; rpt., New York: Dover, 1971), 146, 204; content analysis of advertisements in *El Heraldo de México,* 3 May 1916, 2 May 1920, and 13 Sept. 1925, and *La Opinión,* 6 April 1930 and 5 May 1935.

42. Heller Committee and Panunzio, *Earn and Live,* 52–53.

43. Scott L. Bottles, *Los Angeles and the Automobile: The Making of the Modern City* (Berkeley: Univ. of California Press, 1987), 33–41.

44. See Rodolfo F. Acuña, *A Community under Siege: A Chronicle of Chicanos East of the Los Angeles River, 1945–1975* (Los Angeles: Chicano Studies Research Center Publications, Univ. of California, Los Angeles, 1984), for a discussion of this control in the post–World War II period.

Part Four
Ambivalent Americanism

1. Pablo Guerrero, Letter to Los Angeles County, 28 May 1934, Box 4E, 40.31/340.39, Los Angeles County Board of Supervisors files.

2. Alfred Barela, Letter to the Honorable Arthur S. Guerin, 21 May 1943,

and Arthur S. Guerin, Letter to Mr. Alfred Barela, 2 June 1943, Julian Nava collection, Urban Archives Center, California State University, Northridge.

3. "They Must Know: A One Act Play," *Mexican Voice* (Jan.–Feb. 1939), 15; *Mexican Voice* (Jan.–Feb. 1939), 7.

Chapter 10
The Dilemma of Repatriation

1. *La Opinión*, 30 and 31 Jan. 1931; Los Angeles City Directories, 1927–35, show that the Olazábal family was new to Belvedere in 1931, and that Mr. Olazábal was able to remain employed as a grocer through the most difficult years of the Depression.

2. *La Opinión*, 30 and 31 Jan. 1931.

3. The exact number of Mexicans who returned to their native land from Los Angeles in this period is difficult to know. Official estimates from organized train departures number below 14,000, but at least as many probably left Los Angeles on their own or with the help of private organizations. Carey McWilliams probably has the highest estimate of 75,000 in *Factories in the Field: The Story of Migratory Farm Labor in California* (Santa Barbara: Peregrine Smith, 1935), 125. More reasonable is the estimate arrived at by Matt Meier and Feliciano Rivera (taken apparently from other figures used by McWilliams in the 1930s) of 35,000 in *The Chicanos: A History of Mexican Americans* (New York: Hill and Wang, 1972), 161. Since most historians estimate the overall Mexican heritage population of Los Angeles in 1930 to be about 100,000, close to one-third left the city during these years.

4. John D. Weaver, *L.A.: El Pueblo Grande* (Los Angeles: Ward Ritchie Press, 1973), 76; Leonard J. Leader, "Los Angeles and the Great Depression" (Ph.D. diss., University of California, Los Angeles, 1972), 1–4.

5. Frances Fox Piven and Richard A. Cloward, *Regulating the Poor: The Functions of Public Welfare* (New York: Vintage, 1971), 49.

6. Leader, "Great Depression," 4–11; William H. Mullins, *The Depression and the Urban West Coast, 1929–1933: Los Angeles, San Francisco, Seattle, and Portland* (Bloomington: Indiana Univ. Press, 1991), 127.

7. Francisco Enrique Balderrama, "En Defensa de la Raza: The Los Angeles Mexican Consulate and Colonia Mexicana during the Great Depression" (Ph.D. diss., University of California, Los Angeles, 1978), 25; Mullins, *The Depression and the Urban West Coast*, 52.

8. Leader, "Great Depression," 6; Gerald Nash, *The American West in the Twentieth Century* (Englewood Cliffs: Prentice-Hall, 1973), 137; Mark Reisler, *By the Sweat of Their Brow: Mexican Immigration in the United States, 1900–1940* (Westport, Conn.: Greenwood, 1976), 228–33.

9. Leader, "Great Depression," 6–7, 9; Jacqueline Rorabeck Kasun, *Some Social Aspects of Business Cycles in the Los Angeles Area, 1920–1950* (Los Angeles: Haynes Foundation, 1954), 32.

10. *La Opinión*, 8 Feb. 1931.

11. Robin Fitzgerald Scott, "The Mexican-American in the Los Angeles Area, 1920–1950: From Acquiescence to Activity" (Ph.D diss., University of Southern California, 1971), 116–17; *La Opinión*, 14 Aug. 1931.

12. *Los Angeles Times*, 13 Jan. 1931; *Los Angeles Daily News*, 17 April 1931; Leader, "Great Depression," pp. 29–30; Scott, "Acquiescence," 13.

13. California, State Relief Administration, *Review of the Activities of the State Relief Administration of California: 1933–1935* (San Francisco: State Printing Office, 1936), 349; California, *Report and Recommendations of the California State Unemployment Commission* (San Francisco: State Printing Office, 1932), 371.

14. Leader, "Great Depression," 32–38; California, *Report and Recommendations,* 371–74. For the fullest account of relief distribution in Los Angeles during this period, see Mullin, *The Depression and the Urban West Coast,* 27–36, 62–71, 98, 126.

15. Analysis of naturalization records by linking individuals to Los Angeles city directories; see note 28 below for further description. Abraham Hoffman, *Unwanted Mexican Americans in the Great Depression: Repatriation Pressures, 1929–1939* (Tucson: Univ. of Arizona Press, 1974), 36–37; interview with Carlos Munatones, conducted by Christine Valenciana, 4 Aug. 1971, OH 748, Oral History Program, California State University at Fullerton.

16. Manuel Gamio, *Mexican Immigration to the United States: A Study of Human Migration and Adjustment* (1930; rpt., New York: Dover, 1971), Appendix V, pp. 224–29.

17. Interview with José Castillo, Biographies and Case Histories II folder, Z-R5, Manuel Gamio collection, Bancroft Library, University of California, Berkeley.

18. Hoffman, *Unwanted,* 36–37, and Appendix D, 174–75.

19. Ibid.; Balderrama, "Defensa," and *In Defense of La Raza: The Los Angeles Mexican Consulate and the Mexican Community, 1929 to 1936* (Tucson: Univ. of Arizona Press, 1982), especially Chap. 2; Mercedes Carreras de Velasco, *Los Mexicanos que devolvió la crisis, 1929–1932* (México, D.F.: Secretaría de Relaciones Exteriores, 1974). For contemporary studies, see James Carl Gilbert, "A Field Study in Mexico of the Mexican Repatriation Movement" (Master's thesis, Univ. of Southern California, 1934); and Paul S. Taylor, *A Spanish-Mexican Peasant Community: Arandas in Jalisco, Mexico,* Ibero-Americana, 4 (Berkeley: Univ. of California Press, 1933).

20. Hoffman, *Unwanted,* 39–41, 63.

21. Ibid., 43–63; *La Opinión,* 27 and 28 Feb. 1931.

22. Hoffman, *Unwanted,* 64–65.

23. Ibid., 68–71.

24. Carey McWilliams, "Getting Rid of the Mexican," *American Mercury* 28 (1933), 322–24.

25. Balderrama, *Defense,* 17–22; *La Opinión,* 6 Feb. and 8 March 1931.

26. Interview with Maria Bustos Jefferson, conducted by Christine Valenciana, 11 Sept. 1971, OH 1300, Oral History Program, California State University at Fullerton. See Chapter 5 for a full description of the reasons behind the advocacy for repatriation by the Mexican consulate.

27. Hoffman, *Unwanted,* 84–87; Balderrama, *Defense,* 42.

28. I created a subsample of individuals from my larger group who were living in Los Angeles in 1927–28 at which time they took out first or second papers for naturalization. Linking this information to Los Angeles city directories, I was able to locate 171 of these individuals and their families. Over the next ten years, 62, or 36 percent, disappeared from the directories, apparently moving out of the city. Although I currently have no way of telling whether, in fact, these individuals returned to Mexico, the numbers do suggest that a sub-

stantial proportion of even these Mexicans were affected by the repatriation campaign.

29. Hoffman, *Unwanted,* 88. For an example of a family struggling over whether or not to return to Mexico in 1931, see interview with Mrs. Carmen Landeros, conducted by Christine Valenciana, 7 Aug. 1971, OH 745, Oral History Program, California State University at Fullerton.

30. Lucas Lucio, organizer of three repatriation trips, in Balderrama, *Defense,* 28. See also interview with Allen A. Hunter, conducted by Christine Valenciana, 22 Aug. 1971, OH 744, Oral History Program, California State University at Fullerton.

31. Interview with Maria Bustos Jefferson and interview with Allen A. Hunter, Oral History Program, California State University at Fullerton.

32. Interview record No. 17 (in Spanish) from Gilbert, "Field Study," 146.

33. Emory Bogardus, *The Mexican in the United States* (Los Angeles: Univ. of Southern California Press, 1934), 91. Even the small percentage that settled in agricultural colonies differed with the Mexican government over where those colonies should be located. Most repatriates organized themselves for agriculture endeavors in the northern states, while Mexican government officials wanted colonies set up in the tropical and depopulated south; see Carreras de Velasco, *Devolvió,* 117.

34. Taylor, *Arandas*; Gilbert, "Field Study."

35. Gilbert, "Field Study," 137–42; Bogardus, *Mexican in the United States,* 93–94; interview with Theresa Martinez Southard, conducted by Christine Valenciana, 1 Sept. 1971, OH 753, Oral History Program, California State University at Fullerton.

36. Gilbert, "Field Study," 137–42; Bogardus, *Mexican in the United States,* 93–94; interview with Theresa Martinez Southard.

37. Gilbert, "Field Study," 136–37.

38. Ann L. Craig, *The First Agraristas: An Oral History of a Mexican Agrarian Reform Movement* (Berkeley: Univ. of California Press, 1983), 92–94, 178–82, 211.

39. *La Opinión,* 11, 19, 20 Feb. and 29 April 1933.

40. *La Opinión,* 2 Nov. 1931, 6 and 20 March, 28 May 1932, 19 Jan. 1933.

41. Gilbert found that the majority of repatriates he interviewed in 1934 wanted to return to the United States, for example. "Field Study," 164.

42. Pablo Guerrero, Letter to Los Angeles County, 28 May 1934, Box 4E, 40.31/340.39, Los Angeles County Board of Supervisors files.

43. See Hoffman, *Unwanted,* 91.

44. Ibid., 105–9, 172–73.

45. Evaluation of the characteristics of a sample of 171 individuals present in the city during 1927–28, and their movement out of the city over the next decade. The initial sample is taken from naturalization records, and compared with Los Angeles city directories for the period 1927 through 1935.

46. *La Opinión,* 11 April and 28 May 1932; undated report on Mexican immigrants by DeCourcey to Phillip J. Robinson, 6, Federal Writers Project collection, National Archives, Washington, D.C.; Earl E. Jensen, Superintendent of Charities, Letter to Los Angeles Board of Supervisors, 15 Feb. 1934, p. 3, Board of Supervisors Minutes, vol. 196, p. 233. Paul Taylor also notes the role of adverse reports from Mexico leading to a sharp decline in repatriation in *Mexican*

Labor in the United States: Migration Statistics, IV (Berkeley: Univ. of California Press, 1934), 24.

47. *La Opinión,* 11 March 1933.

48. Jensen to Board of Supervisors, 2; Hoffman, *Unwanted,* 105, 108; Analysis of Naturalization documents, National Archives, Laguna Niguel, California.

49. William R. Brock, *Welfare, Democracy, and the New Deal* (Cambridge, Eng.: Cambridge Univ. Press, 1988), 152–54, 238–43, 339.

50. Although 172 relief cases, totaling 664 individuals, left on this April 1934 train, this effort still failed to raise the number of interested families to pre–New Deal levels. Jensen to Board of Supervisors, 15 Feb. 1934; Hoffman, *Unwanted,* 107–9.

51. Interview with John Anson Ford, conducted by Christine Valenciana, 4 Sept. 1971, OH 759, Oral History Program, California State University at Fullerton; "Repatriation of the 1930s: Mrs. Emilia Castañeda de Valencia," interviewed by Christine Valenciana, 8 Sept. 1971, California State University, Fullerton, Oral History Program: Mexican American Project, 12–13; 16–17. See also Albert Camarillo, *Chicanos in California: A History of Mexican Americans in California* (San Francisco: Boyd and Fraser, 1984), 49–51.

52. "Mrs. Emilia Castañeda de Valencia," 1, 6–7, 11; Analysis of Naturalization records, National Archives, Laguna Niguel, and Los Angeles city directories.

53. G. A. Clements, Memorandum to A. G. Arnoll, 20 July 1933, MS 118, Geo. P. Clements Papers, Department of Special Collections, University of California, Los Angeles; Ronald W. Lopez, "The El Monte Berry Strike of 1933," *Aztlán* 1 (1970), 109–10.

54. Donald L. Zelman, "Mexican Migrants and Relief in Depression California: Grower Reaction to Public Relief Policies as They Affected Mexican Migration," *Journal of Mexican-American History* 5 (1975), 6.

55. Quoted in Zelman, "Grower Reaction," 9.

56. Ibid., 9–13.

57. Porfirio J. Miranda, "Perceptions of Locus of Control among Three Multi-Generation Chicano/Mexican Families" (Ph.D. diss., University of California, Los Angeles, 1978), 215.

58. José Hernandez Alvarez, "A Demographic Profile of the Mexican Immigration to the United States, 1910–1950," *Journal of Inter-American Affairs* 25 (1983), 479.

59. W. W. Robinson, *Los Angeles from the Days of the Pueblo: A Brief History and a Guide to the Plaza Area* (San Francisco: California Historical Society, 1981), 95–102. The painting of a mural by socialist David Alfaro Siqueros produced such an uproar when completed that city officials had it whitewashed rather than depict a historic past not in tune with tourism and promotion.

Chapter 11
Forging a New Politics of Opposition

1. Marie Louise Wuesthoff, "An Inquiry into the Activities of the International Ladies' Garment Workers' Union in Los Angeles" (Master's thesis, University of Southern California, 1938), 70–78; John Laslett and Mary Tyler, *The ILGWU in Los Angeles, 1907–1988* (Inglewood, Calif.: Ten Star Press, 1989),

31–35. Rose Pesotta tries to paint a much more positive account of this meeting in *Bread upon the Waters* (New York: Dodd, Mead, 1945), 41.

2. Pesotta, *Bread upon the Waters,* chaps. 2, 3, 4, and 5. For a fuller scholarly account of Pesotta's life, see Elaine Leeder, "The Gentle Warrior: Rose Pesotta, Anarchist and Labor Organizer" (Ph.D. thesis, Cornell University, 1985).

3. Francisco E. Balderrama, *In Defense of La Raza: The Los Angeles Mexican Consulate and the Mexican Community, 1929 to 1936,* (Tucson: Univ. of Arizona Press, 1982), 115.

4. For a demographic analysis of the transformation of the Mexican-origin population in the twentieth century and the particular role of the 1930s, see Thomas D. Boswell, "The Growth and Proportional Distribution of the Mexican Stock Population of the United States: 1910–1970," *Mississippi Geographer* 7:1 (Spring 1979), 57–76. For figures on Los Angeles, see U.S. Bureau of the Census, *Population of Spanish Mother Tongue: 1940,* Series P-15, No. 1, 9 June 1942, p. 3.

5. See Richard Amado Garcia, "The Making of the Mexican-American Mind, San Antonio, Texas, 1929–1941: A Social and Intellectual History of an Ethnic Community" (Ph.D. diss., University of California, Irvine, 1980), 571–606. See also Richard A. Garcia, "Class, Consciousness, and Ideology—The Mexican Community of San Antonio, Texas: 1930–1940," *Aztlán* 9 (1979), 42–43, 53–57.

6. Luis Leobardo Arroyo, "Industrial Unionism and the Los Angeles Furniture Industry, 1918–1954" (Ph.D. diss., University of California, Los Angeles, 1979); Douglas G. Monroy, "Mexicanos in Los Angeles, 1930–1941: An Ethnic Group in Relation to Class Forces" (Ph.D. diss., University of California, Los Angeles, 1978), 124–25, 138–39; Vicki L. Ruiz, *Cannery Women/Cannery Lives: Mexican Women, Unionization, and the California Food Processing Industry, 1930–1950* (Albuquerque: Univ. of New Mexico Press, 1987). See also Albert Camarillo, *Chicanos in California: A History of Mexican Americans in California* (San Francisco: Boyd & Fraser, 1984), 52–57.

7. See Juan Gómez-Quiñones, "The First Steps: Chicano Labor Conflict and Organizing, 1900–1920," *Aztlán* 3:1 (Spring 1972) and Luis Leobardo Arroyo, "Mexican Workers and American Unions: The Los Angeles AFL, 1890–1933," Chicano Political Economy Collective Working Paper Series #107 (Berkeley, 1981).

8. Arroyo, "Mexican Workers," 10, 12–17.

9. Devra Anne Weber, "The Struggle for Stability and Control in the Cotton Fields of California: Class Relations in Agriculture, 1919–1942" (Ph.D. diss., University of California, Los Angeles, 1986), 128–29; Gómez-Quiñones, "First Steps," 13–50, and *Sembradores: Ricardo Flores Magón y El Partido Liberal Mexicano: A Eulogy and Critique* (Los Angeles: UCLA Chicano Studies Research Center Publications, 1973); W. Dirk Raat, *Revoltosos: Mexico's Rebels in the United States, 1903–1923* (College Station: Texas A&M Univ. Press, 1981), 13–64. See Colin M. MacLachlan, *Anarchism and the Mexican Revolution: The Political Trials of Ricardo Flores Magón in the United States* (Berkeley: Univ. of California Press, 1991), for the latest account of PLM activity north of the border.

10. Raat, *Revoltosos,* 40–42, 56–62; Salvatore Salerno, *Red Nov., Black Nov.: Culture and Community in the Industrial Workers of the World* (Albany: State Univ. of New York Press, 1989), 28–30, 147–51; James R. Green, *Grass-roots*

Socialism: Radical Movements in the Southwest, 1895–1943 (Baton Rouge: Louisiana State Univ. Press, 1978), 176–85.

11. See Juan Gómez-Quiñones, "Piedras contra la Luna, México en Aztlán y Aztlán en México: Chicano-Mexican Relations and the Mexican Consulates, 1900–1920," *Contemporary Mexico: Papers of the IV International Congress of Mexican History,* James W. Wilkie, Michael C. Meyer, and Edna Monzón de Wilkie, eds. (Berkeley: Univ. of California Press, 1976), 494–527; Edward Joseph Escobar, "Chicano Protest and the Law: Law Enforcement Responses to Chicano Activism in Los Angeles, 1850–1936" (Ph.D. diss., University of California, Riverside, 1983), 130–81.

12. Arroyo, "Mexican Workers," 20–23. See Gregg Andrews, *Shoulder to Shoulder?: The American Federation of Labor, the United States, and the Mexican Revolution, 1910–1924* (Berkeley: Univ. of California Press, 1991), 99–168, for an excellent discussion of the AFL's complicated relationship with Mexican labor unions.

13. "Memorandum" from the Confederacion de Sociedades Mexicanas, IV-339–12, Secretaria de Relaciones Exteriores, Mexico City; Rodolfo Acuña, *Occupied America: A History of Chicanos,* 3rd ed. (New York: Harper and Row, 1988), 183; Balderrama, *Defense,* 91–93. My own analysis of these developments parallels that of Mario T. García, as evidenced by his letter to the editor in *Labor History* 24 (1984), 152–54, a critique of Douglas Monroy, "Anarquismo y Comunismo: Mexican Radicalism and the Communist Party in Los Angeles during the 1930s," *Labor History* 24 (Winter 1983), 34–59.

14. Clementina Durón, "Mexican Women and Labor Conflict in Los Angeles: The ILGWU Dressmakers' Strike of 1933," *Aztlán* 15:1 (Spring 1984), 146.

15. Paul S. Taylor, "Mexican Women in Los Angeles Industry in 1928," *Aztlán* 11:1 (Spring 1980), 102–14; Ruiz, *Cannery Women,* 14.

16. Pesotta, *Bread upon the Waters,* 1–3, 11, 19, 40.

17. Ibid., 21.

18. Letter from Rose Pesotta to David Dubinsky, 25 Sept. 1933, David Dubinsky Papers, ILGWU Archives, Cornell University, Ithaca; ibid., 24–25, 37, 43.

19. Ralph Chaplin, "Why I Wrote 'Solidarity Forever,'" *American West* (Jan. 1968); "Ralph H. Chaplin (1887–1961)," *Encyclopedia of the American Left,* Mari Jo Buhle, Paul Buhle, and Dan Georgakas, eds. (New York: Garland, 1990), 127.

20. Pesotta, *Bread upon the Waters,* 48. These songs could also provide unity behind bars in jail; see 52.

21. Ibid., 50–51.

22. Ibid., 32.

23. See Laslett and Tyler, *ILGWU,* 35, and Pesotta, *Bread upon the Waters,* 54–59, for the conclusion of the strike.

24. Pesotta, *Bread upon the Waters,* 26.

25. For examples of male Mexican American organizers in Los Angeles, see Monroy, "Mexicanos," and Arroyo, "Furniture." For an excellent study of Americanism among ethnic workers during the 1930s (in this case, French-Canadians in Massachusetts) see Gary Gerstle, *Working-Class Americanism: The Politics of Labor in a Textile City, 1914–1960* (Cambridge, Eng.: Cambridge Univ. Press,

1989). To contrast this working-class Americanism with that expressed by the Mexican American middle class, see Richard A. Garcia, *Rise of the Mexican American Middle Class: San Antonio, 1929–1941* (College Station: Texas A&M Univ. Press, 1991).

26. Pesotta, *Bread upon the Waters,* 332–33.

27. Ronald W. Lopez, "The El Monte Berry Strike of 1933," *Aztlán* 1:1 (Spring 1970), 101–2; Weber, "Organizing," 319–26; Dorothy Healey and Maurice Isserman, *Dorothy Healey Remembers: A Life in the American Communist Party* (New York: Oxford Univ. Press, 1990), 42–58; American Social History Project, *Who Built America?: Working People & the Nation's Economy, Politics, Culture & Society,* vol. 2 (New York: Pantheon Books, 1992), 357–59.

28. Lopez, "El Monte," 103–5.

29. Ibid., 105–9; *La Opinión,* 21 and 22 June 1933.

30. Acuña, *Occupied America,* 210–17; Balderrama, *Defense,* 93–106; Devra Anne Weber, "The Organizing of Mexicano Agricultural Workers: Imperial Valley and Los Angeles, 1928–34, an Oral History Approach," *Aztlán* 3:2 (Fall 1972), 326–28.

31. *Western Worker,* 19 and 26 June, 17 and 31 July, 7 Aug. 1933.

32. *Western Worker,* 13 and 20 Aug. 1933.

33. *Western Worker,* 4 Dec. 1933.

34. Healey, *Remembers,* 45–46.

35. *Western Worker,* 24 July 1936; Weber, "Organizing," 330–31.

36. *La Opinión,* 28 Oct. 1933.

37. Ibid.; *La Opinión* editorial, 30 Oct. 1933.

38. *La Opinión,* 1 Nov. 1933.

39. Ibid.

40. See Lizabeth Cohen, *Making a New Deal: Industrial Workers in Chicago, 1919–1939* (Cambridge, Eng.: Cambridge Univ. Press, 1990), and Gerstle, *Working-Class Americanism,* for a fuller discussion of these issues in different settings during the decade.

41. See Robin D. G. Kelley, *Hammer and Hoe: Alabama Communists during the Great Depression* (Chapel Hill: Univ. of North Carolina Press, 1990), esp. chap. 5, for a similar interpretation of the role of Communist party organizers in the South during the same period.

42. James R. Green, *The World of the Worker: Labor in Twentieth-Century America* (New York: Hill and Wang, 1980), 133–73; American Social History Project, *Who Built America?,* 373–80.

43. Arroyo, "Furniture," 12–14, 62–69.

44. Ibid., 47–51; *Western Worker,* 4 June, 2 and 16 July 1934.

45. Monroy, "Mexicanos," 120; Arroyo, "Furniture," 69–74, 79–96.

46. Arroyo, "Furniture," 119–51, and "Chicano Participation in Organized Labor; The CIO in Los Angeles, 1938–1950. An Extended Research Note." *Aztlán* 6:2 (Summer 1975), 284–86.

47. Monroy, "Mexicanos," 128–32.

48. Arroyo, "Participation," 280–83.

49. See Ruiz, *Cannery Women,* especially chap. 2.

50. Quoted in Vicki L. Ruiz, "A Promise Fulfilled: Mexican Cannery Workers in Southern California," *Pacific Historian* 30:2 (Summer 1986), 55.

51. Ruiz, *Cannery Women,* 69–75; Healey, *Remembers,* 65–69.

52. Mark Naison, *Communists in Harlem during the Depression* (Urbana:

Univ. of Illinois Press, 1983), 169–72. See *Western Worker,* 13 Aug. 1936, for an account of Earl Browder's campaign swing through California.

53. Healey, *Remembers,* 36–37.

54. David Roediger, "Where Communism Was Black: A Review of *Hammer and Hoe: Alabama Communists during the Great Depression* by Robin D. G. Kelley," *American Quarterly* 44:1 (March 1992), 124.

55. Ruiz, *Cannery Women,* 74–78, 83; Healey, *Remembers,* 69–70, 75–76.

56. Ruiz, *Cannery Women,* 99; Mario T. García, *Mexican Americans: Leadership, Ideology, & Identity, 1930–1960* (New Haven: Yale Univ. Press, 1989), 146; Camarillo, *California,* 58; Albert Camarillo, "Mexicans and Europeans in American Cities: Some Comparative Perspectives, 1900–1940," in *From "Melting Pot" to Multiculturalism: The Evolution of Ethnic Relations in the United States and Canada,* Valeria Gennaro Lerda, ed. (Rome, Italy: Bulzoni Editore, 1990), 251–52.

57. Garcia, *Mexican Americans,* 146–47; Camarillo, *California,* 58–60; Camarillo, "Cities," 252.

58. Ruiz, *Cannery Women,* 78–81; Healey, *Remembers,* 70.

59. Quoted in Camarillo, *California,* 60.

60. García, *Mexican Americans,* 147.

61. "First National Congress of the Mexican & Spanish American People of the United States: Reporte del Comite de Credenciales," 1–4, Ernesto Galarza Papers, M224, Box 13, Folder 9, Stanford University, Department of Special Collections; Camarillo, *California,* 58–64; Camarillo, "Cities," 253–54; Acuña, *Occupied America,* 229–31, 235–39.

62. Naison, *Communists in Harlem,* 178.

63. Ibid., 177–85. Mario García makes the same assessment of El Congreso's relationship to the Popular Front strategy of the American Communist party in *Mexican Americans,* 153–55.

64. "First National Congress," 7, and "Resolutions Adopted by the Second Convention of the Spanish Speaking People's Congress of California, Dec. 9 and 10, 1939," 6, Galarza Papers, Stanford University. I would like to thank Robin Kelley for pointing out this sign for determining "Popular Front" status. See also García, *Mexican Americans,* 165.

Unlike El Congreso, which had just begun its most active period in 1939, the Nazi-Soviet pact seems to have curtailed the political activity of the National Negro Congress substantially by alienating black intellectuals who were committed to the fight against fascism. See Naison, *Communists in Harlem,* 287–93.

65. "First National Congress," 7, and "Resolutions Adopted . . . ," 6.

66. García, *Mexican Americans,* 155–56.

67. Ibid., 156–57.

68. First National Congress, "Digest of Proceedings," 5, Galarza Papers, Stanford University; Camarillo, "Cities," 256; García, *Mexican Americans,* 163–64.

69. Camarillo, "Cities," 255, 257–58.

70. "Resolutions Adopted . . . ," 5.

71. "Digest of Proceedings"; Camarillo, "Cities," 255–56, 258. See also David G. Gutierrez, *Walls and Mirrors: Mexican Americans, Mexican Immigrants, and the Politics of Ethnicity in the American Southwest* (Berkeley: Univ. of California Press, forthcoming), chap. 3, for a thorough examination of El Congreso's position on Mexican immigrants and American citizenship.

72. "Call! To the Fourth State Convention of the Spanish Speaking Peoples

Congress of California," Manuel Ruiz Papers, Box 5, Folder 12, Department of Special Collections, Stanford University, Stanford.

73. Ibid. See also García, *Mexican Americans,* 165–66.

74. Camarillo, "Cities," 259–61; García, *Mexican Americans,* 170–73.

75. Short biographies of members of the Coordinating Council of Latin American Youth, Manuel Ruiz, Jr., papers, Box 3, Folder 16.

76. Manuel Ruiz, Jr., papers, MS #295, Box 16, Folder 2.

77. García, *Mexican Americans,* 173.

78. Ruiz, *Cannery Women,* 116.

79. Address delivered by Luisa Moreno Bemis to the 12th Annual Convention, California CIO Council, 15 Oct. 1949, pp. 3–4.

Chapter 12
The Rise of the Second Generation

1. Alfred Barela, Letter to the Honorable Arthur S. Guerin, 21 May 1943, and Arthur S. Guerin, Letter to Mr. Alfred Barela, 2 June 1943, Julian Nava collection, Urban Archives Center, California State University, Northridge.

2. Barela to Guerin, 21 May 1943, Nava collection.

3. This is not the case for other urban areas of the Southwest, where issues common to the second generation emerged before the onset of the Great Depression. The permanency of Mexican settlement encouraged the development of second-generation organizations such as the Orden de Hijos de America in south Texas, for example, and in southern Arizona, the Alianza Hispano Americana was organized as early as 1892.

4. See Richard Amado Garcia, "The Making of the Mexican-American Mind, San Antonio, Texas, 1929–1941: A Social and Intellectual History of an Ethnic Community" (Ph.D. diss., University of California, Irvine, 1980); Guadalupe San Miguel, Jr., *"Let All of Them Take Heed": Mexican Americans and the Campaign for Educational Equality in Texas, 1910–1981* (Austin: Univ. of Texas Press, 1987); Carl Allsup, *The American G.I. Forum: Origins and Evolution* (Austin: Center for Mexican American Studies, Univ. of Texas, 1982); José Amaro Hernández, *Mutual Aid for Survival: The Case of the Mexican American* (Malabar, Fla.: Robert E. Krieger, 1983), chaps. 2 and 3.

5. Richard Garcia, "The Making of the Mexican American Mind," in *History, Culture and Society: Chicano Studies in the 1980s* (Ypsilanti: Bilingual Press, 1983), 79, 88; O. Douglas Weeks, "The League of United Latin American Citizens: A Texas-Mexican Organization," *Political and Social Science Quarterly* 10 (1929), 257–78.

6. José Hernández Alvárez, "A Demographic Profile of the Mexican Immigration to the United States, 1910–1950," *Journal of Inter-American Studies* 25 (1983), 472–73.

7. The group also produced many individuals who would later be involved in other Chicano organizations in southern California. These individuals include Bert Corona, a community activist whose career in Chicano politics spans six decades, and Richard Ibañez, a civil rights attorney and later a judge.

8. See Chapter 6 and Ricardo Romo, *East Los Angeles: History of a Barrio* (Austin: Univ. of Texas Press, 1983), 142–48 for a description of this activity.

9. "Leaders Meet at Pomona 'Y,' " *Mexican Voice* (July 1938), 5.

10. See Rodolfo Alvárez, "Psycho-historical and Socioeconomic Develop-

ment of the Chicano Community in the United States," *Social Science Quarterly* 53 (1973), 931–36; Miguel David Tirado, "Mexican American Community Political Organization: The Key to Chicano Political Power," *Aztlán: Chicano Journal of the Social Sciences and the Arts* 1 (1970), 53–78; Mario T. García, "Americans All: The Mexican American Generation and the Politics of Wartime Los Angeles, 1941–45," in *The Mexican American Experience: An Interdisciplinary Anthology,* Rodolfo O. de la Garza et al., eds. (Austin: Univ. of Texas Press, 1985), 201–12, for examples of this tendency.

11. José Rodríguez, "The Value of Education," *Mexican Voice* (July 1938), 7.

12. James William Cameron, "The History of Mexican Public Education in Los Angeles, 1910–1930" (Ph.D. diss., University of Southern California, 1976), 147–49; Herman Buckner, "A Study of Pupil Elimination and Failure among Mexicans" (Master's thesis, University of Southern California, 1935), 37–38; Charles Withers, "Problems of Mexican Boys" (Master's thesis, University of Southern California, 1942), 110.

13. José Rodriguez, "Thrift," *Mexican Voice* (Jan. 1940), 14.

14. Editor Félix Gutiérrez, writing under the pen name "Manuel de la Raza," "Nosotros," *Mexican Voice* (Aug. 1938), 10–11.

15. "Editor's Note," *Mexican Voice* (July 1938), 6.

16. "Personalities," *Mexican Voice* (Jan. 1940), 7–8.

17. "Stephen Reyes Answers a Question," *Mexican Voice* (Sept. 1938), 10–11.

18. See Hernández, *Mutual Aid,* for examples of self-help mutualist organizations in Chicano communities in the nineteenth and early twentieth centuries.

19. Mary Peters, "The Segregation of Mexican American Children in the Elementary Schools of California, Its Legal and Administrative Aspects" (Master's thesis, University of California, Los Angeles, 1948), 35; William A. Farmer, "The Influence of Segregation of Mexican and American School Children upon the Development of Social Attitudes" (Master's thesis, University of Southern California, 1937), 7; Annie Reynolds, *The Education of Spanish Speaking Children in Five Southwestern States,* United States Department of the Interior, Office of Education Bulletin 1933:11 (Washington, D.C.: GPO, 1933), 10; Charles M. Wollenberg, *All Deliberate Speed: Segregation and Exclusion in California Schools, 1855–1975* (Berkeley: Univ. of California Press, 1976), 112, 116, 118.

20. Wollenberg, *All Deliberate Speed,* 110–11; *Mexicans in California,* Report of Governor C.C. Young's Mexican Fact-Finding Committee (1930; rpt., San Francisco: R and E Research Associates, 1970), 53; Paul S. Taylor, *Mexican Labor in the United States,* I (Berkeley: Univ. of California Press, 1930), 18, 265–84; Judith Rosenberg Raftery, "The Invention of Modern Urban Schooling: Los Angeles, 1885–1941" (Ph.D. diss., University of California, Los Angeles, 1984), 206.

21. Raftery, "Modern Urban Schooling," 92–95, 173, 220–21; Cameron, "Mexican Public Education," 153–65; Gilbert G. González, "Racism, Education, and the Mexican Community in Los Angeles, 1920–30," *Societas* 4 (1974), 287–301.

22. Cameron, "Mexican Public Education," 162–63.

23. "Club News—Monrovia," *Mexican Voice* (Spring 1939), 16.

24. "Confidence," *Mexican Voice* (Feb. 1940), 15.

25. Mary Martínez, "An Expression," *Mexican Voice* (Jan. 1940), 12.

26. "Memo No. I," Papers of the Supreme Council of the Mexican American Movement, Urban Archives Center, California State University, Northridge.

27. Paul K. Conkin, *The New Deal,* 2nd ed. (Arlington Heights, Ill.: AHM Publishing, 1975), 2.

28. "Racial Minorities Survey—Mexican," interviewed by Eustace L. Williams, 6 May 1937, Federal Writers Project collection, Department of Special Collections, University of California, Los Angeles.

29. Analysis of Naturalization documents, National Archives, Laguna Niguel, California.

30. Rodolfo Acuña, *Occupied America: The Chicano's Struggle Toward Liberation,* 1st ed. (San Francisco: Canfield, 1972), 276; Beatrice Griffith, *American Me* (Boston: Houghton Mifflin, 1948), 290. For the relationship between blacks and Roosevelt's Democratic party, see Harvard Sitkoff, *A New Deal for Blacks: The Emergence of Civil Rights as a National Issue* (New York: Oxford Univ. Press, 1978).

31. See "Advisory Board" in Mexican American Movement file, Box 14, Folder 7, MS #224, Ernesto Galarza collection, Department of Special Collections, Stanford University.

32. See Alice McGrath collection, MS #107, Box 4, Department of Special Collections, University of California, Los Angeles.

33. Robin Fitzgerald Scott, "The Mexican-American in the Los Angeles Area, 1920–1950: From Acquiescence to Activity" (Ph.D. diss., University of Southern California, 1971), 125; Douglas G. Monroy, "Mexicanos in Los Angeles, 1930–1941: An Ethnic Group in Relation to Class Forces" (Ph.D. diss., University of California, Los Angeles, 1978), 156; Gerald D. Nash, *The American West in the Twentieth Century: A Short History of an Urban Oasis* (Englewood Cliffs: Prentice-Hall, 1973), 171.

34. Scott, "Acquiescence," 125; Nash, *American West,* 171–72.

35. Monroy, "Mexicanos," 189.

36. *Mexican Voice* (Sept. 1938), 3–5.

37. Rebecca Muñoz, "Horizons," *Mexican Voice* (July 1939), 1.

38. "They Must Know: A One Act Play," *Mexican Voice* (Jan.–Feb. 1939), 15; *Mexican Voice* (Jan.–Feb. 1939), 7.

39. Dora Ibañez, "Diferencia en la Esfera de Acción de los Padres y sus Hijos en este País," *Mexican Voice* (Aug. 1939), 4–5.

40. Manuel Ceja, "Are We Proud of Being Mexicans?," *Mexican Voice* (Aug. 1938), 9.

41. Paul Coronel, "Social Conditions of the Mexican People in General," *Mexican Voice* (Jan.–Feb. 1939), 3–5.

42. Stuart Cosgrove, "The Zoot-Suit and Style Warfare," *History Workshop Journal* 18 (1984), 78.

43. Cosgrove, "Style Warfare," 78–80; Steve Chibnall, "Whistle and Zoot: The Changing Meaning of a Suit of Clothes," *History Workshop Journal* 20 (1985), 56–63; Mauricio Mazón, *The Zoot-Suit Riots: The Psychology of Symbolic Annihilation* (Austin: Univ. of Texas Press, 1984), 5–8.

44. *Report on Juvenile Delinquency in Wartime,* issued by the District Attorney's Office in Los Angeles, 1943, as referred to by Beatrice Griffith, *American Me* (Boston: Houghton Mifflin, 1948), 45.

45. Vicki Ruiz, " 'Star Struck': Acculturation, Adolescence and the Mexican American Woman, 1920–1940," unpublished paper presented at "Vision and

Representation of History in the Cultural Production of Hispanics in the United States," Sorbonne, Paris, France, 12–14 May 1987, pp. 10–12.

46. Interview with Margarita Salazar McSweyn, in Sherna Berger Gluck, *Rosie the Riveter Revisted: Women, the War and Social Change* (New York: NAL, 1987), 84–85.

47. Rodolfo Acuña, *Occupied America,* 3rd ed. (New York: Harper and Row, 1988), 254–55; Mazón, *Zoot-Suit Riots,* 19–23. See also Alice McGrath papers and Guy Endore papers, Department of Special Collections, University of California, Los Angeles.

48. Box 4, Folder 4, McGrath papers.

49. "Report and Recommendations of Citizens Committee," by Joseph T. McGucken, Auxiliary Bishop of Los Angeles, Chairman, p. 3, Julian Nava collection, Urban Affairs Institute, California State University, Northridge. Sociologist Emory Bogardus estimated between 45 and 70 Chicano gangs in 1943, in "Gangs of Mexican-American Youth," *Sociology and Social Research* 28 (1943), 55–56.

50. Charles Sheldon Thompson, "Democracy in Evolution: The Natural History of a Boys' Club Group" (G. Howland Shaw, 1940), in "Church of All Nations: Boys Work, 1939–40" subject file, Container #58, G. Bromley Oxnam collection, Manuscript Division, Library of Congress, Washington, D.C. See also Joan W. Moore, *Homeboys: Gangs, Drugs, and Prison in the Barrios of Los Angeles* (Philadelphia: Temple Univ. Press, 1978), 55–74.

51. See Mazón, *Zoot-Suit Riots,* especially chap. 5, for the best description of events leading up to the riots to date.

52. From speech given by Manuel Ruiz entitled "Los Angeles' Mexican Minority," 29 Nov. 1944, Manuel Ruiz collection, Department of Special Collections, Stanford University.

53. Octavio Paz, *The Labyrinth of Solitude: Life and Thought in Mexico,* Lysander Kemp, trans. (1950; New York: Grove Press, 1961), 13.

54. Ibid., 14–17.

55. Los Angeles *Times,* 11 June 1943.

Conclusion

1. Octavio Paz, *The Labyrinth of Solitude: Life and Thought in Mexico,* Lysander Kemp, trans. (1950; New York: Grove Press, 1961), 12–13.

2. See Los Angeles *Times,* 18 Dec. 1988.

3. Zeferino Ramirez, "Origins of the Ramirez Family, from What I Remember," Jean Ramirez, trans., Beatrice R. Palomarez, ed. (Unpublished manuscript, 1983), 39. *Belvedere Citizen.*

Appendix

1. See, for example, Helen Walker, "Mexican Immigrants and American Citizenship," *Sociology and Social Research* 13 (1928–29), 465–71; Leo Grebler, "The Naturalization of Mexican Immigrants in the United States," *International Migration Review* 1 (1966), 17–32; Camarillo, *Changing,* 161.

2. See Romo, *East Los Angeles*; Castillo, "Mexican Barrio"; and Monroy, "Mexicanos."

3. See Walker, "Citizenship," 466–67.

Bibliography

Government Publications

California. Commission of Immigration and Housing. *The Home Teacher, Immigrant Education Leaflet No. 5.* San Francisco: State Printing Office, 1916.

———. Commission of Immigration and Housing. *A Discussion of Methods of Teaching English to Adult Foreigners, with a Report on Los Angeles County.* Sacramento: State Printing Office, 1917.

———. Commission of Immigration and Housing. *Report of an Experiment Made in Los Angeles in the Summer of 1917 for the Americanization of Foreign-Born Women.* Sacramento: State Printing Office, 1917.

———. Commission of Immigration and Housing. *Primer for Foreign-Speaking Women, Part II.* Compiled by Amanda Matthews Chase. Sacramento. State Printing Office, 1918.

———. Commission of Immigration and Housing. *Annual Report.* Sacramento: State Printing Office, 1927.

———. State Relief Administration. *Report and Recommendations of the California State Unemployment Commission.* San Francisco: State Printing Office, 1932.

———. State Relief Administration. *Review of the Activities of the State Relief Administration of California: 1933–1935.* San Francisco: State Printing Office, 1936.

Mexico, Instituto Nacional de Estadística, Geografía e Informática. *Resumen del censo general de habitantes de 30 de noviembre de 1921.* Mexico, D.F.: Talleres Gráficos de la Nación, 1928. *Estadísticas Históricas de México,* Tomo I. Mexico, D.F.: Instituto Nacional de Estadística, Geografía e Informática, 1985.

———. *Estadísticas Históricas de México,* Segunda Edición. Mexico, D.F.: Instituto Nacional de Estadística, Geografía e Informática, 1990.

U.S. Bureau of the Census. *Fifteenth Census of the United States: 1930, Population: Special Report on Foreign-Born White Families by Country of Birth of Head with an Appendix Giving Statistics for Mexican, Indian, Chinese, and Japanese Families.* Washington, D.C.: GPO, 1933.

———. *Population of Spanish Mother Tongue: 1940,* Series P-15, No. 1. Washington, D.C.: GPO, 9 June 1942.

U.S. Congress. House. Committee on Immigration and Naturalization. *Hearings on Seasonal Agricultural Laborers from Mexico.* 69th Cong., 1st sess. Washington, D.C.: GPO, 1926.

———. Committee on Immigration and Naturalization. *Hearings on Temporary Admission of Illiterate Mexican Laborers.* 69th Cong., 1st sess. Washington, D.C.: GPO, 1926.

U.S. Department of Interior. Office of Education. *The Education of Spanish Speaking Children in Five Southwestern States.* By Annie Reynolds. Washington, D.C., GPO, 1933.

U.S. Department of Labor. *Mexican Labor in the United States.* By Victor S. Clark. Bulletin, No. 78. Washington, D.C.: GPO, 1908.

———. *Annual Report of the Commissioner-General of Immigration.* Washington, D.C.: GPO, 1918.

———. *Annual Report of the Commissioner-General of Immigration.* Washington, D.C.: GPO, 1920.

———. Bureau of Labor Statistics. "Mexican Families in Los Angeles." *Money Disbursements of Wage Earner and Clerical Workers in Five Cities in the Pacific Region, 1934–36.* Bulletin 639. Washington, D.C.: GPO, 1939.

Books

Acuña, Rodolfo F. *Occupied America: The Chicano's Struggle toward Liberation.* 1st ed. San Francisco: Canfield Press, 1972.

———. *A Community Under Siege: A Chronicle of Chicanos East of the Los Angeles River.* Los Angeles: Chicano Studies Research Center Publications, University of California, Los Angeles, 1984.

———. *Occupied America: A History of Chicanos.* 3rd ed. New York: Harper and Row, 1988.

Allsup, Carl. *The American G.I. Forum: Origins and Evolution.* Austin: Center for Mexican American Studies, University of Texas, 1982.

American Social History Project. *Who Built America?: Working People & the Nation's Economy, Politics, Culture & Society,* vol. 2. New York: Pantheon Books, 1992.

Anderson, Benedict. *Imagined Communities: Reflections on the Origin and Spread of Nationalism.* Rev. ed. London: Verso, 1983, 1991.

Andrews, Gregg. *Shoulder to Shoulder?: The American Federation of Labor, the United States, and the Mexican Revolution, 1910–1924.* Berkeley: Univ. of California Press, 1991.

Anzaldúa, Gloria. *Borderlands/La Frontera: The New Mestiza.* San Francisco: Spinster/Aunt Lute, 1987.

Atherton, Lewis. *Main Street on the Middle Border.* Bloomington: Indiana Univ. Press, 1984.

Balderrama, Francisco E. *In Defense of La Raza: The Los Angeles Mexican Consul-*

ate and the Mexican Community, 1929 to 1936. Tucson: Univ. of Arizona Press, 1982.

Baldwin, Deborah J. *Protestants and the Mexican Revolution: Missionaries, Ministers, and Social Change*. Urbana: Univ. of Illinois Press, 1990.

Balibar, Etienne, and Immanuel Wallerstein. *Race, Nation, Class: Ambiguous Identities*. London: Verso, 1991.

Barrera, Mario. *Race and Class in the Southwest: A Theory of Racial Inequality*. Notre Dame: Univ. of Notre Dame Press, 1979.

Barton, Josef J. *Peasants and Strangers: Italians, Rumanians, and Slovaks in an American City, 1890–1950*. Cambridge, Mass.: Harvard Univ. Press, 1975.

Beezley, William H. *Judas at the Jockey Club and Other Episodes of Porfirian Mexico*. Lincoln: Univ. of Nebraska Press, 1987.

Blassingame, John W. *The Slave Community: Plantation Life in the Antebellum South*. New York: Oxford Univ. Press, 1972.

Bodnar, John. *The Transplanted: A History of Immigrants in Urban America*. Bloomington: Indiana Univ. Press, 1985.

Bogardus, Emory S. *The Mexican in the United States*. Los Angeles: Univ. of Southern California Press, 1934.

Bottles, Scott. *Los Angeles and the Automobile: The Making of the Modern City*. Berkeley: Univ. of California Press, 1987.

Brackenridge, R. Douglas, and Francisco O. Garcia-Treto. *Iglesia Presbiteriana: A History of Presbyterians and Mexican Americans in the Southwest*. San Antonio: Trinity Univ. Press, 1974.

Bresette, Linna E. *Mexicans in the United States: A Report of a Brief Survey*. Washington, D.C.: National Catholic Welfare Conference, 1929.

Briggs, John W. *An Italian Passage: Immigrants to Three American Cities, 1890–1930*. New Haven: Yale Univ. Press, 1978.

Brilliant, Ashleigh. *The Great Car Craze: How Southern California Collided with the Automobile in the 1920s*. Santa Barbara: Woodbridge Press, 1989.

Brock, William R. *Welfare, Democracy, and the New Deal*. New York: Cambridge Univ. Press, 1988.

Bronder, Saul. *Social Justice and Church Authority: The Public Life of Archbishop Robert E. Lucey*. Philadelphia: Temple Univ. Press, 1982.

Buhle, Mari Jo, Paul Buhle, and Dan Georgakas, eds. *Encyclopedia of the American Left*. New York: Garland, 1990.

Bukowczyk, John J. *And My Children Did Not Know Me: A History of the Polish-Americans*. Bloomington: Indiana Univ. Press, 1987.

Cahan, Abraham. *The Rise of David Levinsky*. 1917; rpt. New York: Harper and Row, 1960.

Camarillo, Albert. *Chicanos in a Changing Society: From Mexican Pueblos to American Barrios in Santa Barbara and Southern California, 1848–1930*. Cambridge, Mass.: Harvard Univ. Press, 1979.

———. *Chicanos in California: A History of Mexican Americans in California*. San Francisco: Boyd and Fraser, 1984.

Cardoso, Lawrence A. *Mexican Emigration to the United States, 1897–1931: Socio-economic Patterns.* Tucson: Univ. of Arizona Press, 1980.

Carr, Larry. *More Fabulous Faces: The Evolution and Metamorphosis of Dolores del Río, Myrna Loy, Carole Lombard, Bette Davis, and Katharine Hepburn.* Garden City, N.Y.: Doubleday, 1979.

Carreras de Velasco, Mercedes. *Los Mexicanos que devolvió la crisis, 1929–1932.* México, D.F.: Secretaría de Relaciones Exteriores, 1974.

Carrillo, Leo. *The California I Love.* Englewood Cliffs, N.J.: Prentice-Hall, 1961.

Chan, Sucheng. *Asian Americans: An Interpretive History.* Boston: Twayne, 1991.

Chavez, John R. *The Lost Land: The Chicano Image of the Southwest.* Albuquerque: Univ. of New Mexico Press, 1984.

Cinel, Dino. *From Italy to San Francisco: The Immigrant Experience.* Stanford, Calif.: Stanford Univ. Press, 1982.

Clifford, James, and George E. Marcus, eds. *Writing Culture: The Poetics and Politics of Ethnography.* Berkeley: Univ. of California Press, 1986.

Coatsworth, John H. *Growth Against Development: The Economic Impact of Railroads in Porfirian Mexico.* Dekalb: Northern Illinois Univ. Press, 1981.

Cohen, Lizabeth. *Making a New Deal: Industrial Workers in Chicago, 1919–1939.* Cambridge: Cambridge Univ. Press, 1990.

Conkin, Paul K. *The New Deal.* 2nd ed. Arlington Heights, Ill.: AHM, 1975.

Craig, Ann L. *The First Agraristas: An Oral History of a Mexican Agrarian Reform Movement.* Berkeley: Univ. of California Press, 1983.

Cubberly, Ellwood P. *Changing Conceptions of Education.* Boston: Houghton Mifflin, 1909.

Daniels, Roger. *The Politics of Prejudice: The Anti-Japanese Movement in California and the Struggle for Japanese Exclusion.* Berkeley: Univ. of California Press, 1962.

Davis, Allen F. *Spearheads for Reform: The Social Settlements and the Progressive Movement, 1890–1914.* New York: Oxford Univ. Press, 1967.

Davis, Mike. *City of Quartz: Excavating the Future in Los Angeles.* London: Verso, 1990.

de la Colina, Rafael. *Una vida de hechos.* Mexico City: Secretaría de Relaciones Exteriores, Archivo Histórico Diplomático Mexicano: Serie Testimonios 1, 1989.

De León, Arnoldo. *The Tejano Community, 1836–1900.* Albuquerque: Univ. of New Mexico Press, 1982.

Degler, Carl. *At Odds: Women and the Family in America from the Revolution to the Present.* New York: Oxford Univ. Press, 1980.

D'Emilio, John, and Estelle B. Freedman. *Intimate Matters: A History of Sexuality in America.* New York: Harper and Row, 1988.

Deutsch, Sarah. *No Separate Refuge: Culture, Class, and Gender on an Anglo-Hispanic Frontier in the American Southwest, 1880–1940.* New York: Oxford Univ. Press, 1987.

Diner, Hasia. *Erin's Daughters in America: Irish Immigrant Women in the Nineteenth Century.* Baltimore: Johns Hopkins Univ. Press, 1983.

Elkins, Stanley M. *Slavery: A Problem in American Institutional and Intellectual Life*. Chicago: Univ. of Chicago Press, 1959.

Ellis, Pearl Idelia. *Americanization through Homemaking*. Los Angeles: Wetzel, 1929.

Fábila, Alfonso. *El problema de la emigración de obreros y campesinos mexicanos*. México: Talleres Gráficos de la Nación, 1929.

Fogelson, Robert M. *The Fragmented Metropolis: Los Angeles, 1850–1930*. Cambridge: Harvard Univ. Press, 1967.

Foster, George M. *Tzintzuntzan: Mexican Peasants in a Changing World*. Rev. ed. New York: Elsevier, 1979.

Fowler, Gene, and Bill Crawford. *Border Radio*. Austin: Texas Monthly Press, 1987.

Friedrich, Paul. *Agrarian Revolt in a Mexican Village*. 2nd ed. Chicago: Univ. of Chicago Press, 1970, 1977.

Friedricks, William B. *Henry E. Huntington and the Creation of Southern California*. Columbus: Ohio State Univ. Press, 1992.

Fromm, Erich, and Michael Maccoby. *Social Character in a Mexican Village: A Sociopsychoanalytic Study*. Englewood Cliffs, N.J.: Prentice-Hall, 1970.

Galarza, Ernesto. *Barrio Boy*. Notre Dame: Univ. of Notre Dame Press, 1971

Gamio, Manuel. *La Población del Valle de Teotihuacán*. Mexico City: Talleres Gráficas de la Nación, 1922.

———. *Introduccion, sintesis y conclusion a la poblacion del Valle de Teotihuacan*. Mexico: Secretaria de Educacion Publica, 1924.

———. *Mexican Immigration to the United States: A Study of Human Migration and Adjustment*. 1930; rpt. New York: Dover, 1971.

———. *The Life Story of the Mexican Immigrant*. 1931; rpt. New York: Dover, 1971.

———. *Forjando patria*. 2nd ed. Mexico: Porrua, 1960.

García, Mario T. *Desert Immigrants: The Mexicans of El Paso, 1880–1920*. New Haven: Yale Univ. Press, 1981.

———. *Mexican Americans: Leadership, Ideology, & Identity, 1930–1960*. New Haven: Yale Univ. Press, 1989.

García, Richard A. *Rise of the Mexican Middle Class: San Antonio, 1929–1941*. College Station: Texas A & M Univ. Press, 1991.

Geertz, Clifford. *The Interpretation of Cultures*. New York: Basic Books, 1973.

Geijerstam, Claes af. *Popular Music in Mexico*. Albuquerque: Univ. of New Mexico Press, 1976.

Genovese, Eugene D. *Roll, Jordan, Roll: The World the Slaves Made*. New York: Pantheon, 1974.

Gerstle, Gary. *Working-Class Americanism: The Politics of Labor in a Textile City, 1914–1960*. Cambridge: Cambridge Univ. Press, 1989.

Gil, Carlos B. *Life in Provincial Mexico: National and Regional History Seen from Mascota, Jalisco, 1867–1972*. Los Angeles: UCLA Latin American Studies Center Publications, 1983.

Glazer, Nathan, and Daniel P. Moynihan, eds. *Ethnicity: Theory and Experience*. Cambridge, Mass.: Harvard Univ. Press, 1975.

Gluck, Sherna Berger. *Rosie the Riveter Revisited: Women, the War and Social Change*. New York: New American Library, 1987.

Gómez-Quiñones, Juan. *Sembradores: Ricardo Flores Magón y El Partido Liberal Mexicano: A Eulogy and Critique*. UCLA Chicano Studies Research Center Publications, 1973.

——. *Chicano Politics: Reality and Promise, 1940–1990*. Albuquerque: Univ. of New Mexico Press, 1990.

González, Luis. *San José de Gracia: Mexican Village in Transition*. John Upton, trans. Austin: Univ. of Texas Press, 1972.

Gordon, Milton. *Assimilation in American Life: The Role of Race, Religion, and National Origins*. New York: Oxford Univ. Press, 1964.

Gramsci, Antonio. *Selections from the Prison Notebooks*. Quintin Hoare and Geoffrey Newell Smith, eds. and trans. New York: International, 1971.

Grebler, Leo, Joan W. Moore, and Ralph C. Guzman. *The Mexican American People*. New York: Free Press, 1970.

Green, James R. *Grass-roots Socialism: Radical Movements in the Southwest, 1895–1943*. Baton Rouge: Louisiana State Univ. Press, 1978.

——. *The World of the Worker: Labor in Twentieth-Century America*. New York: Hill and Wang, 1980.

Griffith, Beatrice. *American Me*. Boston: Houghton Mifflin, 1948.

Griswold del Castillo, Richard. *The Los Angeles Barrio, 1850–1890: A Social History*. Berkeley: Univ. of California Press, 1979.

——. *La Familia: Chicano Families in the Urban Southwest, 1848 to the Present*. Notre Dame: Univ. of Notre Dame, 1984.

Gutierrez, David G. *Walls and Mirrors: Mexican Americans, Mexican Immigrants, and the Politics of Ethnicity in the American Southwest*. Berkeley: Univ. of California Press, forthcoming.

Gutierrez, Felix F., and Jorge Reina Schement. *Spanish-Language Radio in the Southwestern United States*. Austin: Center for Mexican American Studies, Univ. of Texas, 1979.

Gutman, Herbert C. *The Black Family in Slavery and Freedom, 1750–1925*. New York: Pantheon, 1976.

——. *Work, Culture, and Society in Industrializing America: Essays in American Working-Class and Social History*. New York: Vintage–Random House, 1977.

Handlin, Oscar. *The American People in the Twentieth Century*. Cambridge, Mass.: Harvard Univ. Press, 1954.

——. *The Newcomers: Negroes and Puerto Ricans in a Changing Metropolis*. Cambridge, Mass.: Harvard Univ. Press, 1959.

——. *The Uprooted: The Epic Story of the Great Migrations That Made the American People*. 2nd ed., enlarged. 1951; Boston: Little, Brown, 1973.

Healey, Dorothy, and Maurice Isserman. *Dorothy Healey Remembers: A Life in the American Communist Party*. New York: Oxford Univ. Press, 1990.

The Heller Committee for Research in Social Economics of the Univ. of California and Constantine Panunzio. *How Mexicans Earn and Live: A Study of the Incomes and Expenditures of One Hundred Mexican Families in San Diego,*

California. Cost of Living Studies, 5. Berkeley: Univ. of California Press, 1933.

Henstell, Bruce. *Sunshine and Wealth: Los Angeles in the Twenties and Thirties*. San Francisco: Chronicle Books, 1984.

Hernández, José Amaro. *Mutual Aid for Survival: The Case of the Mexican American*. Malabar, Fla.: Robert E. Krieger, 1983.

Hewitt de Alcantara, Cynthia. *Anthropological Perspectives on Rural Mexico*. London: Routledge & Kegan Paul, 1984.

Higham, John. *Strangers in the Land: Patterns of American Nativism, 1860–1925*. New York: Atheneum, 1965.

Hobsbawm, Eric, and Terence Ranger, eds. *The Invention of Tradition*. Cambridge, Eng.: Cambridge Univ. Press, 1983.

Hoffman, Abraham. *Unwanted Mexican Americans in the Great Depression: Repatriation Pressures, 1929–1939*. Tucson: Univ. of Arizona Press, 1974.

Holland, Clifton L. *The Religious Dimension in Hispanic Los Angeles, A Protestant Case Study*. South Pasadena: William Carey Library, 1974.

Interchurch Movement of North America. *The Mexican in Los Angeles: Los Angeles City Survey*. 1920; rpt. San Francisco: R and E Research Associates, 1970.

Jaher, Frederic Cople. *The Urban Establishment: Upper Strata in Boston, New York, Charleston, Chicago and Los Angeles*. Urbana: Univ. of Illinois Press, 1982.

Jensen, Richard J. *The Winning of the Midwest: Social and Political Conflict, 1888–96*. Chicago: Univ. of Chicago Press, 1971.

Kasun, Jacqueline Rorabeck. *Some Social Aspects of Business Cycles in the Los Angeles Area, 1920–1950*. Los Angeles: Haynes Foundation, 1954.

Keefe, Susan E., and Amado Padilla. *Chicano Ethnicity*. Albuquerque: Univ. of New Mexico Press, 1987.

Kelley, Robin D.G. *Hammer and Hoe: Alabama Communists during the Great Depression*. Chapel Hill: Univ. of North Carolina Press, 1990.

Kerber, Linda. *Women of the Republic: Intellect and Ideology in Revolutionary America*. Chapel Hill: Univ. of North Carolina Press, 1980.

Kessner, Thomas. *The Golden Door: Italian and Jewish Immigrant Mobility in New York City, 1880–1915*. New York: Oxford Univ. Press, 1977.

Kleppner, Paul. *The Cross of Culture: A Social Analysis of Midwestern Politics, 1850–1900*. New York: Free Press, 1970.

Krauze, Enrique. *Caudillos culturales en la revolución mexicana*. Mexico: Siglo veintiuno editores, 1976.

LeMay, Michael C. *From Open Door to Dutch Door: An Analysis of U.S. Immigration Policy Since 1820*. New York: Praeger, 1987.

Laslett, John, and Mary Tyler. *The ILGWU in Los Angeles, 1907–1988*. Inglewood, Calif.: Ten Star Press, 1989.

Lavender, David. *California: Land of New Beginnings*. Lincoln: Univ. of Nebraska Press, 1972.

Levine, Lawrence W. *Black Culture and Black Consciousness: Afro-American Folk Thought from Slavery to Freedom*. New York: Oxford Univ. Press, 1977.

Lewis, Oscar. *A Death in the Sánchez Family*. New York: Random House, 1969.

Limerick, Patricia Nelson. *Legacy of Conquest: The Unbroken Past of the American West*. New York: W.W. Norton, 1987.

Lipsitz, George. *Time Passages: Collective Memory and American Popular Culture*. Minneapolis: Univ. of Minnesota Press, 1990.

MacLachlan, Colin M. *Anarchism and the Mexican Revolution: The Political Trials of Ricardo Flores Magón in the United States*. Berkeley: Univ. of California Press, 1991.

McWilliams, Carey. *Factories in the Field: The Story of Migratory Farm Labor in California*. Santa Barbara: Peregrine Smith, 1935.

———. *Southern California: An Island on the Land*. New York: Duell, Sloan & Pearce, 1946.

Marchand, Roland. *Advertising the American Dream: Making Way for Modernity, 1920–1940*. Berkeley: Univ. of California Press, 1985.

Martínez, Oscar. *Border Boom Town: Ciudad Juarez since 1848*. Austin: Univ. of Texas Press, 1975.

Massey, Douglas, Rafael Alarcón, Jorge Durand, and Humberto González. *Return to Aztlan: The Social Process of International Migration from Western Mexico*. Berkeley: Univ. of California Press, 1987.

May, Elaine Tyler. *Great Expectations: Marriage and Divorce in Post-Victorian America*. Chicago: Univ. of Chicago Press, 1980.

May, Lary. *Screening Out the Past: The Birth of Mass Culture and the Motion Picture Industry*. Chicago: Univ. of Chicago Press, 1980.

Mazón, Mauricio. *The Zoot-Suit Riots: The Psychology of Symbolic Annihilation*. Austin: Univ. of Texas Press, 1984.

Meier, Matt, and Feliciano Rivera. *The Chicanos: A History of Mexican Americans*. New York: Hill and Wang, 1972.

Mendoza, Vicente T. *El Corrido Mexicano*. México, D.F.: Fondo de Cultura Económica, 1974.

Mexicans in California. Report of Governor C.C. Young's Mexican Fact-Finding Committee. 1930; rpt. San Francisco: R and E Research Associates, 1970.

Meyer, Michael C., and William L. Sherman. *The Course of Mexican History*. 2nd ed. New York: Oxford Univ. Press, 1983.

Mirandé, Alfredo, and Evangelina Enríquez. *La Chicana: The Mexican-American Woman*. Chicago: Univ. of Chicago Press, 1979.

Modell, John. *The Economics and Politics of Racial Accommodation: The Japanese of Los Angeles, 1900–1942*. Urbana: Univ. of Illinois Press, 1977.

Moldow, Gloria Melnick. *Women Doctors in Gilded-Age Washington: Race, Gender, and Professionalization*. Urbana: Univ. of Illinois Press, 1987.

Monroy, Douglas. *Thrown among Strangers: The Making of Mexican Culture in Frontier California*. Berkeley: Univ. of California Press, 1990.

Montejano, David. *Anglos and Mexicans in the Making of Texas, 1836–1986*. Austin: Univ. of Texas Press, 1987.

Moore, Joan W. *Homeboys: Gangs, Drugs, and Prison in the Barrios of Los Angeles*. Philadelphia: Temple Univ. Press, 1978.

Mora, Carl J. *Mexican Cinema: Reflections of a Society, 1896–1980*. Berkeley: Univ. of California Press, 1982.

Morantz-Sanchez, Regina Markell. *Sympathy and Science: Women Physicians in American Medicine*. New York: Oxford Univ. Press, 1985.

Mosqueda, Lawrence J. *Chicanos, Catholicism and Political Ideology*. Lanham, Md.: Univ. Press of America, 1986.

Mowry, George E. *The California Progressives*. Berkeley: Univ. of California Press, 1951.

Mullins, William H. *The Depression and the Urban West Coast, 1929–1933: Los Angeles, San Francisco, Seattle, and Portland*. Bloomington: Indiana Univ. Press, 1991.

Murguía, Edward. *Chicano Intermarriage: A Theoretical and Empirical Study*. San Antonio: Trinity Univ. Press, 1982.

Naison, Mark. *Communists in Harlem during the Depression*. Urbana: Univ. of Illinois Press, 1983.

Nash, Gerald. *The American West in the Twentieth Century: A Short History of An Urban Oasis*. Englewood Cliffs, N.J.: Prentice-Hall, 1973.

Nelson, Cynthia. *The Waiting Village: Social Change in Rural Mexico*. Boston: Little, Brown, 1971.

Olin, Spencer C. *California's Prodigal Sons: Hiram Johnson and the Progressives, 1911–1917*. Berkeley: Univ. of California Press, 1968.

Oliva, Juan Diego Razo. *Rebeldes Populares del Bajío (Hazanas, Tragedias y Corridos, 1910–1927)*. Mexico City: Editorial Katun, 1983.

O'Malley, Ilene V. *The Myth of the Revolution: Hero Cults and the Institutionalization of the Mexican State, 1920–1940*. Westport, Conn.: Greenwood Press, 1986.

Omi, Michael, and Howard Winant. *Racial Formation in the United States: From the 1960s to the 1980s*. New York: Routledge & Kegan Paul, 1986.

Oxnam, G. Bromley. *The Mexican in Los Angeles*. Los Angeles: Interchurch World Movement of North America, 1920.

Paredes, Américo. *A Texas-Mexican Cancionero: Folksongs of the Lower Border*. Urbana: Univ. of Illinois Press, 1976.

Patric, Gladys. *A Study of the Housing and Social Conditions in the Ann Street District of Los Angeles, California*. Los Angeles: n.p., 1918.

Paz, Octavio. *The Labyrinth of Solitude: Life and Thought in Mexico*. Lysander Kemp, trans. 1950; New York: Grove Press, 1961.

Peiss, Kathy. *Cheap Amusements: Working Women and Leisure in Turn-of-the-Century New York*. Philadelphia: Temple Univ. Press, 1986.

Peña, Manuel. *The Texas-Mexican Conjunto: History of a Working-Class Music*. Austin: Univ. of Texas Press, 1985.

Perkins, Clifford Alan. *Border Patrol: With the U.S. Immigration Service on the Mexican Boundary, 1910–1954*. El Paso: Texas Western Press, 1978.

Perry, Louis B., and Richard S. Perry. *A History of the Los Angeles Labor Movement*. Berkeley: Univ. of California Press, 1963.

Pesotta, Rose. *Bread upon the Waters*. New York: Dodd, Mead, 1944.

Pitt, Leonard. *The Decline of the Californios: A Social History of the Spanish-Speaking Californians, 1846–1890*. Berkeley: Univ. of California Press, 1966.

Piven, Frances Fox, and Richard A. Cloward. *Regulating the Poor: The Functions of Public Welfare*. New York: Vintage, 1971.

Pletcher, David M. *Rails, Mines, and Progress: Seven American Promoters in Mexico, 1867–1911*. Ithaca, N.Y.: Cornell Univ. Press, 1958.

Portes, Alejandro, and Robert L. Bach. *Latin Journey: Cuban and Mexican Immigrants in the United States*. Berkeley: Univ. of California Press, 1985.

Quirk, Robert E. *The Mexican Revolution and the Catholic Church, 1910–1929*. Bloomington: Indiana Univ. Press, 1973.

Raat, W. Dirk. *Revoltosos: Mexico's Rebels in the United States, 1903–1923*. College Station: Texas A&M Univ. Press, 1981.

Rawick, George P. *From Sundown to Sunup: The Making of the Black Community*. Westport, Conn.: Greenwood Press, 1972.

Redfield, Robert. *Tepoztlán, A Mexican Village: A Study in Folk Life*. Chicago: Univ. of Chicago Press, 1930.

Reisler, Mark. *By the Sweat of their Brow: Mexican Immigrant Labor in the United States, 1900–1940*. Westport, Conn.: Greenwood Press, 1976.

Rigdon, Susan M. *The Culture Facade: Art, Science, and Politics in the Work of Oscar Lewis*. Urbana: Univ. of Illinois Press, 1988.

Robinson, W.W. *Los Angeles from the Days of the Pueblo: A Brief History and Guide to the Plaza Area*. San Francisco: California Historical Society, 1981.

Rogin, Michael, and John L. Shover. *Political Change in California: Critical Elections and Social Movements, 1890–1966*. Westport, Conn.: Greenwood Press, 1970.

Romanucci-Ross, Lola. *Conflict, Violence, and Morality in a Mexican Village*. Palo Alto, Calif.: National Press Books, 1973.

Romo, Ricardo. *East Los Angeles: History of a Barrio*. Austin: Univ. of Texas Press, 1983.

Rosaldo, Renato. *Culture and Truth: The Remaking of Social Analysis*. Boston: Beacon Press, 1989.

Ruiz, Ramón Eduardo. *The Great Rebellion: Mexico, 1905–1924*. New York: Norton Press, 1980.

Ruiz, Vicki. *Cannery Women/Cannery Lives: Mexican Women, Unionization, and the California Food Processing Industry, 1930–1950*. Albuquerque: Univ. of New Mexico Press, 1987.

Salerno, Salvatore. *Red November, Black November: Culture and Community in the Industrial Workers of the World*. Albany: State Univ. of New York Press, 1989.

San Miguel, Guadalupe. *"Let All of Them Take Heed": Mexican Americans and the Campaign for Educational Equality in Texas, 1910–1981*. Austin: Univ. of Texas Press, 1987.

Scharf, Lois. *To Work and to Wed: Female Employment, Feminism, and the Great Depression*. Westport, Conn.: Greenwood Press, 1980.

Schmidt, Henry C. *The Roots of Lo México: Self and Society in Mexican Thought, 1900–1934*. College Station: Texas A&M Univ. Press, 1978.

Shapiro, Laura. *Perfection Salad: Women and Cooking at the Turn of the Century*. New York: Farrar, Straus and Giroux, 1986.

Shevky, Eshref, and Molly Lewin. *Your Neighborhood: A Social Profile of Los Angeles*. Los Angeles: Haynes Foundation, 1949.

Simmons, Merle E. *The Mexican Corrido as a Source for Interpretive Study of Modern Mexico (1870–1950)*. Bloomington: Indiana Univ. Press, 1957.

Singleton, Gregory H. *Religion in the City of Angels: American Protestant Culture and Urbanization, Los Angeles, 1850–1930*. Ann Arbor: UMI Research Press, 1979.

Sitkoff, Harvard. *A New Deal for Blacks: The Emergence of Civil Rights as a National Issue*. New York: Oxford Univ. Press, 1978.

Sollors, Werner. *Beyond Ethnicity: Consent and Descent in American Culture*. New York: Oxford Univ. Press, 1986.

Soto, Antonio R. *The Chicano and the Church*. San Francisco: Martel Associates, 1973.

Starr, Kevin. *Americans and the California Dream: 1850–1915*. New York: Oxford Univ. Perss, 1973.

———. *Inventing the Dream: California through the Progressive Era*. New York: Oxford Univ. Press, 1985.

———. *Material Dreams: Southern California through the 1920s*. New York: Oxford Univ. Press, 1990.

Steinberg, Stephen. *The Ethnic Myth: Race, Ethnicity, and Class in America*. Boston: Beacon Press, 1982.

Stowell, Jay S. *The Near Side of the Mexican Question*. New York: George H. Doran, 1921.

Takaki, Ronald, ed. *From Different Shores: Perspectives on Race and Ethnicity in America*. New York: Oxford Univ. Press, 1987.

———. *Strangers from a Different Shore: A History of Asian Americans*. New York: Penguin Books, 1990.

Taylor, Paul S. *Mexican Labor in the United States: Migration Statistics*. Vol. 1. Publications in Economics, 6. Berkeley: Univ. of California Press, 1930.

———. *Mexican Labor in the United States: Chicago and the Calumet Region*. Berkeley: Univ. of California Press, 1932.

———. *A Spanish-Mexican Peasant Community: Arandas in Jalisco, Mexico*. Ibero-Americana, 4. Berkeley: Univ. of California Press, 1933.

Thernstrom, Stephan. *Poverty and Progress: Social Mobility in a Nineteenth Century City*. Cambridge, Mass.: Harvard Univ. Press, 1964.

———. *The Other Bostonians: Poverty and Progress in the American Metropolis, 1880–1970*. Cambridge, Mass.: Harvard Univ. Press, 1973.

Tilly, Louise A., and Joan W. Scott. *Women, Work, and Family*. New York: Holt, Rinehart and Winston, 1978.

Tolbert, Emory J. *The UNIA and Black Los Angeles: Ideology and Community in the American Garvey Movement*. Los Angeles: Center for Afro-American Studies, Univ. of California, Los Angeles, 1980.

Tullis, F. LaMond. *Mormons in Mexico: The Dynamics of Faith and Culture*. Logan: Utah State Univ. Press, 1987.

Vaughan, Mary Kay. *The State, Education, and Social Class in Mexico, 1880–1928.* DeKalb: Northern Illinois Univ. Press, 1982.

Vorspan, Max, and Lloyd P. Gartner. *History of the Jews in Los Angeles.* San Marino: Huntington Library, 1970.

Walsh, Mary Roth. *"Doctors Wanted: No Women Need Apply": Sexual Barriers in the Medical Profession, 1835–1975.* New Haven: Yale Univ. Press, 1977.

Weaver, John D. *L.A.: El Pueblo Grande.* Los Angeles: Ward Ritchie Press, 1973.

Wiebe, Robert H. *The Search for Order, 1877–1920.* New York: Hill and Wang, 1967.

Williams, Raymond. *The Country and the City.* New York: Oxford Univ. Press, 1973.

Wollenberg, Charles M. *All Deliberate Speed: Segregation and Exclusion in California Schools, 1855–1975.* Berkeley: Univ. of California Press, 1976.

Worster, Donald. *Rivers of Empire: Water, Aridity, and the Growth of the American West.* New York: Pantheon Books, 1985.

Yans-McLaughlin, Virginia. *Family and Community: Italian Immigrants in Buffalo, 1880–1930.* Ithaca: Cornell Univ. Press, 1971.

Zunz, Olivier. *The Changing Face of Inequality: Urbanization, Industrial Development, and Immigrants in Detroit, 1880–1920.* Chicago: Univ. of Chicago Press, 1982.

Articles

Aiken, Charles S. "The Land of Tomorrow." *Sunset Magazine* 22 (June 1909), 509–10.

Alarcón, Norma. "Chicano Feminism: In the Tracks of 'the' Native Women." *Cultural Studies* 4:3 (October 1990), 248–56.

Alvarez, José Hernández. "A Demographic Profile of the Mexican Immigration to the United States, 1910–1950." *Journal of Inter-American Affairs* 25 (1983), 920–42.

Alvarez, Rodolfo. "The Psycho-historical and Socioeconomic Development of the Chicano Community in the United States." *Social Science Quarterly* 53 (1973), 920–42.

Alvirez, David, and Frank D. Bean. "The Mexican American Family." In *Ethnic Families in America.* Eds. Charles H. Mindel and Robert W. Haberstein. New York: Elsevier, 1976, pp. 269–92.

Arroyo, Luis Leobardo. "Chicano Participation in Organized Labor; The CIO in Los Angeles, 1938–1950; An Extended Research Note." *Aztlán* 6:2 (Summer 1975), 277–303.

———. "Mexican Workers and American Unions: The Los Angeles AFL, 1890–1933." Chicano Political Economy Collective Working Paper Series #107. Berkeley, 1981.

Atl, Dr. "Obregón y el principio de la renovación social," in *Obregón, Aspectos de Su Vida.* México: Editorial "Cultura," 1935, pp. 65–86.

Bastian, Jean-Pierre. "Metodismo y clase obrera durante el Porfiriato." *Historia Mexicana* 33 (1983), 39–71.

Beach, Frank L. "The Effects of the Westward Movement on California's Growth and Development, 1900–1920." *International Migration Review* 3 (1969), 20–35.

Bernard, William S. "A History of U.S. Immigration Policy." In *Immigration: Dimensions of Ethnicity*. Ed. Reed Ueda. Cambridge: Belknap Press, 1982, pp. 87–99.

Bodeen, De Witt. "Dolores Del Rio: Was the First Mexican of Family to Act in Hollywood." *Films in Review* 18 (1967), 266–83.

Bogardus, Emory. "Gangs of Mexican-American Youth." *Sociology and Social Research* 28 (1943), 55–66.

Boswell, Thomas D. "The Growth and Proportional Distribution of the Mexican Stock Population of the United States: 1910–1970." *The Mississippi Geographer* 7:1 (Spring 1979), 57–76.

Brading, David A. "Manuel Gamio and Official Indigenismo in Mexico." *Bulletin of Latin America Research* 7:1 (1988), 75–89.

Camarillo, Albert. "Mexicans and Europeans in American Cities: Some Comparative Perspectives, 1900–1940." In *From "Melting Pot" to Multiculturalism: The Evolution of Ethnic Relations in the United States and Canada*. Ed. Valeria Gennaro Lerda. Rome, Italy: Bulzoni Editore, 1990, pp. 237–62.

Cardoso, Lawrence A. "Protestant Missionaries and the Mexican: An Environmentalist Image of Cultural Differences." *New Scholar* 7 (1978), 223–36.

Castañeda, Daniel. "La Música y la Revolución Mexicana." *Boletín Latin-Americano de Música* 5 (1941), 447–48.

Chang, Gordon H. "Asian Americans and the Writing of Their History." *Radical History Review* 53 (1992), 105–14.

Chaplin, Ralph. "Why I Wrote 'Solidarity Forever.' " *The American West* 5:1 (Jan. 1968), 18–27, 73.

Chibnall, Steve. "Whistle and Zoot: The Changing Meaning of a Suit of Clothes." *History Workshop Journal* 20 (1985), 56–81.

Cornelius, Wayne. "Immigration, Mexican Development Policy, and the Future of U.S.-Mexican Relations." In *Mexico and the United States*. Ed. R. McBride. Englewood Cliffs, N.J.: Prentice-Hall, 1981, pp. 104–27.

Corwin, Arthur F. "Early Mexican Labor Migration: A Frontier Sketch, 1848–1900." In *Immigrants—and Immigrants: Perspectives on Mexican Labor Migration to the United States*. Ed. Arthur F. Corwin. Westport, Conn.: Greenwood Press, 1979, pp. 25–37.

Cosgrove, Stuart. "The Zoot Suit and Style Warfare." *History Workshop Journal* 18 (1984), 77–91.

Cushing, Sumner W. "The Distribution of Population in Mexico." *The Geographical Review* 11 (1921), 227–42.

Deutsch, Sarah. "Women and Intercultural Relations: The Case of Hispanic New Mexico and Colorado." *Signs: Journal of Women in Culture and Society* 12 (1987), 719–39.

DeGraff, Lawrence B. "The City of Black Angels: Emergence of the Los Angeles Ghetto, 1890–1930." *Pacific Historical Review* 39 (1970), 333–36.

Duron, Clementina. "Mexican Women and Labor Conflict in Los Angeles: The ILGWU Dressmakers' Strike of 1933." *Aztlán: International Journal of Chicano Studies Research* 15 (1984), 145–61.

Flores, Juan, and George Yudice. "Living Borders/Buscando America: Languages of Latino Self-Formation." *Social Text* 8:2 (1990), 57–84.

Gamio, Manuel. "Empiricism of Latin-American Governments and the Empiricism of Their Relations with the United States." *Mexican Review* 3:5 (Aug. 1919), 3–21.

García, Mario T. "Americanization and the Mexican Immigrant, 1880–1930." *The Journal of Ethnic Studies* 6 (1978), 19–34.

———. "The Chicana in American History: The Mexican Women of El Paso, 1880–1920—A Case Study." *Pacific Historical Review* 49 (1980), 315–38.

———. "La Familia: The Mexican Immigrant Family, 1900–1930." In *Work, Family, Sex Roles, Language: The National Association of Chicano Studies, Selected Papers 1979*. Eds. Mario Barrera, Alberto Camarillo, and Francisco Hernández. Berkeley: Tonatiuh-Quinto Sol International, 1980, pp. 117–39.

———. "Communications." *Labor History* 24 (1984), 152–54.

———. "Americans All: The Mexican American Generation and the Politics of Wartime Los Angeles, 1941–45." In *The Mexican American Experience: An Interdisciplinary Anthology*. Ed. Rodolfo O. de la Garza et al. Austin: Univ. of Texas Press, 1985, pp. 201–12.

———. "La Frontera: The Border as Symbol and Reality in Mexican-American Thought." *Mexican Estudios/Estudios Mexicanos* 1:2 (Summer 1985), 195–225.

Garcia, Richard Amado. "Class, Consciousness, and Ideology—The Mexican Community of San Antonio, Texas: 1930–1940." *Aztlán* 9 (1978), 23–70.

———. "The Making of the Mexican American Mind." In *History, Culture and Society: Chicano Studies in the 1980s*. Ypsilanti, Mich.: Bilingual Press, 1983, pp. 67–93.

Gómez-Quiñones, Juan. "The First Steps: Chicano Labor Conflict and Organizing, 1900–1920," *Aztlan* 3:1 (Spring 1972), 13–49.

———. "On Culture." *Revista Chicano-Riquena* 5:2 (1977), 29–47.

———. "Piedras contra la Luna, México en Aztlán y Aztlán en México: Chicano-Mexican Relations and the Mexican Consulates, 1900–1920." In *Contemporary Mexico: Papers of the IV International Congress of Mexican History*. Eds. James W. Wilkie, Michael C. Meyer, and Edna Monzón de Wilkie. Berkeley: Univ. of California Press, 1976, pp. 494–527.

Gómez-Sicre, José. "Dolores del Rio." *Américas* 19 (1967), 8–17.

González, Gilbert G. "Factors Relating to Property Ownership of Chicanos in Lincoln Heights Los Angeles." *Aztlán: International Journal of Chicano Studies Research* 2 (1971), 107–44.

———. "Racism, Education, and the Mexican Community in Los Angeles, 1920–30." *Societas* 4 (1974), 287–301.

Gratton, Brian, F. Arturo Rosales, and Hans DeBano. "A Sample of the

Mexican-American Population in 1940." *Historical Methods* 21 (1988), 80–87.

Grebler, Leo. "The Naturalization of Mexican Immigrants in the United States." *International Migration Review* 1 (1966), 17–32.

Griswold del Castillo, Richard. "Patriarchy and the Status of Women in the Late Nineteenth-Century Southwest." In *The Mexican and Mexican American Experience in the 19th Century.* Ed. Jaime E. Rodríguez O. Tempe, Arizona: Bilingual Press, 1989, pp. 85–99.

———. "Southern California Chicano History: Regional Origins and National Critique." *Aztlán* 19:1 (Spring 1988–90), 109–24.

Guendelman, Sylvia, and Auristela Perez-Itriago. "Double Lives: The Changing Role of Women in Seasonal Migration." *Women's Studies* 13 (1987), 249–71.

Hall, Stuart. "Cultural Identity and Diaspora." In *Identity: Community, Culture, Difference.* Ed. Jonathan Rutherford. London: Lawrence & Wishart, 1990, pp. 222–37.

Hernández, Marita, and Robert Montemayor. "Mexico to U.S.—A Cultural Odyssey," *Los Angeles Times,* 24 July 1983, pp. 1, 18–21.

Johnson, Richard. "What Is Cultural Studies Anyway?" *Social Text* 16 (1986–87), 38–80.

Kanellos, Nicolás. "Two Centuries of Hispanic Theatre in the Southwest." *Revista Chicano-Riquena* 11 (1983), 17–36.

———. "An Overview of Hispanic Theatre in the United States." In *Hispanic Theatre in the United States.* Ed. Nicolás Kanellos. Houston: Arte Publico Press, 1984, pp. 7–13.

Kantor, Harvey. "Choosing a Vocation: The Origins and Transformation of Vocational Guidance in California, 1910–1930." *History of Education Quarterly* 26 (1986), 351–75.

Katz, Friedrich. "Labor Conditions on Haciendas in Porfirian Mexico: Some Trends and Tendencies." *Hispanic American Historical Review* 54 (1974), 1–47.

Kemper, Robert V., and Anya Peterson Royce. "Mexican Urbanization since 1821: A Macro-Historical Approach." *Urban Anthropology* 8 (1979), 267–89.

Knight, Alan. "Racism, Revolution, and *Indigenismo:* Mexico, 1910–1940." In *The Idea of Race in Latin America, 1870–1940.* Ed. Richard Graham. Austin: Univ. of Texas Press, 1990. pp. 71–113.

"L.A. Tenement Problems and the Bubonic-Pneumonic Plagues." Municipal League of Los Angeles *Bulletin,* 16 Feb. 1925, pp. 2–3.

Layton, Edwin. "The Better America Federation: A Case Study of Superpatriotism." *Pacific Historical Review* 30 (1961), 137–47.

Lopez, Ronald W. "The El Monte Berry Strike of 1933." *Aztlán: International Journal of Chicano Studies Research* 1 (1970), 101–114.

Luria, Daniel. "Wealth, Capital, and Power: The Social Meaning of Home Ownership." *The Journal of Interdisciplinary History* 7 (1976), 261–82.

McDowell, John Holmes. "The Corrido of Greater Mexico as Discourse, Music and Event." In *"And Other Neighborly Names": Social Process and Cultural Image in Texas Folklore*. Eds. Richard Bauman and Robert D. Abrahams. Austin: Univ. of Texas Press, 1981, pp. 44–75.

McWilliams, Carey. "Getting Rid of the Mexican." *American Mercury* 28 (March 1933), 322–24.

Martínez, Oscar J. "On the Size of the Chicano Population: New Estimates, 1850–1900." *Aztlan: International Journal of Chicano Studies Research,* 6 (1975), 43–67.

Medeiros, Francine. "*La Opinión,* A Mexican Exile Newspaper: A Content Analysis of Its First Years, 1926–1929." *Aztlan* 11:1 (1980), 68–75.

Meyerowitz, Joanne. "Women and Migration: Autonomous Female Migrants to Chicago, 1880–1930." *Journal of Urban History* 13 (1987), 147–68.

Milkman, Ruth. "Women's Work and the Economic Crisis: Some Lessons from the Great Depression." *Review of Radical Political Economics* 8 (1976), 73–97.

Mirandé, Alfredo. "The Chicano Family: A Reanalysis of Conflicting Views." *Journal of Marriage and the Family* 39 (1977), 747–58.

Modell, John, and Tamara Haraven. "Urbanization and the Malleable Household: An Examination of Boarding and Lodging in American Families." In *The American Family in Social-Historical Perspective*. Ed. Michael Gordon. 2nd ed. New York: St. Martin's Press, 1978, pp. 51–68.

Monroy, Douglas. "La Costura en Los Angeles, 1933–1939: The ILGWU and the Politics of Domination." In *Mexican Women in the United States: Struggle Past and Present*. Magdalena Mora and Adelaida R. del Castillo, eds. Los Angeles: Chicano Studies Research Center Publications, 1980, pp. 171–78.

———. "Like Swallows at the Old Mission: Mexicans and the Racial Politics of Growth in Los Angeles in the Interwar Period." *Western Historical Quarterly* 14 (1983), 435–58.

———. "Anarquismo y Comunismo: Mexican Radicalism and the Communist Party in Los Angeles during the 1930s." *Labor History* 24 (Winter 1983), 34–59.

Montiel, Miguel. "The Social Science Myth of the Mexican American Family." *El Grito: A Journal of Contemporary Mexican American Thought* 3 (1970), 56–63.

Moss, Leonard W., and Walter H. Thomson. "The South Italian Family: Literature and Observation." *Human Organization* 18 (1959), 35–41.

Murillo, Nathan. "The Mexican American Family." In *Chicanos: Social and Psychological Perspectives*. Eds. Nathaniel N. Wagner and Marsha J. Haug. St. Louis: C.V. Mosby, 1971, pp. 97–108.

Nelson, Howard J. "The Vernon Area, California—A Study of the Political Factor in Urban Geography." *Annals of the Association of American Geographers* 42 (1952), 177–91.

Oxnam, G. Bromley. "The Mexican in Los Angeles from the Standpoint of the Religious Forces of the City." *Annals of the American Academy* 93 (1921), 130–33.

————. "Los Angeles: A City of Many Nations." *California Christian Advocate,* 18 Feb. 1926, p. 14.

Panuzio, Constantine. "Intermarriage in Los Angeles, 1924–1933." *American Journal of Sociology* 48 (1942), 690–701.

Paredes, Américo. "The Ancestry of Mexico's Corridos: A Matter of Definitions." *Journal of American Folklore* 76 (1963), 231–35.

Patterson, Tim. "Notes on the Historical Application of Marxist Cultural Theory." *Science & Society* 39 (1975), 257–91.

Potter, David M. "The Quest for National Character." In *The Reconstruction of American History.* Ed. John Higham. New York: Hutchinson, 1962, pp. 197–220.

Rabinowitz, Howard N. "Race, Ethnicity, and Cultural Pluralism in American History." In *Ordinary People and Everyday Life: Perspectives on the New Social History.* Eds. James B. Gardner and George Rollie Adams. Nashville: American Association for State and Local History, 1983, pp. 23–49.

Rischin, Moses. "Immigration, Migration, and Minorities in California: A Reassessment." *Pacific Historical Review,* 41 (1972), 71–132.

Roberts, Kenneth L. "The Docile Mexican." *Saturday Evening Post,* 10 March 1928, p. 43.

Roediger, David. "Where Communism Was Black: A Review of *Hammer and Hoe: Alabama Communists During the Great Depression* by Robin D.G. Kelley." *American Quarterly* 44:1 (March 1992), 123–28.

Romano, Octavio Ignacio V. "The Anthropology and Sociology of the Mexican-Americans: The Distortion of Mexican-American History." *El Grito* 2:1 (Fall 1968), 13–26.

Romo, Ricardo. "Responses to Mexican Immigration, 1910–1930." *Aztlán* 6 (1975), 173–90.

————. "The Urbanization of Southwestern Chicanos in the Early Twentieth Century." *New Scholar* 5 (1977), 183–207.

Rosales, Francisco Arturo. "Shifting Self Perceptions and Ethnic Consciousness among Mexicans in Houston, 1908–1946." *Aztlán* 16:1–2 (1985), 71–94.

———— and Daniel T. Simon. "Mexican Immigrant Experience in the Urban Midwest: East Chicago, Indiana, 1919–1945." In *Forging a Community: The Latino Experience in Northwest Indiana, 1919–1975.* Eds. James B. Lane and Edward J. Escobar. Chicago: Cattails Press, 1987, pp. 137–60.

Rose, Kenneth D. " 'Dry' Los Angeles and Its Liquor Problems in 1924." *Southern California Quarterly* 69 (1987), 55–70.

Rouse, Roger. "Mexican Migration and the Social Space of Postmodernism." *Diaspora* 1:1 (Spring 1991), 8–23.

Ruiz, Vicki L. "A Promise Fufilled: Mexican Cannery Workers in Southern California." *Pacific Historian* 30:2 (Summer 1986), 50–61.

Saldivar, José David. "The Limits of Cultural Studies." *American Literary History* 2:2 (Summer 1990), 251–66.

Saragoza, Alex M. "The Conceptualization of the History of the Chicano Family." In *The State of Chicano Research in Family, Labor and Migration Stud-*

ies. Eds. Armando Valdez, Albert Camarillo, and Tomás Almaguer. Stanford: Stanford Center for Chicano Research, 1983, pp. 111–38.

———. "Recent Chicano Historiography: An Interpretive Essay." *Aztlán* 19:1 (Spring 1988–90), 1–77.

Simmons, Merle E. "The Ancestry of Mexico Corridos." *Journal of American Folklore* 76 (1963), 1–15.

Skirius, John. "Vasconcelos and *México de Afuera.*" *Aztlán* 7:3 (Fall 1976), 479–97.

Smith, Peter H. "La política dentro de la Revolución: El congreso constituyente de 1916–1917." *Historia Mexicana* 22 (1973), 363–95.

Smith, Timothy L. "Religion and Ethnicity in America." *American Historical Review* 83:5 (Dec. 1978), 1155–185.

Soto, Antonio R. "The Church in California and the Chicano: A Sociological Analysis." *Grito del Sol* 4 (1979), 47–74.

Taylor, Paul S. "Mexican Women in Los Angeles Industry in 1928." *Aztlán* 11 (1980), 99–131.

Tirado, Miguel David. "Mexican American Community Political Organization: The Key to Chicano Political Power." *Aztlán: Chicano Journal of Social Sciences and the Arts* 1 (1970), 58–78.

Treutlein, Theodore E. "Los Angeles, California: The Question of the City's Original Spanish Name." *Southern California Quarterly* LV (Spring 1973), 1–7.

Treviño, Roberto R. "Prensa y patria: The Spanish-Language Press and the Biculturation of the Tejano Middle Class, 1920–1940." *Western Historical Quarterly* (Nov. 1991), 451–72.

Turner, Timothy. "Where Folks Are Folks." *Los Angeles Times Illustrated Magazine,* 14 Sept. 1924, p. 7.

Vaughn, Mary Kay. "Primary Education and Literacy in Nineteenth-Century Mexico: Research Trends, 1968–1988." *Latin American Research Review* 25:1 (1990), 31–66.

Vecoli, Rudolph J. "Contadini in Chicago: A Critique of *The Uprooted.*" *Journal of American History* 51 (1964), 404–17.

Walker, Franklin. "Pacific Electric." In *Los Angeles: Biography of a City.* Eds. John and LaRae Caughey. Berkeley: Univ. of California Press, 1976, pp. 218–22.

Walker, Helen. "Mexican Immigrants as Laborers." *Sociology and Social Research,* 13 (1928–29), 55–62.

———. "Mexican Immigrants and American Citizenship." *Sociology and Social Research* 13 (1928–29), 465–71.

Weber, Devra Anne. "The Organizing of Mexicano Agricultural Workers: Imperial Valley and Los Angeles, 1928–34, An Oral History Approach." *Aztlán* 3:2 (Fall 1972), 307–47.

Weber, Francis J. "Irish-Born Champion of the Mexican-Americans." *California Historical Society Quarterly* 49 (1970), 233–50.

Weeks, O. Douglas. "The League of United Latin American Citizens: A Texas-

Mexican Organization." *Political and Social Science Quarterly* 10 (1929), 257–78.

White, Richard. "Race Relations in the American West." *American Quarterly* 38 (1986), 396–416.

Ybarra-Frausto, Tomás. "I Can Still Hear the Applause: La Farándula Chicana: Carpas y Tandas de Variedad." In *Hispanic Theatre in the United States.* Ed. Nicolas Kanellos. Houston: Arte Publico Press, 1984, pp. 45–60.

Young, Pauline V. "The Russian Molokan Community in Los Angeles." *American Journal of Sociology* 35 (1929), 393–402.

Zelman, Donald L. "Mexican Migrants and Relief in Depression California: Grower Reaction to Public Relief Policies as They Affected Mexican Migration." *Journal of Mexican-American History* 5 (1975), 1–23.

Dissertations, Theses, and Unpublished Papers

Arroyo, Luis Leobardo. "Industrial Unionism and the Los Angeles Furniture Industry, 1918–1954." Ph.D. diss., University of California, Los Angeles, 1979.

Balderrama, Francisco Enrique. "En Defensa de la Raza: The Los Angeles Mexican Consulate and Colonia Mexicana during the Great Depression." Ph.D. diss., University of California, Los Angeles, 1978.

Bond, J. Max. "The Negro in Los Angeles." Ph.D. diss., University of Southern California, 1936.

Buckner, Herman. "A Study of Pupil Elimination and Failure among Mexicans." Master's thesis, University of Southern California, 1935.

Burke, Rev. Dennis J. "The History of the Confraternity of Christian Doctrine in the Diocese of Los Angeles, 1922–1936." Master's thesis, Catholic University of America, 1965.

Cameron, James William. "The History of Mexican Public Education in Los Angeles, 1910–1930." Ph.D. diss., University of Southern California, 1976.

Castillo, Pedro G. "The Making of a Mexican Barrio: Los Angeles, 1890–1920." Ph.D. diss., University of California, Santa Barbara, 1979.

Chambers, M. Lorena. "Urban Reform: The Effect of Social Work on Mexicans in Los Angeles, 1910–1936." Senior honors thesis. UCLA, 1989.

Culp, Alice Bessie. "A Case Study of the Living Conditions of Thirty-five Mexican Families of Los Angeles with Special Reference to Mexican Children." Master's thesis, University of Southern California, 1921.

Deutsch, Sarah. "Culture, Class, and Gender: Chicanas and Chicanos in Colorado and New Mexico, 1900–1940." Ph.D. diss., Yale University, 1985.

Escobar, Edward Joseph. "Chicano Protest and the Law: Law Enforcement Responses to Chicano Activism in Los Angeles, 1850–1936." Ph.D. diss., University of California, Riverside, 1983.

Farmer, William A. "The Influence of Segregation of Mexican and American School Children on the Development of Social Attitudes." Master's thesis, University of Southern California, 1937.

Fellows, Lloyd. "Economic Aspects of the Mexican Rural Population in California with Special Emphasis on the Need for Mexican Labor in California." Master's thesis, University of Southern California, 1929.

Foster, Nellie. "The *Corrido:* A Mexican Culture Trait Persisting in Southern California." Master's thesis, University of Southern California, 1939.

García, Richard Amado. "The Making of the Mexican-American Mind, San Antonio, Texas, 1929–1941: A Social and Intellectual History of an Ethnic Community." Ph.D. diss., University of California, Irvine, 1980.

Gilbert, James Carl. "A Field Study in Mexico of the Mexican Repatriation Movement." Master's thesis, University of Southern California, 1934.

Griswold del Castillo, Richard. "La Raza Hispano Americana: The Emergence of an Urban Culture among the Spanish Speaking of Los Angeles." Ph.D. diss., University of California, Los Angeles, 1974.

Guerin-Gonzáles, Camille. "Cycles of Immigration and Repatriation: Mexican Farm Workers in California Industrial Agriculture, 1900–1940." Ph.D. diss., University of California, Riverside, 1985.

Gutiérrez, Ramón Arturo. "Marriage, Sex and the Family: Social Change in Colonial New Mexico, 1690–1846." Ph.D. diss., University of Wisconsin-Madison, 1980.

Herman, David G. "Neighbors on the Golden Mountain: The Americanization of Immigrants in California. Public Instruction as an Agency of Ethnic Assimilation, 1850 to 1933." Ph.D. diss., University of California, Berkeley, 1981.

Kienle, John E. "Housing Conditions among the Mexican Population of Los Angeles." Master's thesis, University of Southern California, 1912.

Kirschner, Olive P. "The Italian in Los Angeles." Master's thesis, University of Southern California, 1920.

Lanigan, Mary. "Second Generation Mexicans in Belvedere." Master's thesis, University of Southern California, 1932.

Leader, Leonard J. "Los Angeles and the Great Depression." Ph.D. diss., University of California, Los Angeles, 1972.

Leeder, Elaine. "The Gentle Warrior: Rose Pesotta, Anarchist and Labor Organizer." Ph.D. diss., Cornell University, 1985.

Lipsitz, George. "*Con Safos:* Can Cultural Studies Read the Writings on the Walls?" Unpublished paper, Jan. 1991.

Locke, Neil M. "The Lodging Houses of Los Angeles." Master's thesis, University of Southern California, 1913.

Loza, Stephen Joseph. "The Musical Life of the Mexican/Chicano People in Los Angeles, 1945–1985: A Study in Maintenance, Change, and Adaptation." Ph.D. diss., University of California, Los Angeles, 1985.

McEuen, William W. "A Survey of the Mexicans in Los Angeles." Master's thesis, University of Southern California, 1914.

Monroy, Douglas G. "Mexicanos in Los Angeles, 1930–1941: An Ethnic Group in Relation to Class Forces." Ph.D. diss., University of California, Los Angeles, 1978.

Ortegon, Samuel M. "Mexican Religious Population of Los Angeles." Master's thesis, University of Southern California, 1932.

Palomarez, Beatrice R. "Escuela Mexico." Unpublished manuscript, 1990.

Parlee, Lorena May. "Porfirio Díaz, Railroads, and Development in Northern Mexico: A Study of Government Policy Toward the Central and Nacional Railroads, 1876–1910." Ph.D. diss., University of California, San Diego, 1981.

Peters, Mary. "The Segregation of Mexican American Children in the Elementary Schools of California, Its Legal and Administrative Aspects." Master's thesis, University of California, Los Angeles, 1948.

Porfirio, J. Miranda. "Perceptions of Locus of Control among Three Multi-Generation Chicano/Mexican Families." Ph.D. diss., University of California, Los Angeles, 1978.

Privett, Stephan A., S.J. "Robert E. Lucey: Evangelization and Cathesis among Hispanic Catholics." Ph.D. diss., Catholic University of America, 1985.

Raftery, Judith Rosenberg. "The Invention of Modern Urban Schooling: Los Angeles, 1885–1941." Ph.D. diss., University of California, Los Angeles, 1984.

Ramírez, Zeferino. "Origins of the Ramírez Family, from What I Remember." Unpublished manuscript, 1983.

Ruiz, Vicki L. " 'Star Struck': Acculturation, Adolescence and the Mexican American Woman, 1920–1940." Unpublished paper presented at the Sorbonne conference, "Vision and Representation of History in the Cultural Production of Hispanics in the United States." Paris, France, 12–14 May, 1987.

Sánchez, George J. "Becoming Mexican American: Ethnicity and Acculturation in Chicano Los Angeles, 1900–1943." Ph.D. diss., Stanford University, 1989.

Schmidt, Arthur P. "The Social and Economic Effect of the Railroad in Puebla and Veracruz, Mexico, 1867–1911." Ph.D. diss., Indiana University, 1973.

Scott, Robin Fitzgerald. "The Mexican-American in the Los Angeles Area, 1920–1950: From Acquiescence to Activity." Ph.D. diss., University of Southern California, 1971.

Smith, Clara Gertrude. "The Development of the Mexican People in the Community of Watts, California." Master's thesis, University of Southern California, 1933.

Vargas, Zaragoza. "Mexican Auto Workers at Ford Motor Company, 1918–1933." Ph.D. diss., University of Michigan, 1984.

Weber, Devra Anne. "The Struggle for Stability and Control in the Cotton Fields of California: Class Relations in Agriculture, 1919–1942." Ph.D. diss., University of California, Los Angeles, 1986.

Westhoff, Marie Louise. "An Inquiry into the Activities of the International Ladies' Garment Workers' Union in Los Angeles." Master's thesis. University of Southern California, 1938.

White, Alfred. "The Apperceptive Mass of Foreigners as Applied to Americanization: The Mexican Group." Master's thesis, University of California, Berkeley, 1923.

Withers, Charles. "Problems of Mexican Boys." Master's thesis, University of Southern California, 1942.

Ybarra, Lea. "Conjugal Role Relationships in the Chicano Family." Ph.D. diss., University of California, Berkeley, 1977.

Zinn, Maxine Baca. "Marital Roles, Marital Power and Ethnicity: A Study of Changing Families." Ph.D. diss., University of Oregon, 1978.

Newspapers

El Heraldo de México
Los Angeles *Daily News*
Los Angeles *Examiner*
Los Angeles *Times*
Mexican Voice
La Opinión
The Tidings
Western Worker

Manuscript Collections and Documents

Emory S. Bogardus collection, University Archives, University of Southern California.

Catholic Welfare Bureau collection, Archdiocese of Los Angeles Archives, San Fernando, California.

George P. Clements collection, MS 118, Department of Special Collections, University Research Library, University of California, Los Angeles.

Declarations of Intention and Petitions for Naturalization, Record Group 21, National Archives, Laguna Niguel, California.

David Dubinsky collection, International Ladies' Garment Workers' Union papers, Cornell University, Ithaca.

Guy Endore collection, Department of Special Collections, University Research Library, University of California, Los Angeles.

Federal Writers Project collection, MS 306, Department of Special Collections, University Research Library, University of California, Los Angeles.

Federal Writers Project collection, National Archives, Washington, D.C.

Foreign consulate records for Los Angeles, Secretaría de Relaciones Exteriores, Mexico City, D.F., Mexico.

Ernesto Galarza collection, MS 224, Department of Special Collections, Stanford University.

Manuel Gamio collection, Z-R5, Bancroft Library, University of California, Berkeley.

Los Angeles County Board of Supervisors files and minutes, County Hall of Records, Los Angeles, California.

Robert E. Lucey collection, Catholic Archives at San Antonio, Chancery Office, San Antonio, Texas.

Alice McGrath collection, Department of Special Collections, University Research Library, University of California, Los Angeles.

Mexican American Movement, Papers of the Supreme Council, Urban Archives Center, California State University, Northridge.

Julian Nava collection, Urban Archives Center, California State University, Northridge.

Oral History Program: Mexican American Project, California State University, Fullerton.

G. Bromley Oxnam collection, MS 58, Manuscript Division, Library of Congress, Washington, D.C.

Manuel Ruiz Jr., collection, MS 295, Department of Special Collections, Stanford University, Stanford.

Paul S. Taylor collection, 74/187c, Bancroft Library, University of California, Berkeley.

U.S. Department of State, Records Relating to the Internal Affairs of Mexico, 1910–1929, Record Group 59, National Archives, Washington, D.C.

U.S. Immigration and Naturalization Service Records, Record Group 85, National Archives, Washington, D.C.

U.S. Mediation and Conciliation Service Records, Record Group 280, National Archives, Washington, D.C.

YMCA Room, Whittier College Library.

Miscellaneous

Annual Catholic Welfare Bureau Reports, 1925–32.

Associated Catholic Charities Reports, 1919–23.

Ballad of an Unsung Hero. San Diego: Cinewest/KPBS, 1983

González, Pedro J. Videocassette tape interviews and transcripts for *Ballad of an Unsung Hero.* San Diego: Cinewest/KPBS, 1983. Audio-visual Department, Stanford University.

Los Angeles City Directories, 1915–40

Palomarez, Beatrice R., Pico Rivera, California, 3 January 1991. Interview conducted by George J. Sánchez.

Sonnichsen, Philip. *Texas-Mexican Border Music, Vols. 2 & 3: Corridos Parts 1 & 2.* Arhoolie Records, 1975.

Index

351